Quantum Physics and Cosmology

SCIENCES

Sociology, Ethics and Epistemology of Sciences,
Field Directors – Bernard Reber and Jim Dratwa

Epistemology of Physics and Earth Sciences,
Subject Head – Hervé Zwirn

Quantum Physics and Cosmology

The Mysteries of the Infinitely Small and the Infinitely Large

Coordinated by
Hervé Zwirn

iSTE

WILEY

First published 2026 in Great Britain and the United States by ISTE Ltd and John Wiley & Sons, Inc.

ISTE Ltd
27-37 St George's Road
London SW19 4EU
UK

www.iste.co.uk

John Wiley & Sons, Inc.
111 River Street
Hoboken, NJ 07030
USA

www.wiley.com

Library of Congress Control Number: 2025945969

British Library Cataloguing-in-Publication Data
A CIP record for this book is available from the British Library
ISBN 978-1-78945-233-4

ERC code:
PE2 Fundamental Constituents of Matter
PE2_5 Classical and quantum physics of gravitational interactions
PE9 Universe Sciences
PE9_9 Cosmology and large-scale structure, dark matter, dark energy

The manufacturer's authorized representative according to the EU General Product Safety Regulation is Wiley-VCH GmbH, Boschstr. 12, 69469 Weinheim, Germany, e-mail: Product_Safety@wiley.com.

Contents

Introduction

Hervé ZWIRN[1,2]
[1] *Institut d'Histoire et de Philosophie des Sciences et des Techniques, Paris, France*
[2] *École Normale Supérieure de Paris-Saclay, Centre Borelli, France*

In recent years, contemporary physics has made spectacular progress in both infinitely large and infinitely small realms. The aim of this book is to share with the reader some of these discoveries, presented from a generalist and sometimes philosophical angle. The aim is not to present things in a technical way, which is only accessible to specialists, but rather to adapt the level so that the chapters are comprehensible, at least for the most part, by most readers with scientific knowledge at the first-year university level.

The chosen style of presentation is comparable to that of high-caliber popular magazines such as Scientific American (Pour la Science). I would like to thank the authors who have contributed to this volume and have made the educational effort to present their discipline in a way that is accessible to as many readers as possible. However, not all of the chapters in this book are equal in terms of ease of reading; some require slightly more effort from readers than others. This is mainly due to the diversity of the scientific fields covered.

Some chapters present recent observations that shake up what we thought we knew, such as how the universe was formed and evolved. Other chapters look at scientific questions that are currently unresolved. Today's leading scientists are currently so hard pressed to solve them that they are torn between whether answers can be found within the framework of current theories or whether, on the contrary, it will be necessary to completely revolutionize our way of thinking in order to do so.

Understandably, the effort required to understand each subject will have to be adapted to its intrinsic difficulty.

I would particularly like to thank Nathalie Palanque-Delabrouille for her careful proofreading and correction of the chapters in the cosmology section.

The subjects covered are quantum physics and cosmology, which make up the two main subjects of this book. Of course, the aim is not to be exhaustive – that would take more than one volume. Instead, we will focus on certain aspects of these disciplines that have recently undergone major advances or that are simply of particular interest. Many topics will not be covered here due to insufficient space, but readers who have been captivated by the stimulating chapters in this book will be able to continue their journey through the many sources available, both in specialized books and on the Internet.

In 2025, we celebrated the centenary of the discovery of quantum mechanics. Quantum mechanics describes the behavior of atoms and elementary particles and is one of the most fertile and best corroborated physical theories. First and foremost, it is fertile because it explains many phenomena that classical physics has failed to understand, such as the stability of matter, the color of bodies, black-body radiation, the photoelectric effect, the behavior of semiconductors, superfluidity and superconductivity. Without it, many of our everyday objects would never have been invented, starting with our smartphones and computers. Second, it is fertile because it is the basis of quantum field theory, which is the foundation on which the standard model of particle physics rests today. This model's success no longer needs to be demonstrated, including the prediction of the existence of the W and Z intermediate bosons (mediators of the electroweak interaction) and the Higgs boson (involved in the mechanism that gives particles their mass), all of which were discovered at CERN. This is corroborated by the fact that, since its creation, it has been subjected to a staggering number of tests in all kinds of circumstances and has never been contradicted by any experiment.

Paradoxically, however, despite this success, quantum mechanics is still subject to fierce debate in regard to understanding what it tells us about the world. Physicists all agree on how to use its equations to predict the outcome of performed experiments but disagree on the ontological significance of formalism. What does quantum mechanics tell us about reality? Unlike classical physics, which can be understood directly as describing the world as it is, quantum mechanics poses serious conceptual problems when we attempt to approach it in the same way. Of course, quantum mechanics puts forward many strange properties, such as indeterminism, the superposition of states and the incompatibility of certain properties. However, the most serious problems are the so-called measurement problem, for which no consensual solution has yet been found. The various proposals for solving this problem result in what are known as “interpretations”, which often give incompatible views of the world. The problem is that, to date, there has been no experimental evidence to enable us to settle between them.

To set the scene and enable the reader to situate the debates in question, Hervé Zwirn’s overview chapter (Chapter 1) briefly presents the problem of measurement, the reasons for these debates and the main interpretations to which they have given rise.

Thomas Ryckman analyzes the position of Niels Bohr, who, along with Werner Heisenberg, is considered the main author of the so-called “Copenhagen interpretation”. This interpretation, which for a long time has remained the overwhelming majority view among physicists and is still highly influential today, is not monolithic. The positions of Heisenberg and Bohr differ in many subtle respects. However, beyond these divergences, Bohr’s thought itself has often been described as obscure, and it is not easy to obtain a clear idea of what Bohr thought about reality. This is the subject of Chapter 2.

Lev Vaidman described the multiple worlds interpretation (MWI), which was first invented by Hugh Everett in 1957. This interpretation, which was later popularized by DeWitt and Graham in 1973, has been widely mentioned for its extraordinary character, which has led some to describe it as science fiction. Indeed, the concept of parallel worlds inevitably brings to mind films of this nature. However, as it may seem, this way of solving the problem of measurement is not absurd and is supported by many advocates. Nevertheless, there are several ways of understanding Everett’s position which are not compatible with each other. Lev Vaidman describes his personal version here (Chapter 3) after recalling the basic elements common to all versions.

Jeffrey Barrett focuses on how to make sense of the concept of probability in Everett’s interpretation. Indeed, one of the main criticisms of this interpretation is that it is difficult to reconcile with the usual probabilistic approach to quantum mechanics. Since all possible events occur during a measurement, giving rise to as many distinct worlds as there are possible outcomes, it seems that the notion of probability disappears. Various attempts to solve this problem have been proposed but have failed to win the support of physicists. Jeffrey Barrett (Chapter 4) analyzes some of these attempts in detail and arrives at a conclusion in favor of one of them.

Pragmatism is one possible approach to the problems of interpretation; this was originally coined by Charles Peirce, John Dewey and William James. When applied to quantum physics, the wave function that is supposed to describe the state of a system in quantum mechanics does not actually have this role. More generally, for pragmatists, a scientific theory does not describe the world as it is but is intended to guide an agent in estimating the probabilities of a given outcome occurring in the course of the agent's experiments. In Chapter 5, Richard Healey presents his pragmatist position and how it solves the measurement problem and avoids introducing non-locality into Einstein–Podolski–Rosen-type experiments.

It seems that if we stick to the standard formalism of quantum mechanics without modifying it, it is difficult to solve the problem of measurement without making the observer play a central role. This is the approach taken by Hervé Zwirn, who, in Chapter 6, presents convivial solipsism (ConSol). This interpretation, which he has developed, considers that no modification of the system's physical state takes place during a measurement. In this view, a measurement is essentially an act of perception by an observer. This way of looking at measurement solves the problem posed in the traditional framework and provides a number of solutions to other problems, such as non-locality.

ConSol is part of a large family of interpretations that David Schmid, Yile Ying and Matthew Leifer describe as "Copenhagenish", in that they are related to, but distinct from, the Copenhagen interpretation. These interpretations have in common that they consider the result an observer obtains from a measurement to be unique. They also consider that quantum mechanics applies to all physical systems on any scale, quantum mechanics is a complete theory that cannot be supplemented by hidden variables and the state vector of a system does not directly represent its physical state. In Chapter 7, the authors compare different interpretations of this type and analyze how they explain Wigner's friend experiment (a brief description of which can be found in Chapter 1).

At the other end of the dimensional spectrum, in the infinitely large realm, the profound questions raised by recent discoveries have multiplied in recent years. The discovery that the expansion of the universe was not slowing but rather accelerating came as a great surprise. This acceleration was first observed in 1998, when astronomers noticed that some supernovae (stars exploding at the end of their lives) appeared to be farther away than expected, suggesting that the universe was expanding faster than previously thought. To explain this phenomenon, astronomers turned to a mysterious form of energy known as "dark energy", which makes up approximately 70% of the universe. Dark energy represents one of the greatest challenges of modern physics. Although its existence is widely accepted owing to its observable influence on the expansion of the universe, its true nature remains a mystery. In Chapter 8, Nathalie Palanque-Delabrouille explains the current state of thought on this subject.

Dark matter is another great mystery of modern cosmology. The introduction of dark matter stems from the observation that galaxies spin on themselves at a speed that would disintegrate them if they contained only the mass of their visible matter, whose gravitation would not be sufficient to hold them together. Dark matter is thought to make up 25% of our universe, which means that with 70% dark energy, the ordinary matter we know weighs in at just 5% of the total. Many hypotheses have been proposed concerning the nature of this dark matter, but none of these hypotheses, which are based on known matter particles, have been satisfactory. It would seem, therefore, that dark matter is made up of unknown exotic particles, unless the alternative hypothesis of a change in the law of gravitation is finally accepted. Françoise Combes explains the latest advances in this field in Chapter 9. Nevertheless, we already know that 95% of what makes up the universe is totally unknown to us!

In 2015, the first detection of gravitational waves was performed, followed by the merger of binary neutron stars in 2017. Since then, gravitational wave astronomy has rapidly progressed, producing many scientific results in astrophysics, cosmology and fundamental physics. This new way of exploring the distant universe is fascinating because it gives us access to events that we would not be able to see using our usual means of observation on the basis of electromagnetic waves. The LIGO–Virgo–KAGRA international gravitational wave network has already observed more than 200 gravitational wave sources, and plans are underway to improve current detectors and extend their astrophysical reach. In a presentation (Chapter 10) by Matteo Barsuglia and Simone Mastrogiovanni, they describe the importance of the results already obtained and the plans for the next two decades.

Bruno Mansoulié (Chapter 11) analyzes the problem posed by vacuum energy. This question, linked to dark energy and the cosmological constant, is one of the most intriguing in modern physics. The value of this energy when calculated in general relativity via the cosmological constant is 10^{120} times smaller than when calculated in the formalism of quantum field theory. This discrepancy is a record of the disparity of predictions that two theories can make. We know that we lack a unifying theory of gravity and quantum physics, but here we are in the presence of something outrageous. How can we reconcile the two predictions? No one currently knows the answer to this question, and this chapter presents the highly prospective avenues explored today.

In Chapter 12, Marc Lachieze-Rey analyzes the notion of time in relativity and shows us, through the presentation of various relativistic temporal effects, that time, understood in the usual sense – namely, classical time – does not exist. The only thing that exists is the duration of each story.

One of the greatest challenges facing modern physics is to unify the two theoretical pillars of quantum mechanics and general relativity. Numerous attempts have been made to this end, and several theories are currently in competition, none of which have yet been deemed satisfactory. The very question of the need to quantify gravity (in other words, to treat it quantum mechanically) is controversial. It would be possible for gravity to be a classical force, albeit existing in a quantum mode. However, arguments of symmetry and homogeneity of theories (and even aesthetics) tend to push us in the direction of considering that gravity must be quantum, like other forces. Olival Freire and Thiago Hartz recount the history of quantum gravity thought experiments (Chapter 13) from the beginning of quantum mechanics to the present day.

In Chapter 14, Baptiste Le Bihan, Emilia Margoni and Annica Vieser take a more epistemological look at the concept of possibilities in classical physics, general relativity and quantum physics. These possibilities can be kinematic or dynamic. The question is as follows: how can we clearly define the difference between these two types of possibility, and do they all represent genuine physical possibilities?

These chapters show just how much progress modern physics has made in recent years (we have undoubtedly learned more in the last hundred years than in the entire history of mankind) while at the same time raising new questions at least as numerous as the discoveries. Will unanswered questions today, such as the mystery of dark energy and dark matter, vacuum energy, the problem of measurement, the absence of a unifying theory between gravity and quantum mechanics, etc., turn out to be temporary difficulties that will be resolved within the framework of current theories, or will they play the same role as the two clouds evoked by Lord Kelvin at the end of the 19th century (the constancy of the speed of light incompatible with the ether, and the spectrum of black-body radiation), which led to the two greatest revolutions in physics, namely, Einstein's relativity and quantum mechanics? Only time will tell.

1

A Brief Introduction to Quantum Mechanics and a Review of Interpretations

Hervé ZWIRN[1,2]

[1]*Institut d'Histoire et de Philosophie des Sciences et des Techniques, Paris, France*
[2]*École Normale Supérieure de Paris-Saclay, Centre Borelli, France*

To set the scene for the various chapters of this book's analyses of interpretations of quantum mechanics, it is useful to gain an overview of why these interpretations have been proposed, how they relate to each other and what criticisms have been made against them. This is the purpose of this introductory chapter[1].

We might say that the main reason these interpretations were invented was to solve the measurement problem, which, from the very beginning of quantum mechanics, gave rise to much questioning and heated debate. To understand this problem, however, we need to know a little about the formalism used in quantum mechanics.

Let us begin by giving the necessary rudiments. Then, let us look at where this problem comes from and finally present the various solutions that have been found.

---

1 Since this chapter references the various chapters of the book, we will refrain from providing references here.

Quantum Physics and Cosmology,
coordinated by Hervé ZWIRN.

1.1. Quantum formalism

1.1.1. *States and observations*

This is not intended to be a complete course in quantum mechanics, which is well beyond the scope of this chapter. We simply give a few points regarding quantum formalism, which the reader should take at face value. Readers who do not have sufficient mathematical knowledge to fully understand the meaning of the objects presented here will have to be content, while retaining the vocabulary associated with certain quantum concepts, such as the state vector, observable or eigenvalue. This will be sufficient to understand the rest of the presentation.

Quantum mechanics is based on a number of rules and concepts that can seem confusing or even arbitrary when first encountered. This is the experience that all students have when they discover this discipline. The correct attitude to adopt at the outset is to see it as a kind of apprenticeship to the rules of a new game. The motivation for this is that this new game represents the most fruitful and best verified physical theory in the history of physics. As such, these rules are not a purely gratuitous game.

Quantum mechanics describes the states of systems by means of a vector in a vector space over the field of complex numbers. This space, with its special properties, is called a Hilbert space and is denoted as $\mathcal{H}$. The state vector is denoted as $|\psi\rangle$, according to a notation coined by the physicist Paul Dirac. For reasons linked to the measurement operation (which we present below), the length (or modulus) of such a vector must be normalized to 1. If the position of a system is x, its state is denoted as $|x\rangle$, and its length is assumed to be 1.

One of the fundamental principles of quantum mechanics is the principle of superposition, according to which any linear combination of possible state vectors is itself a possible state vector. Therefore, if $|\psi\rangle$ and $|\phi\rangle$ are two possible states of a system, the state $a|\psi\rangle + b|\phi\rangle$ (with $|a|^2 + |b|^2 = 1$) will also be a possible state. This means that if x and y are two possible positions for a system, the state $^1/_{\sqrt{2}}|x\rangle + ^1/_{\sqrt{2}}|y\rangle$ will also be a possible state. Here, we encounter a first difficulty: how do we interpret such a so-called "superposed" state, which has no equivalent in classical physics? What is the position of the system when it is in such a state? We answer this question below.

In a Hilbert space, we can define the scalar product of two vectors, which associates two vectors with a complex number. This is expressed as: $\langle\phi|\psi\rangle = \alpha$. Two vectors are said to be orthogonal when their scalar product is zero.

Any physical quantity, such as position, velocity and energy is associated with a Hermitian operator, called an observable. When the image of a vector by an operator is a multiple of the initial vector, the vector is said to be an eigenvector of the operator. If A is an operator and $|\psi\rangle$ is a vector, we say that $|\psi\rangle$ is an eigenvector of A with eigenvalue λ if A $|\psi\rangle = \lambda|\psi\rangle$.

Hermitian operators (observables) have special properties: if an operator A is Hermitian, then:

– the eigenvalues of A are real;

– eigenvectors corresponding to distinct eigenvalues are orthogonal;

– there is a basis of the Hilbert space formed by the eigenvectors of A.

1.1.2. *Evolution of the state of a system without and with measurement*

In the case where a system is allowed to evolve without being measured (i.e. without any experiment to determine the value of a property), the quantum formalism tells us that its state $|\psi\rangle$ evolves according to Schrödinger's equation:

$$i\hbar\frac{d|\Psi\rangle}{dt} = H|\Psi\rangle$$

The most important thing to remember about this equation is that the evolution thus described is linear and unitary (meaning, in figurative language, that it preserves the length of the vectors, known as their modulus). This means that any linear combination of vectors will be transformed by this equation into a linear combination of the same length.

The situation is different if a measurement is made on the system. Each physical property of system S is associated with an observable A. Let $|v_1\rangle, |v_2\rangle, \ldots, |v_n\rangle$ be the eigenvector of A, associated with the eigenvalues $\lambda_1, \lambda_2, \ldots, \lambda_n$ and forming a basis of $\mathcal{H}$.

Suppose that S is in the state $|\psi\rangle$. To determine the numerical value of the property in question, we carry out an experiment using a measuring device that is supposed to yield a result corresponding to this measurement. A measurement of S of the property associated with A obeys the following rules (known as the "wave packet collapse" or reduction principle):

– The result of a measurement of the physical property associated with A on system S can only be an eigenvalue of A.

– The probability of obtaining[2] the eigenvalue λ_i is $P(\lambda_i) = |\langle v_i|\psi\rangle|^2$. This is known as Born's rule.

– If the measurement of A on system S in state $|\psi\rangle$ has given the result λ_i, then, immediately after the measurement, the new state of the system is the normalized projection of $|\psi\rangle$ onto the subeigenspace associated with λ_i (i.e. $|v_i\rangle$ in the simple case chosen here of non-degenerate eigenvalues).

Two points can already be made to illustrate the difference between quantum mechanics and classical mechanics. The first is that the values that property A can possess are not arbitrary but limited to the eigenvalues of the observable associated with A. The second is that the result of the measurement cannot generally be predicted with certainty – only probabilistically.

A third important point is that once the measurement has been taken, the state of the system will have changed according to the result obtained.

1.2. The measurement problem

The measurement problem stems from the fact that quantum mechanics contains two rules for calculating the evolution of a system. The first is Schrödinger's equation:

$$i\hbar\frac{d|\Psi\rangle}{dt} = H|\Psi\rangle$$

which is supposed to be used for an isolated system when no measurements are performed on it.

The second rule is the reduction principle, which states that when a measurement of some observable A is made on a system that is initially in a state that is a superposition of eigenstates of A, $|\Psi\rangle = \sum c_i|v_i\rangle$, then after the measurement, if the result is λ_k, an eigenvalue of A, the state $|\Psi\rangle$ is projected onto the eigenvector $|v_k\rangle$ related to this eigenvalue $\sum c_i|v_i\rangle \rightarrow |v_k\rangle$ (or onto the sub-eigenspace of the Hilbert space spanned by the eigenvalue λ_k if λ_k is degenerate).

This would not be a problem if it were possible to clearly define what a measure is. In that case, there would be two rules, each of which would apply in a specific case. However, this is not the case. It would be natural to define a measurement as the interaction of a system with an appropriate measuring device. This is what happens in classical physics: if the temperature of an object is to be measured, for

2 Here, we take the simplest case of non-degenerate eigenvalues.

example, an instrument (a thermometer) will be brought into contact with the object and will give the desired temperature value on its graduated scale.

Let us look at such an experiment. Let us consider system S and device P, which is supposed to measure the value of a certain property A of the system. Prior to the interaction between the system and the device, the state of the system is assumed to be a superposed state of the type $|\Psi_S\rangle = \sum c_i|v_i\rangle$. For the device, it will be in an initial state in which we note that $|P_0\rangle$. Quantum mechanics tells us that the initial state of the overall system made up of system S and the measuring device is then the tensor product[3] of the two states:

$$|\Psi_{SP}\rangle = |\Psi_S\rangle \otimes |P_0\rangle = \sum c_i|v_i\rangle \otimes |P_0\rangle$$

In the traditional way of presenting a measurement (analogous to temperature with a thermometer), once the measuring device and the system have interacted, we look at the display of the measuring device, which provides an accurate result. Therefore, the device will go from the initial state $|P_0\rangle$ to a state $|P_i\rangle$ corresponding to having recorded the result λ_i. We deduce that the measurement of the system's property A yields the result λ_i and, in accordance with the reduction principle, that each of the values λ_i is possible with a probability given by Born's rule.

This is a perfectly natural description of what a measurement is. However, quantum mechanics offers another, equally legitimate way of describing the same experience. To carry out a measurement, the system and the device must interact. Quantum mechanics tells us that such interaction is described by a Hamiltonian of interaction between S and P. As a result, the evolution of the overall system considered as a whole is described by Schrödinger's equation. Applying this equation to the initial state $|\Psi_{SP}\rangle = \sum c_i|v_i\rangle \otimes |P_0\rangle$ produces a final state $|\Psi_{SP}\rangle = \sum c_i|v_i\rangle \otimes |P_i\rangle$.

This final state is what we call an entangled state. This does not correspond to the device having produced a single precise result; in contrast, the device itself is in a state corresponding to the superposed results. This situation is illustrated by the famous Schrödinger's cat experiment. In this experiment, a cat is locked in a box with a vial of poison, which may or may not be broken depending on the disintegration of a radioactive atom. As the atom is in a superposed "disintegrated/not disintegrated" state, this superposition is transmitted to the

3 Readers unfamiliar with these concepts should not be alarmed. Consider that a tensor product of two vectors is a vector that is a sort of juxtaposition of these two vectors.

"broken/not broken" state of the vial and the "dead/alive" state of the cat. Therefore, after a certain time, is the cat dead or alive?

In the same vein, a famous thought experiment was also proposed by Eugène Wigner. He imagined that a friend was locked in a laboratory, where he carried out a measurement of the spin along Oz of a particle initially in a superposed state of this property. The friend had a 50% probability of finding + and a 50% probability of finding – as the result. However, after his measurement, orthodox quantum mechanics tells us that the friend must assign a reduced state to the particle ($|+\rangle$ if he found + as the result and $|-\rangle$ otherwise) and no longer a superposed state. Wigner, who has remained outside the lab, can consider the larger system (lab, friend, particle, measuring apparatus) as a system in itself, on which no measurements have been made. He must therefore attribute to it an entangled state, consisting of the sum of tensor products of the type we mentioned earlier, which, for the system, is totally different from the state assigned to it by his friend. Who is right? Wigner or his friend?

We therefore find ourselves in possession of two descriptions, each just as legitimate as the other, but which in no way give the same result. Of course, in practice, when we use a good measuring device, we obtain a precise result, and we never face a device in a superposed state (any more than with a dead–alive cat). However, this observation is empirical and provides no precise theoretical criteria to explain why, in this case, we need to use the reduction principle and not apply Schrödinger's equation to the overall system. If we want to simplify the situation, we can say that a measurement is nothing other than an interaction between a device and a system, meaning an interaction between two systems, since the measuring device is itself a physical system. The interaction between two physical systems is governed by Schrödinger's equation, and there is no reason to project the superposed state of one of the systems onto a particular eigenvector of the measured observable. The interaction between two electrons, for example, produces no precise observable value. The two electrons will interact, but there will be no measurement. This description is therefore unsuitable. We could then suggest that the measurement comes from the macroscopic nature of the device. However, on the one hand, the microscopic/macroscopic distinction is difficult to make rigorously and, on the other hand, if we assume that quantum mechanics is universal, it would be awkward to restrict its field of application exclusively to microscopic objects. Nevertheless, we shall see that Copenhagen's interpretation contains some of this idea.

In fact, no satisfactory solution to this problem has been proposed within the framework of an unmodified quantum formalism, which is why a considerable number of interpretations are still on the table.

1.3. Interpretations

Quantum mechanics was born in 1925 with two totally different and seemingly contradictory formalizations: Werner Heisenberg's matrix mechanics and Erwin Schrödinger's wave mechanics. Sometime later, John von Neumann and Paul Dirac showed that these two formulations are in fact equivalent and developed a formalism that unifies the two. This is the one used today, which has been briefly presented above. A question arises from the outset: what does the state vector $|\psi\rangle$ (often referred to as the wave function, in reference to wave mechanics) represent? Should it be seen as a real wave propagating in space? Does it represent the actual physical state the system is in, or simply the physicist's knowledge of it? Max Born was the first to understand that it can be understood as a "probability wave" that calculates the probability of obtaining a given value during a measurement. However, as we observed above, it is extremely difficult, if not impossible, to provide a precise and rigorous definition of what a measurement is in the context of quantum formalism. When should we use Schrödinger's equation to describe the evolution of a system, and when is it legitimate to apply the wave packet reduction principle? The many ways of answering these questions have given rise to multiple "interpretations", each associated with a different vision.

1.3.1. *Changes in formalism*

Faced with the problem of measurement, some physicists have attempted to propose modifications to the original formalism of quantum mechanics. This is understandable, given the difficulty of precisely defining what a measurement is. Broadly speaking, there are two classes of such attempts.

The first approach, whose motivation dates back to the very beginnings of quantum mechanics, was first proposed by Louis de Broglie (who played a major role in the early days of quantum mechanics) in 1927 and was developed from 1952 onward by David Bohm. For this reason, it is often referred to as dBB (de Broglie–Bohm) theory. It considers that the standard formalism is incomplete and that it can be completed by assuming that particles are accompanied by a wave that guides them. In this case, the positions (improperly called hidden variables) of the systems have defined properties at all times, contrary to what the standard formalism says. In this theory, the problem of measurement disappears. Moreover, its empirical predictions are exactly the same as those of standard quantum mechanics. However, it is not widely used by physicists, partly because the calculations are more complicated but also because of a number of conceptual problems (such as the difficulty of developing a relativistic version), the details of which we cannot cover here.

The second approach modifies the Schrödinger equation by adding a nonlinear term that explicitly causes the wave function to collapse at random. These theories are sometimes referred to as "spontaneous collapse" theories. The best known of these is the GRW theory developed in 1986 by Ghirardi, Rimini and Weber. The way in which spontaneous collapse is invoked enables several different theories to be created, such as the version developed by Diosi and Penrose, which links collapse to gravitation. Unlike the dBB theory, these theories have empirical consequences that may differ slightly from usual quantum mechanics, such as the non-conservation of energy. This is advantageous because, in principle, it is possible to test these theories and eventually disprove them. However, the experimental effects are slightly beyond the limits of our current detection capabilities, and no conclusive experiments have yet been carried out.

These interpretations will not be studied in this book and are only mentioned here to enable the reader to realize that the problems posed by quantum mechanics have also given rise to attempts to modify this particularly well-confirmed theory.

In what follows, only interpretations that conform to the standard formalism will be discussed.

1.3.2. *The Copenhagen interpretation*

For decades, this has been the prevailing way of understanding quantum mechanics. Even today, many physicists put it forward when asked about their favorite interpretation. Although it is the work of Bohr and Heisenberg, it is not monolithic. There are different descriptions, partly because the two authors did not always agree on certain points but also because of the obscurity of some of Bohr's writings, which evolved in the way he presented it.

In the initial version, according to Bohr, the observation of a quantum object involves an uncontrollable physical interaction with a classical measuring apparatus that affects both systems. Following a famous paper by Einstein, Podolski and Rosen (which we cannot detail here but which will be presented in Chapter 6 on ConSol), Bohr's position evolved toward a more radical break with the view that microscopic systems are classical objects with definite properties. He argued that a measurement is an interaction between the system and a classical macroscopic apparatus, and that the property measured does not belong to the system itself but rather to the whole system and apparatus. He went so far as to say that the very meaning of a property depends on the experimental device used and that speaking of a property of a system without reference to the apparatus used to measure that property is meaningless. Since the physicist who uses a macroscopic device to measure a physical property on a system knows perfectly well what they are doing, there is no ambiguity, and the reduction postulate must be applied. This is where Bohr introduced his famous principle of complementarity, according to which a system can appear as either a particle or a wave, depending on the experiment. This mature version of Bohr's ideas is often considered to eliminate all problems of interpretation in quantum mechanics.

Although this point of view works perfectly well in practice, it does leave open the question of what a macroscopic device is. The microscopic versus macroscopic distinction is not easy to rigorously define. Another difficulty here is that quantum mechanics is assumed to be universally valid and should be applied to any physical system, whether microscopic or macroscopic. In reality, however, a closer analysis of Bohr's position reveals that the classical behavior of the device was not directly related to its macroscopic aspect but rather was related to the use the observer wanted to make of it. Bohr had pragmatic reasons in mind. If the device is used to make an observation on another system, it behaves classically. However, if the observer wants to make a measurement on the device via another device, it has quantum behavior. This interpretation gives the observer an important role (without saying so), even though it claims to be objective. Moreover, in this interpretation, the reduction in the wave packet is a real physical phenomenon. This raises inconsistencies linked to the measurement of intricate particles and the non-locality this leads to.

In Chapter 2, Thomas Ryckman analyzes in detail the Copenhagen interpretation and, above all, Bohr's ideas. These flaws in the Copenhagen interpretation have led physicists to propose alternative interpretations.

1.3.3. *The dualist interpretation of von Neumann, Wigner, London and Bauer*

These researchers reported that, at the end of any measurement, an observer takes note of a precise result and that at this point, we can truly be sure that a measurement has been made. Therefore, the observer clearly plays an essential role in the measurement process. The natural consequence is to assume that the observer's consciousness is responsible for reducing the wave packet and that without a conscious observer being aware of a result, there can be no measurement. To avoid falling back into the measurement problem described above, with the observer's consciousness entangled with the measuring apparatus and the system being measured, we have to postulate that consciousness does not obey quantum mechanics and is therefore not reducible to a physical system. The result is a dualism in which consciousness, although non-physical, nonetheless has physical action on the system by projecting the superposed state toward a proper state of the measured observable. This dualism is considered unacceptable by most physicists, and this interpretation, although proposed by some of the greatest thinkers of the time, has never convinced many physicists.

1.3.4. *Everett's interpretation*

Hugh Everett was concerned with the reduction postulate, which he saw as an inconsistency in von Neumann's formulation of quantum mechanics. Furthermore, he noted that in the Copenhagen interpretation of standard theory, the observer must be treated as an external system. This prevents the Universe from being described as a whole. He therefore proposed dispensing with this postulate altogether. In his interpretation (sometimes referred to as the many-worlds interpretation), there is no collapse, and the universal wave function (including observers) evolves uniquely under the Schrödinger equation.

In Chapters 3 and 4, Lev Vaidman and Jeffrey Barrett analyze some of the specific difficulties of Everett's interpretation, and we limit ourselves here to a very brief description.

In the description of the measurement problem, in an experiment designed to measure the value of a certain property of a system in a superposed state through a measuring device, the use of Schrödinger's equation applied to the set "system, measuring device, observer" results in an entanglement of the system, the device and even the observer. In the simplest case, where the initial superposition of the system concerns two possible values (to be noted as + and –), the state of the system is as follows:

$$|\Psi_S\rangle = \alpha\,|+\rangle_S + \beta\,|-\rangle_S$$

After interaction with the measuring device and the observer, the result is an entangled global wave function of the following form:

$$|\Psi_{SPO}\rangle = \alpha\,|+\rangle_S\,|\uparrow\rangle\,|O_+\rangle + \beta\,|-\rangle_S\,|\downarrow\rangle\,|O_-\rangle$$

where $|\uparrow\rangle$ is the state of the device having recorded the + value and $|O_+\rangle$ the state of the observer having observed the + value (the same applies for –).

We can thus see that the observer is in an entangled state that does not correspond to having observed a definite result. In the absence of the reduction postulate (which Everett rejects), it therefore remains to explain how it is possible for the observer's experience to correspond to the observation of a definite result and not to a superposition of results. To do this, Everett imagines that for each such measurement, the observer splits into two observers, each seeing one of two possible outcomes. There is thus a multiplication of observers (and associated worlds) into a number equal to the number of possible results. In each of these worlds, the observer residing there sees a defined result.

Beyond its surprising aspect, this interpretation raises a number of questions that we do not address here. We leave it to the reader to refer to the relevant chapters in the book.

1.3.5. *Relational interpretation*

This interpretation, developed by Carlo Rovelli (one of the founders of loop quantum gravity theory), holds that the state of a system has no meaning in absolute terms but only in relation to another system. The other system is then called the observer, and it is assumed that any interaction between two systems is equivalent to a measurement of one system by the other. An electron can play the role of an observer in relation to another electron.

The problem here is that there is confusion between "correlation" and "measurement". When two systems interact, a correlation is established between them (or, more precisely, between the values possessed by the observables attached to the systems). However, this is not enough to obtain a measurement. A measurement is not a simple correlation but rather the selection of a single possibility from all those possible in the correlation.

Beyond the absurdity (pointed out by many) of considering that one electron can observe another electron, it is surprising that this interpretation enjoys a certain prestige and is quoted quite often, even though it is based on a rather unforgivable confusion. Recent attempts to resolve these shortcomings have been proposed, but we cannot delve into their description here.

1.3.6. *QBism*

In the above interpretations, the wave function (or state vector) is assumed to refer directly to the actual physical state of the system. Some interpretations challenge this link by assuming that the wave function does not represent the actual state of the system but only refers to our knowledge of it. This type of interpretation is often referred to as epistemic.

QBism is an interpretation of this type. It was originally developed by Fuchs, Schack and Mermin. QBists believe that the primitive concept of experience is the central subject of science. The word "experience" is to be understood here as "personal experience", meaning for each agent what the world has induced in them in the course of their life. This is partly an instrumentalist position, since, for QBists, quantum mechanics is merely a tool enabling any agent to calculate their probabilistic expectations for their future experience from knowledge of the results of their past experience. QBists adopt the subjective interpretation that probabilities represent an agent's degree of belief and, therefore, are unique to that agent. QBism considers that the entities of formalism (wave functions, probabilities) are relative to a particular agent, who uses the formalism to make predictions about the experiments they will carry out. QBists reject the idea that the quantum state of a system is an objective property of that system. It is merely a tool for assigning a subjective probability to the agent's future experience. Quantum mechanics therefore says nothing directly about the "outside world". A measurement (in the usual sense) is just a special case of what QBism calls experience, namely, any action performed by an agent in their external world. A measurement does not reveal a preexisting state of affairs but creates a result for the agent. The way QBists solve the measurement problem is very simple: they assume that the direct internal perception of their private experience is the only phenomenon accessible to an agent that they do not model with quantum mechanics and that the agent's perception is the result of the experience. There is therefore no longer any ambiguity about the use of the reduction postulate, which says nothing about the "real state" of the system being measured but is nothing more than the updating of the agent's state assignment on the basis of their experience. In particular, there is no measurement when there is no agent.

QBism has undergone significant development in recent years but leaves a number of questions about its ontology unanswered. What it tells us about reality and how to understand how an agent creates a result of their own through experience is ambiguous and needs to be clarified. However, it shares a number of points in common with ConSol, developed by Hervé Zwirn, which will be presented in Chapter 6.

1.3.7. *Pragmatism*

The pragmatist interpretation of quantum mechanics is part of a philosophical approach according to which theory does not necessarily describe an objective reality independent of the observer but rather limits itself to providing effective tools for predicting and manipulating observed phenomena. Unlike realist interpretations, which attempt to assign an ontological existence to quantum states, pragmatism sees quantum concepts as useful constructs for organizing our experience and guiding our actions in the world.

This approach stems from the thinking of philosophers such as William James and John Dewey, who emphasized the value of theories in terms of their practical effectiveness rather than their correspondence with an ultimate reality. In quantum mechanics, this means focusing on the results of measurements and the efficiency of calculations rather than on the profound nature of the probability wave or the principle of superposition.

The pragmatist interpretation focuses on the role of the experimenter and on the fact that quantum mechanics does not describe independent objects but rather the relationships between our actions and the results obtained. As such, the concept of the wave function is seen not as a real physical entity but as a mathematical tool for calculating probabilities.

This approach avoids certain paradoxes, such as the Schrödinger cat paradox, by considering that the quantum state is not an intrinsic property of the system but rather a representation of our knowledge and expectations of an experiment. From this point of view, the question of the collapse of the wave function becomes less problematic: it is not an instantaneous physical change but simply an update of the available information.

In summary, the pragmatist interpretation of quantum mechanics rejects any metaphysical speculation on the ultimate nature of reality and insists that theory must be understood as an operative framework that facilitates our interactions with the world. It is part of a broader perspective in which the value of a scientific theory is measured not by its correspondence with absolute reality but by its ability to guide our predictions and experimental interventions. Richard Healey presents his version of pragmatist interpretation in this book (Chapter 5).

1.3.8. *Convivial solipsism*

ConSol is an interpretation I developed following a remark by Bernard d'Espagnat, who was not satisfied with the infinite multiplication of the number of worlds in Everett's interpretation. He therefore set out to solve the problem of measurement by dispensing (like Everett) with the postulate of wave packet reduction while retaining a single world. To this end, the observer was placed at the very heart of the measurement operation. A measurement is not a physical action on the system but rather the selection of a certain "point of view" by the observer's perception, which filters out the different components of a superposed state. Reality then becomes a notion relative to each observer and is no longer common to all, although it is impossible to realize this individually. This will be described in detail in Chapter 6.

1.4. Conclusion

The aim of this short chapter was simply to provide readers with the minimum essential background to enable them to situate themselves in the more advanced chapters that make up the body of this book (in the quantum physics section). It is of course impossible to be exhaustive in this field (several volumes would not suffice), but this brief overview will give a (very small) idea of the diversity of approaches that have been developed to solve the conceptual problems posed by quantum mechanics.

2

Bohrian Realism

Thomas RYCKMAN
Stanford University, USA

Bohr is universally recognized as a progenitor and leading advocate of "complementarity". Less precisely, he is known as the *éminence grise* of the "Copenhagen Interpretation" of quantum mechanics, although recent scholarship has shown that his doctrine is far from univocal, with nearly as many variants as proponents. The intricacies of Bohr's argument, adumbrated in numerous philosophical papers over several decades are underappreciated. He maintained that quantum theory poses a "problem of observation", the solution of which provides "an epistemological lesson with bearings on problems far beyond the domain of physical science". The observational problem stems from Planck's discovery of the quantum of action: how may objectivity be retained despite the brute physical fact of exchange of energy quanta during observation? In broad epistemological consideration, this is "the problem of how objectivity may be retained during the growth of experience beyond the events of daily life". Bohr's solution holds that it is necessary to loosen conventions of language (in particular, the presupposition of a fixed subject/object distinction) familiar with ordinary usage, which is simply assumed in property attribution within classical mechanics. Relaxation of a tacit convention of a fixed subject/object distinction has far reaching consequences concerning fundamental concepts of epistemology, semantics and even metaphysics. In brief, the "epistemological lesson" from quantum theory has occasioned revisions in the notions of "observation", "objectivity", "subject" and "object", "reference", "meaning" and "physical reality". Bohr sought to provide illustrative examples of subject/object fluidity in so-called "complementary phenomena" within disciplines far removed from quantum theory, such as psychology, biology and ethics. My concern here is narrower and less textually explicit. I claim that the implications of the quantum postulate for the description of nature led Bohr implicitly to a two-tiered conception of realism analogous, but not identical, to the Kantian distinction between *phenomena* and *noumena*. In misunderstanding Bohr's often opaque, at times convoluted, argumentative style, many readers conclude that Bohr targets realism *tout court*. Hence, his views may be characterized as neo-positivist, antirealist, instrumentalist or loosey-goosey pragmatist. In what follows, I suggest that these attributions are inaccurate and that Bohr's intent, despite its obscure articulation, is primarily to *identify from within* physical science the boundaries of any "objective" representational realism (one concerning "phenomena", in Bohr's sense of the term, hence an "empirical realism") while also pointing to a realism perhaps only dimly indicated by the phenomena but lying "beyond" them, a realism that by construction is non-objectivist and nonrepresentational. Transcending current empirical access pertains exclusively to the particular "quantal features" of atomic objects revealed in the phenomena, that is, information derived from observation. In conclusion, I consider the legitimacy of positing a "reality" beyond the limits of representational realism.

2.1. The observational problem

Quantum mechanics, according to Bohr, provides "an epistemological lesson with bearings on problems far beyond the domain of physical science".[1] That lesson is contained in drawing out consequences of the "quantum postulate" ("symbolized by Planck's quantum of action")[2], a postulate expressing the "essence" of quantum theory. The term was introduced in Bohr's 1913 theory of the hydrogen atom to refer to the discontinuous series of values of energy in the transition, including emission and adsorption of radiation, between different stationary states.

In quantum mechanics, the postulate is encapsulated in two relations in which Planck's constant fundamentally appears, linking quantities (energy–time, momentum–space) that Bohr will come to term "complementary":

$$E = hf \text{ (Einstein–Planck)}$$

and

$$\boldsymbol{p} = h\sigma := h\frac{1}{\lambda}\hat{\boldsymbol{n}} \text{ (de Broglie)}$$

1 Bohr (1954, pp. 68–69).

2 Bohr (1928, p. 53).

where $\hat{n}$ is a unit vector in the momentum direction. Thus, a determinate energy is associated with a definite frequency, and a given value of momentum is associated with a particular wavelength. It is then immediately clear from the Heisenberg uncertainties that any attribution of a definite value of either energy or momentum on the left-hand side gives rise to a massive underdetermination of the value of the quantity on the other, for example, a definite momentum is associated with a wave of infinite extent. Two epistemological considerations follow. First, Bohr surely is a realist about Planck's constant coupling of both sides in these relations: "Planck's discovery of the so-called quantum of action" is the discovery of a brute fact of nature. Second, this fact of nature, the quantum of action, has created a "problem of observation", one that distinguishes quantum phenomena from those of classical physics and the events of everyday life. Planck's discovery implies a "wholeness" foreign to everyday experience and to classical physics, where it is possible to distinguish sharply between the behavior of objects and the means of observation. Hence, the quantum of action can be ignored. "The indivisibility of the quantum of action [that] requires an essential element of discontinuity in the description of atomic phenomena" reflects a limit, imposed by nature, "on the possibility of speaking about phenomena as existing objectively".[3] This limit gives rise to "the problem of observation". The problem is clearly outlined in the 1927 "Como" lecture:

> On the one hand, the definition of the state of a physical system, as ordinarily understood, claims the elimination of all external disturbances. But in that case, according to the quantum postulate, any observation will be impossible [...] On the other hand, if in order to make observation possible we permit certain interactions with suitable agencies of measurement, not belonging to the system, an unambiguous definition of the state of the system is naturally no longer possible...[4]

This tension between "definition" ("as ordinarily understood") and observation has the implication that "no independent reality in the ordinary physical sense can [...] be ascribed to the phenomena [or] to the agencies of observation" (ibid.). The epistemological/conceptual challenge is then how to bring observation and "unambiguous definition" into harmony. It is essentially a problem of "objective description", or, as Bohr preferred, "objective description of experience". The observational problem sets the stage for a far-reaching "epistemological lesson" with ramifications beyond physical science.

---

3 Bohr (1929a, pp. 107 and 114).

4 Bohr (1928, p. 580).

2.2. Objectivity

> We shall […] be concerned here […] especially with the problem of how objectivity may be retained during the growth of experience beyond the events of daily life.[5]

The observational problem, characterized here, poses another problem: "the problem of how objectivity may be retained during the growth of experience beyond the events of daily life". The familiar definition of objective knowledge as an accurate representation of what is outside the mind presupposes what is meant by "accurate representation". In search of an independent test of "accuracy of representation", the latter term could be unpacked as "objective description of observable facts", but then the definition of objective knowledge becomes woefully circular. Somehow, criteria of "objective description" must be further specified. Bohr argues that any further specification of criteria of objective description requires that we recognize the previous criteria are formulated within a conceptual framework proving "too narrow" to accommodate "the growth of experience". In short, we must realize that:

> All knowledge presents itself with a conceptual framework adapted to account for previous experience and that any such frame may prove too narrow to comprehend new experiences (ibid).

Without prejudice either to Bohr or his critics, we may assume that, by conceptual framework, Bohr intends to address what American philosopher Donald Davidson termed a "conceptual scheme", that is, "a way of organizing experience, a system of categories giving form to the data of sensation, a point of view from which individuals, cultures or periods survey the passing scene".[6]

Admittedly, an extremely broad characterization, Davidson nonetheless invokes the relativization of simultaneity as an illustrative instance of conceptual change. This, of course, is an example repeatedly used by Bohr. As the Newtonian concept of "simultaneity" has been adequate for objective description of experience in prerelativistic physics but not for special relativity, Bohr argued that the conceptual framework of classical physics has proven "too narrow" to accommodate objective description of the phenomena of atomic physics. Hence, the problem of retention of objectivity within this new domain of experience arises.

5 Bohr (1954, p. 67).

6 Davidson (1973–1974, p. 5).

Bohr's proposed solution to the problem of retention of objectivity is twofold: (1) to identify the framework proving "too narrow" and, indeed, the reason for this, and (2) to expand or transform the conceptual framework in such a way as to accommodate objective description of the "new experiences" in atomic physics. It is couched in the language of measurement, observation and interaction. It consists of giving principled reasons to why it is necessary to modify the realist notion of objectivity as the accuracy of representation presumed in classical physics (i.e. correct state assignment of preexisting values to properties independently of measurement), to a novel notion of objectivity, or, as Bohr preferred, of "objective descriptions of experience" in the domain of atomic physics. The innovative step here will be Bohr's relaxation of the rigid subject/object distinction lurking behind, as presupposed, the realist notion of objectivity as the accuracy of representation of objects existing independently of observational interaction and thus independent of the existence of the quantum postulate. This will lead Bohr's redefinition of the term "phenomena".

2.3. Subject and object

> Our present QM formalism is a peculiar mixture describing, in part, the laws of nature and, in part, incomplete human information about nature – all scrambled together by Bohr into an omelette that nobody has seen how to unscramble. However, we think that the unscrambling is a prerequisite for any further advance in basic physical theory…[7]

"Objective description" presupposes or rather "demands the contraposition of subject and object"[8]. Indeed, "differentiation between subject and object" underlies "the general limits of man's capacity to create concepts"[9]. It is an indispensable distinction, although usually only implicit, in ordinary language usage. Familiar speech acts presuppose it and are straightforwardly understood: a subject refers to, speaks "about", attributes of objects by predicating their properties. In simple grammatical subject–predicate form, properties are attributed to objects by a speaker within a common perceptual reality. Canonically, the attention of interlocutors is directed to a commonly perceptible particular object situated in a commonly shared space and time. If I refer to a particular elm tree, I can identify the tree to those in my visual field by its position in relation to readily observable macroscopic states of affairs, "in the northeast corner of the yard" or "behind the garage". This

7 Jaynes (1990, p. 387).

8 Bohr (1960, p. 14).

9 Bohr (1929b, p. 96).

paradigmatic use of everyday language occurs in a commonly shared space and time. It is the basis for a "pictorial" (as Bohr preferred) understanding of predication or attribution, that is, a semantics of word–world relations supposedly necessary for understanding and for a conception of truth of assertions as accuracy of representation.[10]

Classical physics simply assumes this fixed subject/object distinction of everyday language with its word-world semantics:

> While in the mechanical [i.e. classical] conception of nature, the subject-object distinction was fixed, room is provided for a wider description through the recognition that the consequent [sic: consistent] use of our concepts requires a different placing of such a separation.[11]

Bohr stressed that the classical meanings of the kinematical and dynamical concepts "position", "direction of motion", "cause", "effect", etc. are grounded in ordinary usage. On any particular occasion, the meaning of these terms are always contextually bound, yet they retain a universally understood significance as pertaining to objects situated in space and time, at rest or undergoing change in motion on account of perceptible or inferable causes. By default, such objects with their preexisting properties are regarded as isolated or autonomous, without the need to consider means or methods of observation or indeed knowability.[12] Hence, the classical meaning of kinematic and dynamical terms rests on the sharp subject/object distinction, where it is "an assumption inherent in ordinary conventions of language, that it is possible to distinguish sharply between behavior of objects and the means of observation".[13]

---

10 A semantics of "the cat is on the mat" variety, in which terms pick out and are associated with readily perceptible objects and relations, the ontology of what philosophers call "middle-sized dry goods". For an extensive discussion, see Wilson (2006).

11 Bohr (1955, pp. 92–93).

12 Bohr (1958b, p. 59): "The description of ordinary experience presupposes the unrestricted divisibility of the course of the phenomena in space and time and the linking of all steps in an unbroken chain in terms of cause and effect. Ultimately, this viewpoint rests on the fineness of our senses, which for perception demands an interaction with the objects under investigation so small that in ordinary circumstances it is without appreciable influence on the course of events. In the edifice of classical physics, this situation finds its idealized expression in the assumption that the interaction between the object and the tools of observation can be neglected, or, at any rate, compensated for".

13 Bohr (1939, p. 269).

As Bohr also observed, the kinematical and dynamical terms of classical physics are idealizations. The idealizations are evident in that they rest upon the presupposition of a notion of "observation" from everyday usage where a position can be attributed to a classical particle without suggesting interaction with a device that could measure a position with respect to a given reference body. This is possible through the introduction of space and time coordinates. These can be unambiguously defined in a given frame of reference, allowing the circumvention of mention of any particular conscious subject/observer attributing properties, an implicit reference that "infiltrates"[14] everyday language usage. While implicitly following ordinary language usage, the concept of position is quantified through assigning object coordinate values in a given frame. The attribution of a definite position to an object no longer involves the space and time of a specific macroscopic state of affairs but locates the object within a particular frame of spatiotemporal reference (an abstract concept), assigning it a particular *n*-tuple term in that frame.[15] The representational or referential character of ordinary language is retained except that the word–world relation has been idealized to the relational form, the abstract term-world. In this type of word–world correlation, the cognizing, experiencing subject has vanished altogether, with the result that the predicative description of the object meets the objectivity constraint of "accuracy of representation". In summary, the linguistic convention of a fixed subject/object distinction undergirds the attribution of properties to objects. It is common to "events in daily life" and classical physics and is the basis of the conception of objectivity as grounded in accurate representation of the object.

We refer to this conception of objectivity as the accuracy of representation as "representationalist".[16] This concept has several well-known aspects. One is the idea of a semantic characterization of truth familiar in philosophy from Aristotle's *metaphysics* (Γ, 7, 27): "to say of what is that it is; to say of what is not that it is not, it is true". Logician Alfred Tarski constructed his famous *semantical definition* of truth on this account, noting that its "intuitive meaning and general intention" is "quite clear and intelligible".[17] The contemporary guiding metaphor of "correspondence" between

14 Bohr's term (1954, p. 68).

15 Such descriptions are also suited to the demands of relativity since they need make "no explicit reference to any individual observer" (Bohr 1958a, p. 3).

16 Following Rorty (1979, p. 11): Representationalism is "the attempt (which has defined traditional philosophy) to explicate 'rationality' and 'objectivity, in terms of conditions of accurate representation". According to Rorty, this is a "self-deceptive effort to eternalize the normal discourse of the day, and that, since the Greeks, philosophy self-image has been dominated by this attempt". Rorty's extensive diagnosis of the flaws of representationalist epistemology in this text remains unscathed by his own subsequent descent into the morass of postmodern relativism.

17 Tarski (1936, p. 155).

linguistic expression and mind-independent states of affairs is much less transparent. Early logical empiricism[18] sought to make this realist metaphor precise as a univocal coordination, a 1-1 mapping between statements and terms and objects "out there in the world" such that each well-defined linguistic entity has a determinate reference. There are many problems with this conception, not the least of which is that a bijection is a relation obtained between sets but, obviously, only one of these sets, the linguistic one, admits a precise definition. These representational presuppositions of realism are readily apparent in EPR's famous characterization of a "criterion of reality":

> If, without in any way disturbing a system, we can predict with certainty (i.e. with probability equal to unity), the value of a physical quantity, then there exists an element of physical reality corresponding to this physical quantity (original italics).[19]

Bohr immediately recognized these presuppositions that are evident in his initial response to EPR, a short note in *Nature*:

> I should like to point out that the [EPR] criterion [of reality] contains an essential ambiguity when it is applied to problems of quantum mechanics. […] a closer examination reveals that the procedure of measurements has an essential influence on the conditions on which the very definition of the physical quantities in question rests. Since these conditions must be considered as an inherent element of any phenomenon to which the term 'physical reality' can be unambiguously applied, the conclusion of [EPR] does not seem to be justified.[20]

EPR's "element of physical reality" straightforwardly assumes the fixed subject/object distinction presupposed in classical physics, enabling "objective description" independent of observational interaction.[21] The attribution of properties is idealized in two ways: (i) through precise symbolic term–object correlation and (ii) without the need to reference the circumstances or agency through which

18 Schlick (1918) and Reichenbach (1920).

19 Einstein et al. (1935).

20 Bohr (1935a, p. 65).

21 Lest we think that EPR does not align with Einstein's own philosophical views, this representationalist realism is found in other late texts of Einstein countering what he considers Bohr's subjectivist understanding of quantum mechanics (e.g. Einstein 1953, p. 6): "There is such a thing as the 'real' state of a physical system existing independently of any measurement or observation that in principle can be described by the means of expression in physics". The inverted quotes surrounding the term 'real' merely indicate Einstein's caution in assuming such a realism.

attributions of definite values are known. With these idealizations, classical physics retains the concept of an object as an autonomous bearer of determinative properties. Assignment of a physical state to an object or system is understood to provide a consistent and complete list of possessed properties. Observation or measurement simply reveals values of preexisting properties that, if changing in time, nonetheless have definite values at any given time, without regard to measurement or observation. Representationalist objectivity eliminates any taint of subjectivity by obliterating reference to any particular subject. This is emphasized in relativity theory since "no explicit reference is made to any individual observer". Einstein's term "element of reality" implicitly invokes another aspect of representationalist objectivity: in everyday life and in classical physics, representation is of objects *in* space and time. Of course, in Kant's "Transcendental Aesthetic", this is a necessary condition for the possibility of any "object of experience", that is, any object of which we can have knowledge. For Bohr, representation in space and time is the core of "pictorial" or "visualizable" accounts of objective description. Supplemented with causal and dynamical attributes, in classical physics, it is possible to form "a consistent picture of the behavior of the object under investigation". As Bohr argues, this ideal of description "when combination into a single picture is attempted" will appear contradictory if it is extended to the quantum domain. A complete "pictorial" understanding of representationalist descriptions of the causal behavior of objects, as in classical physics, can be retained only in circumstances in which the quantum of action can be neglected.[22]

In summary, the fixed subject/object distinction required for unambiguous definition of the state of a physical system goes hand in hand with a default conception of objectivity and of classical description in classical physics as realist and representationalist, resting upon accurate attribution of definite values to preexisting properties. An objective description presumently has the representational sense taken over from everyday language of the attribution of properties corresponding with an antecedently determinate reality. In fact, objectivity in classical physics can also have another sense, often, but not necessarily, regarded as a consequence of the former. By eliminating mention of any "individual subjective judgment",[23] the expressions of classical physics, expressed by mathematical symbols well defined in the Hamiltonian or other formalisms, can be unambiguously communicated. This weaker conception of objectivity as intersubjectivity will be

22 Bohr (1938, p. 104): "the very ideal of causality, which can aim only at the synthesis of phenomena describable in terms of a behavior of objects independent of the means of observation".

23 Bohr (1960, p. 10): "[...] our task must be to account for [human] experience in a manner independent of individual subjective judgment and therefore objective in the sense that it can be unambiguously communicated in the common human language".

retained by Bohr for quantum theory through redefinition of the notion of "phenomenon".

2.4. Redefining "phenomenon"

> Considering any atomic phenomenon, we must include a specification of the experimental arrangement fixing the conditions of observation, but a sharp distinction must be made between the parts of the total system that forms the apparatus and the part that is the object of observations, and which may exhibit quantal properties.[24]

The existence of the quantum postulate implies that the representationalist account of objective description customary from ordinary life and commonly assumed in classical physics fails for atomic physics. It can no longer be upheld on account of "the impossibility of distinguishing in our customary way between physical phenomena and their observation". The impossibility of strict separation between phenomena and means of observation reveals a "failure of our [customary] forms of perception", a failure connected with forced surrender of the pictorial property realist semantics presumed for objective descriptions of experience in everyday life and in classical physics as well as the rigid subject–object distinction these semantics assume. However, as the human capacity to form concepts relies on the differentiation between the subject and the object,[25] the distinction must be drawn. Accordingly, Bohr argued that there must be a reconciliation of two seemingly incompatible postulates, one stemming from the quantum postulate and the other from the requirement of objective description. The first states that the "wholeness" of the quantum postulate has shown an "impossibility of any sharp separation between the behavior of atomic objects and the interaction with the measuring instruments which serve to define the conditions under which the phenomena appear"[26]. The second postulate, apparently antagonistic, nonetheless requires "a fundamental distinction between the measuring apparatus and the objects under investigation".[27] The latter distinction is required for the attribution of

---

24 Rosenfeld (1963, p. 51).

25 Bohr (1929b, p. 96): "[...] a close connection exists between the failure of our forms of perception, which is founded on the impossibility of a strict separation of phenomena and means of observation, and the general limits of man's capacity to create concepts which have their roots in our differentiation between subject and object".

26 Bohr (1949, pp. 39–40, original italic).

27 Bohr (1958a, p. 3, original italic).

properties to objects and thus for objective description in a sense to be further specified.

Reconciliation is attained by requiring that the term "phenomenon" be redefined in the quantum context such that the description of phenomena is "exclusively to refer to the observations obtained under specified circumstances, including an account of the whole experimental arrangement"[28]. The extension of classical kinematical and dynamical concepts to describe experience in quantum physics mandates, via the aforementioned "impossibility", reference to the conditions of epistemic access to the quantum world. Epistemic access requires reference to a subject, but a Bohrian subject is a specific experimental set up or arrangement, one presuming an *inter alia* state preparation and free choice of agencies of measurement for a particular observable in virtue of which attribution of specific values of these properties to atomic objects can be ascertained. A Bohrian subject is a specific measurement context in which kinematic and dynamical concepts are well defined. Contextualizing the appropriate conditions for the attribution of properties naturally relaxes the fixed subject/object distinction presupposed by representationalist objectivity, yet Bohr insisted over and again that objectivity itself is not at all impaired. In a 1955 letter to Pauli, Bohr claimed that

> [Einstein's] 'ideal of the detached observer' may be retained [...] in quantum physics by recognizing that we are always speaking of well-defined observations obtained under specified conditions.[29]

This is just to say that the "problem of observation" prompted by Planck's discovery of the quantum of action has been solved by contextualization of conditions for objective description. A Bohrian subject meets the objectivity condition inherent in the "ideal of the detached observer" since all subjectivity is avoided by proper attention to circumstances required for well-defined use of elementary physical concepts"[30].

In the objective description of quantum phenomena, the kinematic and dynamical concepts of classical physics acquire unambiguously defined contextualized meanings since:

> An account of the functioning of the measuring instruments is indispensable to the definition of phenomena and we must, so to

28 Bohr (1949, p. 64).

29 Bohr to Pauli, March 2, 1955, in Bohr (1999, p. 573).

30 Bohr (1958a, p. 7).

> speak, distinguish between subject and object in such a way that each single case secures the unambiguous application of the elementary physical concepts used in the description.[31]

"Objective description" of quantum phenomena "can be achieved only by including in an account of the phenomena explicit reference to the experimental conditions", a proviso underscoring that our despite our realist intuitions, our conception of knowledge is inseparable from considerations of the conditions of epistemic access ("our possibilities of inquiry").[32] Consequently, the value of a property attributed to a quantum object as a result of a measurement *cannot be regarded as an intrinsic property of the object itself*; rather, only its behavior manifests observationally when it is placed in a given experimental context and when a result is recorded. This conclusion was pithily expressed in a simple sentence by John Wheeler, who regarded it both as "Bohr's point" in response to Einstein and as "the central point of quantum theory":

> No elementary phenomenon is a phenomenon until it is a registered (observed) phenomenon.[33]

Property attribution requires epistemic accessibility. This is precisely the core message of both of Bohr's 1935 responses to Einstein. The fact that our conception of knowledge is inseparable from considerations of the conditions of epistemic access ("our possibilities of inquiry") is, in fact, the general epistemological lesson taught by quantum physics. In quantum physics, of course, this implies the impossibility of ignoring "the essential wholeness of the quantum phenomena" and hence our inability to unambiguously assign attributes to objects independent of the conditions under which they are observed. A semantic consequence is that terms denoting quantum attributes are homonyms of their more robust counterparts in classical physics that, even if idealized, convey "pictures" of property attribution to objects considered independent of any epistemic access. Within Bohr's extended definition of "phenomenon", these classical terms acquire a restricted, highly contextual, meaning well-defined solely within specific experimental conditions suited for outputting a definite value of the corresponding classical kinematical or dynamical property. The attribution of a given value of a classical kinematic or dynamical quantity to the target system of interest involves "essential ambiguity" unless such attribution necessarily refers to the experimental conditions by which

31 Bohr (1955, p. 91).

32 Bohr (1960, p. 12).

33 Wheeler (1983, p. 184).

that information is obtained.[34] This is the essential claim of "complementarity", together with the accompanying lemmas of exclusivity and exhaustiveness. Complementarity does not do away with "picturability" or representationalism per se. It simply explains *why* information about all observables cannot be combined into "a single picture".[35]

The Bohrian realm of phenomena then marks out from within the context of measurement and experiment with the limits of "objective description", "picturability" and "objective knowledge", all mandating descriptions employing only elementary physical concepts derived from their everyday precursors. In this sense, Bohr can claim that complementarity shows that quantum physics is a "rational generalization" of classical physics. The requisite condition of epistemic access for "objective description" delimits the realm of phenomena as those of an *empirical realism* for quantum physics.

2.5. Realism, without representationalism

> At the start of every scientific enterprise a fundamental postulate is made that nature possesses an objective reality, independent of our sensory perception and of our means of investigation; the object of theory is to give an account of this objective reality.[36]

> There is no quantum world. There is only an abstract quantum physical description. It is wrong to think that the task of physics is to find out how nature is. Physics concerns what we can say about nature.[37]

Arguably, the aim of every natural scientist is to attempt to give accurate account of the way *nature really is*. Regardless of whether Bohr actually stated "there is no

34 Bohr (1949, p. 40): "[…] an essential ambiguity is involved in ascribing conventional physical attributes to atomic objects, as is at once evident […] where we have to do with contrasting pictures, each referring to an essential aspect of empirical evidence".

35 Bohr (1960, p. 12): "evidence obtained by different, mutually exclusive experimental arrangements, may […] at first sight appear contradictory. It is in this situation that the notion of complementarity is called for to provide a frame wide enough to embrace the account of fundamental regularities of nature which cannot be comprehended within a single picture. Indeed, evidence obtained under well-defined experimental conditions – and expressed by adequate use of elementary physical concepts – exhausts in its entirely all information about the atomic objects that can be communicated in common language".

36 d'Espagnat (1983, p. 57), quoting Messiah's *Méchanique Quantique*.

37 Bohr, as quoted in Petersen (1963, p. 12).

quantum world" (as reported by Petersen), Bohr considered "the entire [quantum mechanical] formalism, [whether matrices or wave functions] as a tool for deriving predictions of definite or statistical character".[38] It follows from his revised definition of "phenomenon" that quantum formalism can only be assigned a physical meaning in a specified experimental context. The very idea of the quantum state, which is the supposed representational target of the wave function, has been replaced by a notion that has the necessary reference "to the observations obtained under specified circumstances, including an account of the whole experimental arrangement". As a result, property attribution to phenomena in the quantum domain is completely different than in descriptions of objects of classical mechanics. Attributions of particular properties, for example, q or p, to quantum objects must be understood contextually.

Bohr's view of wave functions may be illustrated by von Neumann's account of quantum measurement. Von Neumann's assumption that the measurement state is the tensor product superposition $|S_i\rangle \otimes |M_i\rangle \otimes |O_i\rangle$ seemingly symbolizes all possible experimental set-ups for the observable physical quantities, each outcome, which is unknown at this stage, as associated with a corresponding apparatus and subject.[39] Thus, it abstracts away from any given experimental setup required to unambiguously define the terms involved in describing the measurement outcome. Obviously, von Neumann's account is not acceptable to Bohr. In fact, in 1938, Bohr responded directly to von Neumann and to the ambiguity of the subject/object distinction in von Neumann's account, that is, as to "where" the so-called "collapse of the wave function" occurs. Bohr's reply predictably maintained that interpretation of von Neumann's formal expression becomes unambiguous once a choice is made regarding the choice of experimental conditions.[40] Objective description of the phenomenon, by definition, certainly omits reference to any individual observer S. Implicit references to a subject occurs only in the particular choice of an observable to measure and in the corresponding experimental setup for which measurement M is a suitable apparatus. In general, for Bohr, superposition expressions are

38 Bohr (1948, p. 314).

39 The tensor product of state vectors for *subject S, measuring apparatus M* and *object system O*. Indices run over all possible subjects, measuring devices and outcomes.

40 Bohr (1938, p. 120). A written record of Bohr's verbal intervention after von Neumann's talk reports: "the duality [Bohr] noticed in the interpretation of the formalism of quantum mechanics was, in [Bohr's] opinion a question of choosing the most adequate description of the experiment. If we decide to include in the enumeration of the exterior conditions all the instruments which must be used for the study of the whole phenomenon, the only arbitrary factor remaining is [...] the free choice of these experimental conditions, and, apart from this freedom, the interpretation of the solution of the problem, concerning the predictions to which the phenomenon we are studying leads, is perfectly unequivocal".

understood as abstractly symbolizing many completely different, possibly mutually exclusive, phenomena, of which an unambiguous account must include a description of all relevant features of the experimental arrangement. By itself, a quantum superposition is merely part of "a mathematical formalism [...] adapted to the measuring results obtainable in the new fields of observation"[41] (Bohr 1946b, p. 125). It cannot be the objective description of any "state" prior to observation, yielding epistemic access.

A prerequisite for any realist theory of the microphysical world must allow observations and measurements to reveal information about the subclassical or quantal aspects of the atomic objects of interest. Bohr argues that quantum theory does yield such information, but what is the status of such information? It is produced only indirectly.[42] It is not yet "picturable" since this condition requires description in terms of elementary physical concepts that are well defined only for phenomena, and for that reason, alone can be communicated in a common language. It is information – conjecture, hypothesis, extrapolation from a formalism that has physical interpretation – about an inferred reality, "beyond" the phenomena, a reality that cannot yet be reliably approached quantitatively, nor as yet can be considered as embedded in a spatiotemporal framework within which elementary physical concepts admit unambiguous definition. Unquestionably, it is a reality that exists for the simple reason that it exerts an influence on phenomena. However, it is not the "reality" of representational realism with its ties to a fixed subject/object distinction, the realism of a mind-independent world whose very conception is due (largely) to a classical metaphysics requiring that distinction, a world of individual objects with real properties and local action. It is a reality somehow allowing for attested quantum phenomena such as statistical violations of the Bell inequalities and the agreeable relativistic prohibition against superluminal causal influences. This reality may yield an explanation for the mysterious appearance of three generations of matter particles in the standard model. It is seemingly implies that the laws of physics are but reflections, distorted no doubt by the conditions of epistemic access involving the quantum of action. However, its undoubted existence suggests that a realism beyond empirical realism can be maintained even after jettisoning an outmoded metaphysical notion of *the real* as comprised of objects "out there" possessing at all times definite values of properties, values that are exactly revealed by measurement or observation. There is little doubt that Bohr believed in such a transcendent reality to the extent that he did not bother to argue for it. Presumably,

41 Bohr (1946b, p. 125).

42 Bohr (1958a, p. 4): "[...] the quantal features of the phenomenon are revealed in the information about the atomic objects derived from the observations".

he thought it was merely common sensical to hold such a belief.[43] Therefore, in numerous places, he contemplated "the exploration of harmonies which cannot be comprehended in the pictorial conceptions adapted to the account of more limited fields of physical experience". Indeed, if the above presentation of Bohr is nearly correct, his philosophical project can be encapsulated precisely in an attempt to show how the discovery of the quantum of action radically transforms epistemology, semantics and metaphysics in our quest to retain objectivity in the growth of human experience beyond the events of everyday life and classical physics.

2.6. References

Bohr, N. (1928). The quantum postulate and the recent development of quantum theory. In *The Philosophical Writings of Niels Bohr. Volume I: Atomic Theory and the Description of Nature*. Ox Bow Press, Woodbridge, CT.

Bohr, N. (1929a). The atomic theory and the fundamental principles underlying the description of nature. In *The Philosophical Writings of Niels Bohr. Volume I: Atomic Theory and the Description of Nature*. Ox Bow Press, Woodbridge, CT.

Bohr, N. (1929b). The quantum of action and the description of nature. In *The Philosophical Writings of Niels Bohr. Volume I: Atomic Theory and the Description of Nature*. Ox Bow Press, Woodbridge, CT.

Bohr, N. (1935a). Quantum mechanics and physical reality. *Nature*, 136, 65. doi: 10.1038/136065a0.

Bohr, N. (1935b). Can quantum-mechanical description of physical reality be considered complete? In *The Philosophical Writings of Niels Bohr. Volume IV: Causality and Complementarity*, Faye, J. and Folse, H. (eds). Ox Bow Press, Woodbridge, CT.

Bohr, N. (1938). The causality problem in quantum physics. In *The Philosophical Writings of Niels Bohr. Volume IV: Causality and Complementarity*, Faye, J. and Folse, H. (eds). Ox Bow Press, Woodbridge, CT.

Bohr, N. (1939). Natural philosophy and human cultures. *Nature*, 143, 268–272. doi: 10.1038/143268a0.

Bohr, N. (1946a). Newton's principles and modern atomic mechanics. In *The Philosophical Writings of Niels Bohr. Volume IV: Causality and Complementarity,* Faye, J. and Folse, H. (eds). Ox Bow Press, Woodbridge, CT.

---

43 Unlike d'Espagnat (1983, 1995), Bohr did not argue for a transcendent reality "behind" the empirical reality of phenomena. Opinions may differ as to how successful are d'Espagnat's arguments for a "veiled reality". My view is that they are merely suggestive. What I have tried to show here is that Bohr implicitly characterizes such a reality as outside of representation by elementary concepts of physics, which are unambiguously defined only in the context of quantum phenomena.

Bohr, N. (1946b). On the problem of measurement in atomic physics. In *The Philosophical Writings of Niels Bohr. Volume IV: Causality and Complementarity*, Faye, J. and Folse, H. (eds). Ox Bow Press, Woodbridge, CT.

Bohr, N. (1948). On the notions of causality and complementarity. *Dialectica*, 2(3–4), 312–319.

Bohr, N. (1949). Discussion with Einstein on epistemological problems in atomic physics. In *The Philosophical Writings of Niels Bohr. Volume II: Essays 1933–1957 on Atomic Physics and Human Knowledge*. Ox Bow Press, Woodbridge, CT.

Bohr, N. (1954). The unity of knowledge. In *The Philosophical Writings of Niels Bohr. Volume II: Essays 1933–1957 on Atomic Physics and Human Knowledge*. Ox Bow Press, Woodbridge, CT.

Bohr, N. (1955). Atomic theory and human knowledge. In *The Philosophical Writings of Niels Bohr. Volume II: Essays 1933–1957 on Atomic Physics and Human Knowledge*. Ox Bow Press, Woodbridge, CT.

Bohr, N. (1958a). Quantum physics and philosophy: Causality and complementarity. In *The Philosophical Writings of Niels Bohr. Volume III: Essays 1958–1962 on Atomic Physics and Human Knowledge*. Ox Bow Press, Woodbridge, CT.

Bohr, N. (1958b). The Rutherford memorial lecture. In *The Philosophical Writings of Niels Bohr. Volume III: Essays 1958–1962 on Atomic Physics and Human Knowledge*. Ox Bow Press, Woodbridge, CT.

Bohr, N. (1960). The unity of human knowledge. In *The Philosophical Writings of Niels Bohr. Volume III: Essays 1958–1962 on Atomic Physics and Human Knowledge*. Ox Bow Press, Woodbridge, CT.

Bohr, N. (1999). *Niels Bohr Collected Works. Part V, Selected Correspondence*. Elsevier, Amsterdam.

Davidson, D. (1973–1974). On the very idea of a conceptual scheme. *Proceedings and Addresses of the American Philosophical Association*, 47, 5–20. doi: 10.2307/3129898.

Einstein, A. (1953). Einleitende Bemerkungen über Grundbegriffe. In *Louis de Broglie: Physicien et Penseur*. Éditions Albin Michel, Paris.

Einstein A., Rosen, N., Podolsky, B. (1935). Can quantum mechanical description of reality be considered complete? *Physical Review*, 47, 777–780. doi: 10.1103/PhysRev.47.777.

d'Espagnat, B. (1983). *In Search of Reality*. Springer, New York/Berlin.

Jaynes, E.T. (1990). Probability in quantum theory. In *Complexity, Entropy and the Physics of Information*, Zurek, W. (ed.). Westview Press, Boulder, CO.

Petersen, A. (1963). The philosophy of Niels Bohr. *Bulletin of the Atomic Scientists*, 19(7), 8–14.

Reichenbach, H. (1920). *Relativitätstheorie und Erkenntnis A Priori*. J. Springer, Berlin.

Rorty, R. (1979). *Philosophy and the Mirror of Nature*. Princeton University Press, Princeton.

Rosenfeld, L. (1963). Niels Bohr's contribution to epistemology. *Physics Today*, 16(10), 47–54.

Schlick, M. (1918). *Allgemeine Erkenntnistheorie*. J. Springer, Berlin.

Tarski, A. (1936). The concept of truth in formalized languages. In *Logic, Semantics, Metamathematics. Papers from 1923–1938*. Clarendon Press, Oxford.

Wheeler, J. (1983). Law without law. In *Quantum Theory and Measurement*, Wheeler, J.A. and Zurek, W. (eds). Princeton University Press, Princeton.

Wilson, M. (2006). *Wandering Significance. An Essay on Conceptual Behavior*. Oxford University Press, New York.

3

The Many-worlds View of Quantum Mechanics

Lev VAIDMAN[1,2]

[1] *School of Physics and Astronomy, Tel-Aviv University, Israel*

[2] *Institute for Quantum Studies, Chapman University, Orange, USA*

A particular version of the many-worlds interpretation (MWI) is presented. I argue that the only ontology of quantum mechanics is the universal wavefunction following unitary deterministic evolution. The other part of the theory postulates connecting experiences of agents in multiple parallel worlds with branches of this wavefunction.

A few decades ago, I spent a day of discussions with David Bohm. I remember telling him that I liked his theory (Bohm 1952) because it is a candidate for an ontological theory that explains everything. David's reply was that this is nonsense: humans are limited, and we will never find such a theory, only a progressively better approximation of it. Many interpretations agree with Bohm's approach which asserts that quantum physics is only capable of telling us how to predict the outcomes of measurements we perform, not what nature is. My colleagues often start teaching students quantum mechanics by saying that they only can explain how to use it, not what it means. By working on foundations of quantum mechanics, starting with my PhD studies with Yakir Aharonov, I feel that we can be more optimistic.

I want to believe that today we understand nature. I do not try to understand why our world happened to be this particular one and not another. I only try to understand how our world works. Beyond predicting the outcomes of the measurements, we also want to explain the process of measurements. The task is to identify, in all that we know, sets of similar situations which can be considered as preparations of measurements, and to find a theory that explains the common patterns of behavior in these situations, the results of measurements. At the end of the 18th century, there was a feeling that science had reached a satisfactory solution for this task. However, this famous quote from Kelvin (1901) shows that this feeling was very wrong:

> The beauty and clearness of the dynamical theory, which asserts that heat and light are modes of motion, is at present obscured by two clouds.

Physics was completely changed by the two clouds that Kelvin was referring to – the theory of relativity and quantum mechanics. Today, we have a much better reason to believe that we understand nature: experimental data and predictions of our current theory agree with unprecedented precision. We do not have a satisfactory quantum explanation of gravity, which the physics community expects, but this fact is irrelevant to explain the experiments that we perform.

In spite of the extraordinary success of science in explaining the results of experiments, the view that today we understand nature is not popular. One sociological reason regards remembering being wrong in 1900 when asserting that we understood nature, when we had no quantum and relativistic theories. Still, maybe the main reason is that, paraphrasing Lord Kelvin, the beauty and clearness of the dynamical theory is at present obscured by one big cloud: the measurement problem.

Quantum mechanic equations describing quantum measurements have terms corresponding to all possible outcomes, but in the laboratory, we always observe only one of them. How do we resolve this difficulty? Physicists were looking for alternative equations that yield one outcome, like in all situations described by classical physics from the time of Lord Kelvin, but without success. Another approach is to accept the equations of evolution of quantum states but add equations that remove all terms appearing in measurement procedures, except the one that corresponds to the observed result. Von Neumann postulated that all other terms disappear without providing any mechanism in mathematical terms. To justify this ad hoc and vaguely defined postulate, he showed that we would not be able to see a difference between a wide range of mechanisms erasing the other terms. Among the physical proposals for such a mechanism (Pearle 1976; Ghirardi et al. 1986; Diósi 1987; Penrose 1996), most also have an ad hoc character. Instead of changing the evolution of quantum states, David Bohm suggested adding the ontology of particle positions (Bohm 1952), which singles out one result for each quantum measurement.

I understand the motivation for having, in theory, only the measurement results we observe in a laboratory, but since all current proposals achieving this are radically different from all other aspects of physics that are tremendously successful in explaining all that we observe, I cannot accept these proposals. The main issue is that they all introduce action at a distance (see Vaidman (2024b)). I prefer to keep physics clear and elegant without non-local actions. This leads me to abandon the von Neumann process of type I, to reject modifications of standard quantum mechanical equations and to avoid adding new ontology like Bohmian positions of particles. There is a price: accepting a multiplicity of worlds corresponding to all measurement outcomes.

The following is an example to demonstrating the MWI and action at a distance of a single-world interpretation. The principle of superposition is well established for microscopic objects like atoms (see Figure 3.1(a)). If there are two stable states of an atom, the ground state $|g\rangle$ and a particular excited state $|e\rangle$, then the atom can be in a superposition state:

$$\frac{1}{\sqrt{2}}(|g\rangle + |e\rangle). \qquad [3.1]$$

The MWI states that the superposition principle is universally valid, not just for microscopic systems, but for cats, people, measuring devices, etc. So, if we have the state of a macroscopic pointer measuring the ground state of the atom, $|G\rangle_{MD}$, and we have the state of a macroscopic pointer measuring the excited state of the atom, $|E\rangle_{MD}$, then the superposition state:

$$\frac{1}{\sqrt{2}}(|E\rangle_{MD} + |G\rangle_{MD}), \qquad [3.2]$$

exists too (see Figure 3.1(b)).

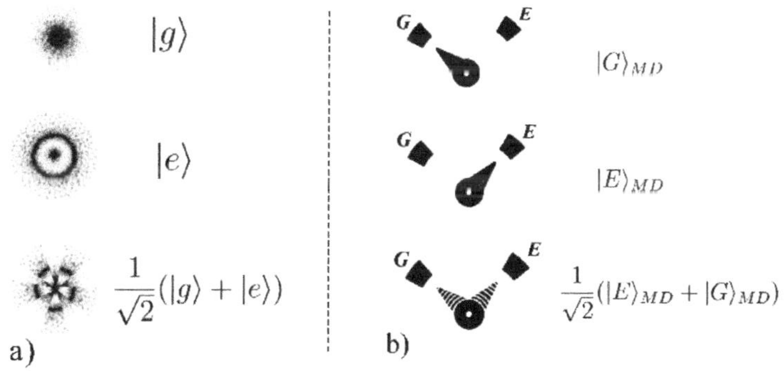

Figure 3.1. *Superposition principle: (a) If an atom can be in state $|g\rangle$ and it can be in state $|e\rangle$, then it can be in state $\frac{1}{\sqrt{2}}(|g\rangle + |e\rangle)$. (b) If a pointer of the measuring device can be in state $|G\rangle_{MD}$ and it can be in state $|E\rangle_{MD}$, then it can be in state $\frac{1}{\sqrt{2}}(|G\rangle_{MD} + |E\rangle_{MD})$. For macroscopic measuring devices, the principle holds in the framework of MWI, but it does not hold in the frameworks of single-world interpretations*

The 2022 Nobel Prize in Physics for the experimental demonstrations of non-local features of quantum mechanics was apparently triggered by "loophole-free" Bell-type experiments with, arguably, the most convincing results in Munich (see Rosenfeld et al. (2017)).

These experiments show that quantum superposition holds for two separated in space (400 m in the Munich experiment) atoms A and B. We can have a state with an atom A in an excited state and an atom B in a ground state, $|e\rangle_A|g\rangle_B$. We can also have an excited atom B and a ground state atom A, $|g\rangle_A|e\rangle_B$. The experiments showed that the entangled state of the atoms, the superposition:

$$\frac{1}{\sqrt{2}}(|e\rangle_A|g\rangle_B + |g\rangle_A|e\rangle_B), \qquad [3.3]$$

exists too (see Figure 3.2).

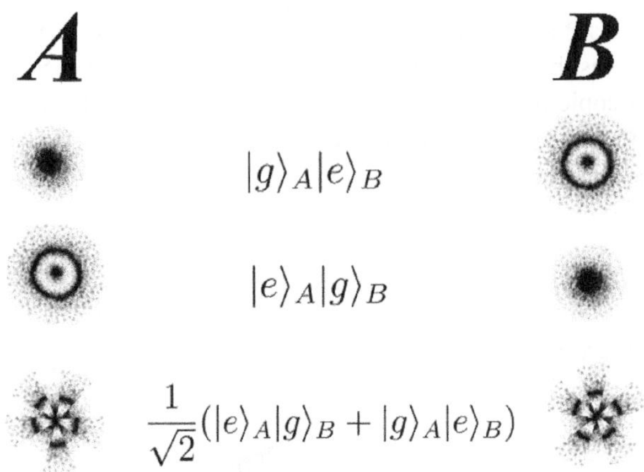

Figure 3.2. *Separated atoms in an entangled state: the superposition principle holds for separated atoms. We have two different states in which one atom is in the ground state and another in excited state. The superposition of these states exists too*

In the framework of MWI, the measurement of the state of atom A of the entangled pair (equation [3.3]) is described in the following way:

$$\frac{1}{\sqrt{2}}|R\rangle_{MD}(|e\rangle_A|g\rangle_B + |g\rangle_A|e\rangle_B) \rightarrow \frac{1}{\sqrt{2}}(|E\rangle_{MD}|e\rangle_A|g\rangle_B + |G\rangle_{MD}|g\rangle_A|e\rangle_B). \qquad [3.4]$$

In view of the experimental results of loophole-free Bell experiments, single-world interpretations cannot reject the existence of the state in equation [3.3] of entangled atoms, but since there is no technology that can test for the interference of macroscopic bodies, these interpretations implicitly postulate that superpositions of macroscopically different macroscopic states like equation [3.2] do not exist. For a single-world interpretation, this postulate is also necessary. Without this postulate, two different readings of the measuring device coexist. This is the coexistence of two histories, the superposition of two worlds. This is the MWI.

The measurement process of an atom is not instantaneous, but it is a local operation at site A and, therefore, the duration of the measurement does not depend on the distance between the sites A and B. In the loophole-free Bell-type experiments, the duration of the measurement of the state of the atom was shorter than the time required for light to go from A to B. If something at site B becomes different between the situation described by equation [3.3] and the right-hand side of equation [3.4], the measurement at A would cause a superluminal change in B, violating the special theory of relativity. However, for MWI, this is not the case (see Figures 3.3(a) and (b)). The atom in B has the same local description in two cases; it is a mixed state described density matrix:

$$\rho_B = \frac{1}{2}\left(|g\rangle\langle g| + |e\rangle\langle e|\right). \qquad [3.5]$$

In single-world interpretations, the process (equation [3.4]) cannot happen, as it leads to a superposition of macroscopically different state of a macroscopic pointer, or it must immediately be followed by the collapse to a quantum state corresponding to one result (see Figures 3.3(c) and (d)). Thus, the time evolution of the atoms and the measuring device measuring atom A is, instead:

$$\frac{1}{\sqrt{2}}|R\rangle_{MD}(|e\rangle_A|g\rangle_B + |g\rangle_A|e\rangle_B) \rightarrow \left\{|E\rangle_{MD}|e\rangle_A|g\rangle_B \text{ or } |G\rangle_{MD}|g\rangle_A|e\rangle_B\right\}. \qquad [3.6]$$

This is action at a distance! If we do not make measurements of the atom in A, we remain with the entangled state in equation [3.3] and the local description of the atom B is given by a mixed state [3.5]. If we perform a measurement in A, the atom in B immediately changes from the mixed state [3.5] to the pure state $|e\rangle_B$ or $|g\rangle_B$.

A local observer will not be able to observe the change. For them, it was (d'Espagnat 2018) an "improper mixture" (equation [3.5]) before the measurement and a "proper mixture" $\{|g\rangle\langle g|$ or $|e\rangle\langle e|\}$ after the measurement. The "proper" mixture is a subjective description of an observer who is ignorant of the complete description of the Universe. The local indistinguishability between proper and improper mixtures ensures the impossibility of superluminal signaling, despite the presence of action at a distance. Proper and improper mixtures are objectively different. Everyone will be ready to place a bet on an atom in an improper mixture because everyone knows the best anyone can do is a random guess. This is not true for an atom in a proper mixture. Since there is a matter of fact about the state of the atom – it is either in the ground state or in the excited state – someone might know, so there is no reason to agree on the bet if it is suggested to you.

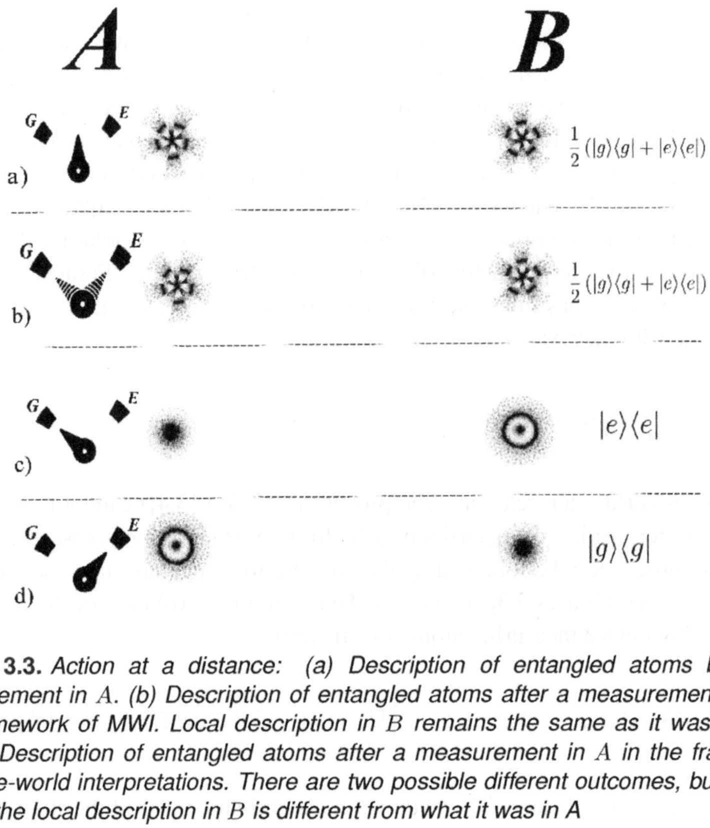

Figure 3.3. *Action at a distance: (a) Description of entangled atoms before a measurement in A. (b) Description of entangled atoms after a measurement in A in the framework of MWI. Local description in B remains the same as it was in A. (c and d) Description of entangled atoms after a measurement in A in the framework of single-world interpretations. There are two possible different outcomes, but in both cases, the local description in B is different from what it was in A*

Abandoning collapse leads to many worlds. However, there are many versions of MWI. In my version, I abandon not only the collapse of the universal wavefunction but also refuse to add *anything else* to my physical theory: no ontological worlds named "universes" (DeWitt 1971), no ontological minds of "many-minds view" (Albert and Loewer 1988) and no ontology of agents as four-dimensional "worms" (Wilhelm 2023). The physical (mathematical) part of my version of MWI postulates even less than the "bare theory" of quantum mechanics (Albert 1992) because I do not postulate the eigenvalue–eigenstate link. Instead, I add a non-mathematical part to the theory that connects the ontology with the experience of sentient beings. In terms of physics, I postulate as given the three-dimensional space (3D), instead of trying to derive it as Carroll and Singh (2019).

My explanation has two very different parts:

1) The ontology, stuff that exists, including laws constraining possible states of stuff and dynamical laws of time evolution of stuff.

2) The prescription of correspondence of the state of stuff with our experiences which we inquire through our sensory organs, sometimes equipped with instruments such as telescopes, microscopes, sonars, etc.

To define part 1), I start with 3D. For Newtonian mechanics, stuff is point-like particles placed in 3D. There are no constraints for the configuration of particles and their velocities (apart from not putting particles on top of each other because of the singularity of local interactions). Newtonian mechanics with Newtonian gravity, which features action at a distance, has to be modified by adding fields to mediate the interactions to avoid action at a distance. The stuff has two forms: particles and fields. However, fields and matter could not be arbitrarily distributed in space. The first pair of Maxwell's equations (the equations without time derivative) are strong constraints on possible configurations of charges and fields. Empirical evidence shows that classical physics is incorrect, so the equations must be modified again. We must be in the framework of quantum mechanics. The particles are now represented by wavefunctions. Relativistic corrections make us move to fields, wave functionals, (maybe strings!) so even part 1), the nature of stuff, has to be modified. However, I do not foresee revolutionary changes in the solution of the measurement problem when we move to the next-level theories that are necessary for analyzing physics at a higher energy scale.

Apart from putting constraints on the configuration of stuff, part 1) also has dynamical laws that specify the evolution of stuff with time. The basic equation is the Shrödinger equation, but of course the exact equation will be its relativistic generalization. I consider the dynamical laws of a theory to be the most important part. The constraints tell us the options for the Universe, given our theory. However, there is one actual universe, and the hypothetical question of what it could have been is of some philosophical interest, but it does not add much to the main question I want to understand: how does the Universe work? In our current memories, there is a large set of situations that can be considered as preparations of measurements together with their results. The regularities of these records are what I want to explain: the connections between preparations and results. These connections are given by dynamical laws. Note that the constraint laws are for the Universe, but they provide constraints in all worlds. The constraint laws shed light on why we have in our memories a particular set of preparations and not any other, which I find to be a less urgent and important question.

The formalism of quantum mechanics, even without many worlds, does not provide a simple, transparent picture corresponding to our experience. Maudlin (2010) strongly opposes the idea that the quantum state defined in the configuration space can describe the reality that we experience in 3D. To avoid changing the standard paradigm of a physical theory that does not have part 2), many physicists try to modify the quantum formalism in such a way that it will lead to a familiar picture of the reality we see. The influential work of Bell (1976) led many to look for local beables that can provide a familiar picture of reality in 3D. Today, it has the name of primitive ontology (Allori et al. 2014), with Bohmian positions of particles as the main example, supported by the fact that the trajectories of the Bohmian particles look very much like the trajectories of classical particles.

However, the formalism of Bohmian mechanics without the postulate of part 2) has an ambiguity. The pilot wave of Bohmian mechanics is the wavefunction of the MWI, which can arguably also provide an explanation of our experience(s). So, there is no unique way to connect the formalism to our experience. In fact, in Bohmian mechanics, there is also a third way that connects the experience with the effective "collapsed" wavefunction of the observer defined by the location of Bohmian positions of systems entangled with the observer (see Vaidman (2024a)).

The Bohmian effective collapsed wavefunction is very similar to the original GRW-Pearle wavefunction. Both differ very little from the von Neumann collapsed wavefunction, the wavefunction explicitly or implicitly considered in textbooks on quantum mechanics. For half a century, roughly until the experiment by Aspect et al. (1981), physicists accepted this wavefunction as the ontology adequately explains well the observed reality. Sometimes, the wavefunction was considered to be part of reality together with the classical physics required to describe measuring devices (see Landau and Lifshitz (2013)). The measurement problem was which dynamical equations, which describe a measurement process, such as equation [3.4], led to a detector that simultaneously shows several different outcomes, such as $|G\rangle_{MD}$ and $|E\rangle_{MD}$. Only the addition of collapse, the von Neumann process I, led to a wavefuntion corresponding to a single outcome $|G\rangle_{MD}$ or $|E\rangle_{MD}$. The eigenvalue–eigenstate link was essentially the postulate of correspondence between ontology and experience. The classical physics connection to experience was accepted without stating it, so the eigenvalue–eigenstate link connected the collapsed wavefunction and experience.

Even if the effectively collapsed Bohmian wavefunction, the GRW and gravitational collapse wavefunctions, and the von Neumann collapsed wavefunction are not identical, they are the same for all practical purposes (FAPP) (Bell 1990). In part 2), I wanted to use this FAPP concept of collapsed wavefunction to connect ontology and experience. If we try to use this concept to define the physical process of collapse, its vagueness is unacceptable; however, in part 2) of the theory, which describes the experience of sentient beings, FAPP is good enough by definition.

Instead of using other interpretations to state the connection in MWI between the branch wavefunction and experience in the corresponding world, I recommend an independent definition. I do not start from the ontology but from the concept of sentient beings. My definition (Vaidman 2021) is:

> A world is the totality of macroscopic objects: stars, cities, people, grains of sand, etc. in a definite classically described state.

"The definitive classically described state" is well defined in the framework of classical physics. We think and communicate using classical concepts, so since "world" is our concept, the definition is rigorous. However, it is not rigorous in the framework of exact sciences. Classical physics is an approximation. Quantum physics, on the other hand, does not have a well-defined counterpart of "definite classically described state".

I have two alternative approaches to make the connection between wavefunction and definite classical state (see Vaidman (2019, 2022)). Both are based on 3D structures. First, an observation: the Heisenberg uncertainty principle – which states that we cannot have a definite position and velocity of an object – does not provide a measurable constraint on macroscopic objects. I can live all my life with an uncertainty of less than a picometer in the position of my body's center of mass. So, position variables of macroscopic bodies can be described as extremely well-localized wave packets.

The first approach to make the connection is to specify the values of macroscopic variables, such as centers of mass, of macroscopic objects. The variable describing my position does not have to be center of mass; it can be, instead, the average of the positions of all of the particles of my body, although the center of mass has an advantage as it can play a role in some conservation laws. To specify the state of my body, I also need to know the relative positions of the macroscopic parts of the body and their orientations. These are also described by well-localized coordinates. There is no entanglement in the world's wavefunction for these macroscopic variables. There are many relative variables with entanglement: relative positions of electrons within atoms, positions of atoms within molecules and possibly positions of molecules within cells. Many of these entangled wavefunctions are definite energy states, such as internal relative states of particles in atoms.

The second option is fully geometrical. There is no need to look for observables that describe the position and state of the body – just the 3D graph of the mass density or the graph of the particle density (mass density is often considered as a primitive ontology (Allori et al. 2014). However, I think this move adds nothing since mass or particle densities are not contingent properties for a given wavefunction).

We identify objects according to their shape in 3D. In fact, the identity of elementary particles and the impossibility of keeping the diachronic identity of elementary particles when they meet particles of the same kind at a beamsplitter change the classical concept of an object. I am not a set of elementary particles that make up my body. I am a particular pattern of the wavefunction of these particles. Of course, it should be the wavefunction of degrees of freedom of the "correct" type. If all electrons in my body were replaced by protons and all protons by electrons, then it would not be me. This modified "me" will have a shape resembling me for an astronomically short time. This Gendanken situation resembles a Boltzmann brain, which I would not consider as my brain, even if it has momentarily the same shape.

The importance of 3D follows from the dynamical property of our Universe belonging to part 1) of our theory. All interactions are local in 3D. The general Hamiltonian formalism of quantum mechanics allows all types of interactions. The locality of interactions is the property required by the theory of relativity. There is a similar property in part 2). In our everyday life, interactions are local in 3D. We read these lines when the book or monitor is near us. I model the experience of sentient beings as memories of preparations of measurements and memories of results of measurements. Here, a famous observation of David Bohm (which was one of the motivations for Bohmian mechanics) is relevant: all measurements end up being measurements of position in 3D; either it is the position of the pointer of our ammeter, or a pixel of the digital readout of another measuring device. The 3D feature in part 2) follows from the property that sentient beings are local and that they have local senses, but of course, the analysis of the evolution that created local sentient beings has its roots in the locality of dynamical interactions of part 1).

Although there is a huge conceptual difference between the ontology of the Bohmian effective collapsed wave function of all particles and the Bohmian positions of all particles, the 3D picture of them is very similar. Bohmian positions do not "sit" at the centers of the wave packets of the particles, but these wave packets are localized so well that the global picture of all of them tells us the same story. Since for me, worlds are classically understandable stories, we have no difference here. Similarly, the von Neumann collapsed wavefunction, as well as the GRW or Diosi–Penrose collapsed wavefunctions, correspond to the same story. The difficulty appears in the MWI because the 3D graph of the mass density of the MWI universal wavefunction (without collapses) does not provide a picture of a familiar world. It is not supposed to provide such a picture, because it corresponds to a multitude of worlds.

I propose to postulate in my theory part 2): the correspondence of the ontology to the experiences of sentient beings. Here, I deviate from the exact sciences: experience is not a precise concept. Other approaches add a new ontology trying to stay as much as possible in the realm of exact sciences; I understand the motivation but do not see a success: a good scientific theory that provides a familiar picture of the world we observe. My alternative approach is to accept existing physics, which is a very successful exact science theory, and to add an invariably vague postulate that connects the ontology of the exact theory to our approximate and vague concepts of experience. This should include a description of a sentient being, which is biological but also robotic; since, if not today, in the near future, robots equipped with artificial intelligence (AI) will be considered FAPP as sentient beings as (or better than) humans. Since we are building robots, we understand their behavior better than we understand ours, so it seems to be an easier task to understand the connection between the "experience" of the robots, which is just records of their sensory devices, and the ontology, the wavefunction.

We consider a robot with a classically designed brain (not a quantum computer). So, our classically defined robot in a world has, by definition, a classically defined sentient state. Part 2) provides a correspondence between the wavefunction of the robot and its sentient state. If we describe the state of the robot using the position and orientation of its macroscopic parts, they all have to be well-localized wave packets. Alternatively, if we define the state of the robot through the 3D map of its mass density, it has to be a precise "sculpture" of the robot.

In a hypothetical universe in which all that exists is a room with the robot and some objects that are not quantum measuring, such as Geiger counters, single-particle detectors, etc., the wavefunction of the room with the robot and the objects provides a good description of the robot and its experiences for a reasonable period of time. However, it is clear that our universe is not like this, both because there are many more objects than those in the room and because there are many quantum devices that performed in the past and are performing now quantum experiments, including natural phenomena like supernovae, which amplify the superposition of some microscopic degrees of freedom to the macroscopic level.

Does the universal wavefunction and the postulate of connection between part of the wavefunction and the experience of a sentient being provide a good explanation of what we see? The MWI is often presented in the following way. There is a universal wavefunction. The decoherence with the environment leads to the emergence of branches of the universal wavefunction that are autonomous and correspond to the experiences of sentient observers "living" in these branches. I would not say that this picture is wrong, but it has not been shown that the emergence program works. In my understanding, this picture is supposed to start with a hot universe at the Big Bang and show how decoherence creates stars, planets, people and robots. Current physics excellently explains the results of all experiments: regularities between preparations and outcomes. It is less successful (maybe yet) in explaining the history of the Universe, starting from its creation. One difficulty is that we only have very partial information about the state of the Universe at the beginning. A lot of data are inaccessible, and, more importantly, most of the data are present in parallel worlds – not in the world to which we have direct access.

I am skeptical of giving a very important role to environmental decoherence. I know that we have to fight decoherence in quantum optics interferometric experiments, such as experiments in which I have been involved when I tried to give meaning to locations of quantum particles within an interferometer (Danan et al. 2013). The usual claim is that decoherence ensures the autonomy of branches. The usefulness of this claim is less clear to me. The linearity of quantum mechanics ensures that each branch in the superposition is autonomous with or without environment.

If we start with a single branch with a robot in the room and no quantum devices, the typical dynamics will keep this single branch as the only one for a long time. Usually, the spread of wavepackets of macroscopic objects is very small. There will be some tails, but we can take a natural approach to disregard branches that have a very small amplitude, that is, they have a very small "existence" (the square of the amplitude of a branch I name the *measure of existence*).

I see one situation in which decoherence might help. If we start with a highly localized wave packet of the position of the robot, due to the uncertainty relation, the wave packet will spread very fast. Now, we can view the particles of the environment as measurement devices, which verify with high frequency that the position of the robot remains localized (not localized with very small uncertainty, just well localized as most macroscopic objects are localized in the room). Then, the quantum Zeno effect will suppress the spreading of the wavefunction. However, we should not expect extreme position localization, because the creation of a high-precision localization of a macroscopic body requires huge energy resources.

If spread is not an issue, then the role of decoherence is just spoiling the interference. However, we do not expect to observe interference of macroscopic bodies anyway. This requires a technologically unthinkable experiment in which sentient beings in two different branches split into identical branches (with different relative phases). I once considered the Gedanken story with a sentient neutron

(Vaidman 1998). The experimentalists perform such experiments with neutrons (Greenberger 1983). In a neutron experiment, decoherence changes the probability of a neutron being in different output ports. In fact, in a tuned interferometer, the neutrons without decoherence might be more surprised that their odds of passing a beamsplitter apparently changed (due to interference with a neutron from another "neutron" word). Anyway, it is not a dramatic change: it only puts a question mark over the validity of a statistical Born rule. The neutron cannot be sure that its behavior was not autonomous and was influenced by the neutron in the other world. Note that considering a neutron as a sentient being is only a Gedanken story (e.g. because it does not have enough internal degrees of freedom to be sentient), and for representative (macroscopic) sentient beings, the interference experiment is not feasible, so the decoherence is not relevant for the absence of interference of macroscopically different worlds.

The concept of decoherence might be useful due to the observation that awareness states of sentient beings are stable under decoherence interactions. In my scheme, the question of why sentient beings (people and robots) are the way they are is not part of the interpretation. (we can investigate this question, and the locality of interactions in 3D will play an important role in this analysis). Given the locality of sentient beings and the locality of the pointer states, the locality of interactions tells us that these states are stable. If $|G\rangle_{MD}$ and $|E\rangle_{MD}$ are two well-localized pointer states in two separate positions, then starting with any one of them, we will continue to keep this state for a long period of time. If, instead, we start with the state in equation [3.2], which is one of the nonlocal states $|\perp\rangle_{MD} = \frac{1}{\sqrt{2}}(|E\rangle_{MD} + |G\rangle_{MD})$, then, due to decoherence, in no time ($t \ll 10^{-10}s$, see Joos and Zeh (1985)) we will get a mixture of $|+\rangle_{MD}$ and $|-\rangle_{MD}$ states. Thus, non-local states cannot serve the purpose of awareness states. However, again, decoherence is just an additional argument, which is not really needed. The complexity of the manipulation of non-local states required for information transfer and information analysis cannot be compared with the complexity of manipulating the local pointer states which are used in the robot's "brains" (and, as far as we understand, in our brains too). So, even if super-technology offered to remove the environmental decoherence, we would probably not use robots with non-local awareness states. An exception is a special task machine, the quantum computer. There, indeed, we make a (big) effort to remove decoherence to allow running special programs that provide quantum supremacy.

I can summarize the discussion of part 2) of the MWI, the correspondence between ontology and experiences of sentient beings, by stating two alternative postulates. To simplify them, I assume that we understand the operation of sensory devices well. This is certainly true for robots (we could not build them without understanding this), and apparently our sensory devices are not conceptually different. Then, instead of describing the state of the brain with particular experience, we can describe the states of macroscopic objects that sentient beings observe.

> An agent experiences a macroscopic object if all macroscopic variables defining the state and position of the object are well-localized wavepackets in the world branch of the wavefunction.

Or, an alternative postulate:

> An agent experiences a macroscopic object if the 3D picture of the density of particles of the object calculated based on the world branch of the wavefunction faithfully represents the object.

Note that the branch of the wavefunction has to keep the pattern of the object for a reasonable time, so that the brain of the sentient being is capable of experiencing the object.

However, these postulates are not enough. Apart from observing classical objects, we witness quantum phenomena. There are superconductivity and superfluidity, which classical physics has difficulty explaining (together with the stability of atoms), but what is really different are quantum measurements. The laws of classical physics cannot explain experiments with identical preparations that might have different outcomes. The randomness observed in quantum experiments and the statistics of the outcomes, which has no roots in the ignorance of some agent about initial conditions, is another kind of experience of sentient beings that requires an explanation. Fortunately, the ontology – the universal wavefunction – has parameters that played no role in explaining the observations of the classical picture. The branches with the same experiments but differing results might have different amplitudes that play no role in describing the 3D patterns or the positions of macroscopic objects. These amplitudes explain the Born rule statistics of the experiments through the second postulate of part 2) (see Vaidman (2021)):

> The probability of self-location in a world with a particular outcome is proportional to the measure of existence of this world.

The measure of the existence of the world is the square of the amplitude of the corresponding branch of the wavefunction. Apart from providing the probability of self-location in a world, the measure of existence of the world characterizes the power of interference of this world if a super-technology were to arrange an interference experiment with other world branches. I explained this in my Gedanken story about a sentient neutron inside a Mach Zehnder interferometer (Vaidman 1998).

I view the postulates of part 2) of the MWI as part of the theory. While the first postulate (in both forms) is very natural (it is a "duck test"), the probability postulate is more substantive. Essentially, it is the counterpart of the Born rule of collapse theory in MWI. Since, in the framework of the MWI, all possible outcomes of quantum measurements are realized, we cannot define the Born rule in a usual form. There is no meaning to the probability of obtaining a particular outcome in a measurement when all outcomes take place. We also cannot ask what outcome a particular agent will see: they will split into multiple agents, with everyone seeing some outcome. The MWI has no diachronic identity as an agent toward the future. So, to state the probability postulate, we need a special arrangement (sleeping pill experiment Vaidman (1998)), which will lead to *different* agents that are not aware of who they are, in which world they are, that is, what the measurement outcome in their world is (although they might know all relevant aspects of the ontology of all worlds).

Note that there are multiple "derivations" of the Born rule in the framework of the MWI (Gleason 1975; Deutsch 1999; Aharonov and Reznik 2002; Wallace 2007; Vaidman 2011; Zurek 2005; Sebens and Carroll 2018). I have argued (Vaidman 2020) that all have some explicit or implicit assumptions beyond the quantum formalism of part 1). Many of these assumptions are very natural; Vaidman (2011) assumed symmetry of space and impossibility of superluminal signaling, but without any assumptions, the Born rule cannot be derived.

The mathematical part describing the constraints and the evolution of the ontological (physical, mathematical) part of the theory, which I named part 1), leaves an important part (the Born rule) undefined. Outside of the framework of the MWI, the Born rule belongs to the physical part of the theory. It is an important postulate, but apparently due to the recent quantum information technology revolution, it has received disproportionate attention. Usually, when we ask the question what the probability is of a particular choice among a set of possibilities, we find many situations in which we have reasons to consider all possibilities equally probable, and then a simple counting provides a good answer. If all branches, actual in MWI, or possible in single-world theories, are similar (their measures of existence must be equal), then simple counting will also provide the correct answer in the quantum realm.

Part 1) of my MWI is ontological, while part 2) is epistemic. Usually, ontological interpretations start with what is and derive an ontic picture that is very close to our experience. The Bohmian positions of atoms of a duck have the shape of the duck, the Bohmian positions of atoms of wings of the duck change in time as flapping wings change their position in time, the oscillation of Bohmian positions of air molecules looks like sound waves, so the ontic picture of Bohmian positions explains our experience. Epistemic interpretations start with experience and do not necessarily try to answer questions about what exists (see the recent review by Barzegar and Oriti (2024)); the theory describes experience in terms of the state of knowledge of sentient agents. In relational approaches to quantum mechanics (Rovelli 1996; Zwirn 2023), quantum theory is a theory of information, where information is subjective to agents. Thus, in Bohmian mechanics, we start with the ontology of the Universe, which includes Bohmian positions of an agent who has their experiences, while in relational approaches, the universal picture does not exist: the experience of an agent is described by their relations to other objects.

Today, a popular (frequently named Oxford) approach to the MWI is finding emerging worlds with their agents with particular experiences by analyzing decoherence in the universal wavefunction. So, it is an ontological approach like the Bohmian one which, however, leads to multiple worlds (instead of one) with agents whose experiences are explained by the theory. This route has to solve a difficult problem that Bohmian mechanics apparently does not have. How do we obtain separate worlds that do not interfere with one another? Bohmian mechanics has one world, so the question usually does not arise, although, in fact, Bohmian mechanics has essentially the same problem. There are "empty" worlds with "empty" wavepackets, which, in principle, can affect the actual Bohmian world. This would happen for a neutron world in a Mach–Zehnder interferometer without the second beam splitter in which neutrons follow "surrealistic trajectories" (Englert et al. 1992) (note that we cannot really get an experimental proof that Bohmian trajectories exist (Vaidman 2024a)). Coming back to Oxford school of MWI, I find that the analysis of evolution from the Big Bang, which leads to emergence of worlds of the type we know, is a difficult problem that requires complicated physical analysis in cosmology, etc. We also have very limited information about the current universal wavefunction because we know very little about parallel worlds. Thus, I do not suggest starting the analysis from the universal wave function.

I observe that there are situations in which the world splitting is clear: when we perform quantum measurements, like creating a single photon and detecting it after it passes a beam splitter. The Geiger counter clicks also split the world, although it is more complicated as we have to understand what the time difference is between the clicks, which makes the worlds different. So, my proposal is to start from our world. My world is not an ontological concept (it exists for me but not for my copies in parallel worlds). Therefore, what I propose is a hybrid approach. I start with the epistemic part 2). My experience and knowledge of my world give me information about the corresponding branch of the wavefunction. My memories of quantum experiments that have been performed in my world also give me some information about a few parallel worlds. I believe in their existence and in the existence of all other worlds about which I have almost no information because I believe that there is an ontological theory for the whole Universe, which evolves according to physical laws of the same type as the laws that explain all experimental data we observe in our world. The laws we know are deterministic (Vaidman 2014) and have no action at distance (Vaidman 2024b). To explain my experience, on top of the ontological story of the unitary evolution of the wavefunction of the Universe, I need again the epistemic part 2) with the almost obvious connection between myself and the wavefunction of my particles in my branch, which looks (and sounds) like me, and a natural probability postulate that has numerous supporting arguments, but no derivation from part 1).

In conclusion, physics explains all phenomena we observe on Earth extremely well. Collapse of a quantum state at measurement is an ugly scar on the beautiful quantum mechanics. The only role of the collapse is to avoid parallel worlds that are not supposed to be seen according to the theory. Without collapse, we have to accept MWI. Apart from the disappointment in understanding that I am not a unique Lev Vaidman, and that there are multiple copies of me in other worlds, MWI allows me to believe that by and large, I understand how the Universe works. For me, parallel worlds are not too high a price to pay for understanding nature.

This work has been supported in part by the Israel Science Foundation Grant No. 2689/23.

References

Aharonov, Y. and Reznik, B. (2002). How macroscopic properties dictate microscopic probabilities. *Physical Review A*, 65(5), 052116.

Albert, D.Z. (1992). *Quantum Mechanics and Experience*. Harvard University Press, Cambridge, MA.

Albert, D.Z. and Loewer, B. (1988). Interpreting the many-worlds interpretation. *Synthese*, 77, 195–213.

Allori, V., Goldstein, S., Tumulka, R., Zanghì, N. (2014). Predictions and primitive ontology in quantum foundations: A study of examples. *The British Journal for the Philosophy of Science*, 65(2), 323–352.

Aspect, A., Grangier, P., Roger, G. (1981). Experimental tests of realistic local theories via Bell's theorem. *Physical Review Letters*, 47, 460.

Barzegar, A. and Oriti, D. (2024). Epistemic–pragmatist interpretations of quantum mechanics: A comparative assessment. *Foundations of Physics*, 54(5), 1–34.

Bell, J.S. (1976). The theory of local beables. In *Quantum Mechanics, High Energy Physics and Accelerators*, Bell, M., Gottfried, K., Veltman, M. (eds). World Scientific, Singapore.

Bell, J.S. (1990). Against "measurement". *Physics World*, 3(8), 33.

Bohm, D. (1952). A suggested interpretation of the quantum theory in terms of "hidden" variables. *Physical Review*, 85, 166.

Carroll, S.M. and Singh, A. (2019). Mad-dog Everettianism: Quantum mechanics at its most minimal. In *What is Fundamental?*, Aguirre, A., Foster, B., Merali, Z. (eds). Springer, Cham.

Danan, A., Farfurnik, D., Bar-Ad, S., Vaidman, L. (2013). Asking photons where they have been. *Physical Review Letters*, 111(24), 240402.

Deutsch, D. (1999). Quantum theory of probability and decisions. *Proceedings of the Royal Society of London. Series A: Mathematical, Physical and Engineering Sciences*, 455(1988), 3129–3137.

DeWitt, B. (1971). The many-universes interpretation of quantum mechanics. In *Foundations of Quantum Mechanics, Proceedings of the International School of Physics "Enrico Fermi"*. Academic Press, New York.

Diósi, L. (1987). A universal master equation for the gravitational violation of quantum mechanics. *Physics Letters A*, 120(8), 377–381.

Englert, B.-G., Scully, M. O., Süssmann, G., Walther, H. (1992). Surrealistic Bohm trajectories. *Zeitschrift für Naturforschung A*, 47, 1175–1186.

d'Espagnat, B. (2018). *Conceptual Foundations of Quantum Mechanics*. CRC Press, Boca Raton.

Ghirardi, G.C., Rimini, A., Weber, T. (1986). Unified dynamics for microscopic and macroscopic systems. *Physical Review D*, 34, 470–491.

Gleason, A.M. (1975). Measures on the closed subspaces of a Hilbert space. In *The Logico-Algebraic Approach to Quantum Mechanics: Volume I: Historical Evolution*, Hooker, C.A. (ed.). Springer, Dordrecht.

Greenberger, D.M. (1983). The neutron interferometer as a device for illustrating the strange behavior of quantum systems. *Reviews of Modern Physics*, 55(4), 875.

Joos, E. and Zeh, H.D. (1985). The emergence of classical properties through interaction with the environment. *Zeitschrift für Physik B Condensed Matter*, 59, 223–243.

Kelvin, L. (1901). Nineteenth century clouds over the dynamical theory of heat and light. *The London, Edinburgh, and Dublin Philosophical Magazine and Journal of Science*, 2(7), 1–40.

Landau, L.D. and Lifshitz, E.M. (2013). *Quantum Mechanics: Non-relativistic Theory*, volume 3. Elsevier, Amsterdam.

Maudlin, T. (2010). Can the world be only wavefunction. In *Many Worlds? Everett, Quantum Theory, & Reality*, Saunders, S., Barrett, J., Kent, A., Wallace, D. (eds). OUP, Oxford.

Pearle, P. (1976). Reduction of the state vector by a nonlinear Schrödinger equation. *Physical Review D*, 13(4), 857.

Penrose, R. (1996). On gravity's role in quantum state reduction. *General Relativity and Gravitation*, 28, 581–600.

Rosenfeld, W., Burchardt, D., Garthoff, R., Redeker, K., Ortegel, N., Rau, M., Weinfurter, H. (2017). Event-ready bell test using entangled atoms simultaneously closing detection and locality loopholes. *Physical Review Letters*, 119, 010402.

Rovelli, C. (1996). Relational quantum mechanics. *International Journal of Theoretical Physics*, 35, 1637–1678.

Sebens, C.T. and Carroll, S.M. (2018). Self-locating uncertainty and the origin of probability in everettian quantum mechanics. *The British Journal for the Philosophy of Science*, 69(1). doi: 10.1093/bjps/axw004.

Vaidman, L. (1998). On schizophrenic experiences of the neutron or why we should believe in the many-worlds interpretation of quantum theory. *International Studies in the Philosophy of Science*, 12(3), 245–261.

Vaidman, L. (2011). Probability in the many-worlds interpretation of quantum mechanics. In *Probability in Physics*, Ben-Menahem, Y. and Hemmo, M. (eds). Springer, Berlin, Heidelberg.

Vaidman, L. (2014). Quantum theory and determinism. *Quantum Studies: Mathematics and Foundations*, 1, 5–38.

Vaidman, L. (2019). Ontology of the wave function and the many-worlds interpretation. In *Quantum Worlds: Perspectives on the Ontology of Quantum Mechanics*, Lombardi, O., Fortin, S., López, C., Holik, F. (eds). Cambridge University Press, Cambridge.

Vaidman, L. (2020). Derivations of the Born rule. In *Quantum, Probability, Logic: The Work and Influence of Itamar Pitowsky*, Hemmo, M. and Shenkar, O. (eds). Springer, Cham.

Vaidman, L. (2021). Many-worlds interpretation of quantum mechanics. In *The Stanford Encyclopedia of Philosophy*, Zalta, E.N. (ed.). Stanford University, Stanford.

Vaidman, L. (2022). Wave function realism and three dimensions. In *Quantum Mechanics and Fundamentality: Naturalizing Quantum Theory between Scientific Realism and Ontological Indeterminacy*, Allori, V. (ed.). Springer, Cham.

Vaidman, L. (2024a). Are there observational differences between Bohmian mechanics and other interpretations? In *Physics and the Nature of Reality: Essays in Memory of Detlef Dürr*, Bassi, A., Goldstein, S., Tumulka, R., Zanghì, N. (eds). Springer, Cham.

Vaidman, L. (2024b). The many-worlds interpretation of quantum mechanics is the only way to avoid action at a distance [Online]. Available at: https://philsci-archive.pitt.edu/id/eprint/24414/.

Wallace, D. (2007). Quantum probability from subjective likelihood: Improving on Deutsch's proof of the probability rule. *Studies In History and Philosophy of Science Part B: Studies In History and Philosophy of Modern Physics*, 38(2), 311–332.

Wilhelm, I. (2023). Centering the Born rule. *Quantum Reports*, 5(1), 311–324.

Zurek, W.H. (2005). Probabilities from entanglement, Born"s rule $p_k = |\psi_k|^2$ from envariance. *Physical Review A – Atomic, Molecular, and Optical Physics*, 71(5), 052105.

Zwirn, H. (2023). Everett's interpretation and convivial solipsism. *Quantum Reports*, 5(1), 267–281.

4

Probability in Everett

Jeffrey A. BARRETT
University of California, Irvine, USA

4.1. Introduction

Hugh Everett III understood the quantum measurement problem as one of the determinate records of physically situated observers.[1] This way of thinking about the measurement problem helps clarify his understanding of quantum probabilities in terms of the relative records of a typical observer. It also provides guidance on how we may extend our account to allow for standard forward-looking probabilities. In this context, we consider a few competing approaches. Some of these do not work at all, but one clearly does. These examples illustrate why obtaining forward-looking probabilities requires us to supplement Everett's theory with an appropriate metaphysical framework.

A situated observer is a physical observer characterized by their physical theory. It is a physically embedded observer, an observer modeled within the world described by the theory. For a theory to explain the records of a situated observer, it must predict physical records on which the observer's experience might plausibly supervene. If so, we say that the theory is *strongly adequate*.

Strong adequacy is a variety of empirical adequacy that meshes well with the explanatory aims of our best physical theories. A theory is empirically adequate only if it makes the right empirical predictions. For a physically situated observer, this

1 Ismael (2007, 2016) has used the notion of a situated agent in discussions of such issues as freewill. The basic idea was also central to Everett's views.

means that the theory must predict physical records that account for their experience. If it does, it is strongly adequate.[2]

Importantly, there is no probability in Everett's relative-state formulation of pure wave mechanics. Rather, he sought to show that the theory predicts that the sequence of measurement records of a typical situated observer will exhibit standard quantum statistics. That it was strongly adequate in this sense is why he took it to provide a satisfactory solution to the quantum measurement problem.

There is good reason, however, to suppose that Everett wanted more than this. At the end of the short version of his PhD thesis, he claimed that "our theory [the relative-state formulation of pure wave mechanics] ultimately justifies the use of the probabilistic interpretation as an aid to making practical predictions" (Everett 1957, p. 196).[3] This suggests that while he held that his formulation of quantum mechanics involved no probabilities, he also took it in some sense to *justify* the forward-looking probabilistic predictions of the usual collapse formulation when it makes coherent predictions.

The *probability problem* is to get from what he showed in his thesis to something that entails standard forward-looking quantum probabilities.

While Everett never explained why he thought we were justified in using the usual probabilistic interpretation to make practical predictions, there is good reason to want a theory that predicts the right forward-looking probabilities for the physical records of a situated observer.[4] After all, the successful forward-looking predictions of the theory have always been the chief empirical virtue of quantum mechanics.

We start with Everett's relative-state formulation of pure wave mechanics, his model of observation, and his argument that the theory is empirically faithful. Then, we consider how we might supplement his theory to obtain forward-looking predictions for the records of situated observers.

---

2 See Barrett (2020, 2021) for discussions of strong adequacy.

3 See Everett (1956, p. 159) for a similar conclusion in the long version of the thesis.

4 More detail regarding what Everett did say on the subject may be helpful. In brief, he held that the justification of quantum probabilities works "in a manner analogous to the methods of statistical mechanics" (Everett 1956, p. 152). That said, while statistical mechanics, like pure wave mechanics, is a deterministic theory where we wish to justify making probabilistic predictions, he never said how he took this to work in that case either. How to get probabilities in statistical mechanics is a topic that remains contentious. One way to put this is that statistical mechanics itself requires something like what we will call a ⋆-principle later. See Albert (2000) and Frigg and Werndl (2024) for introductions.

4.2. How Everett understood the measurement problem

Everett took the quantum measurement problem to provide a clear account of the experience of physically situated observers. In particular, he took a solution to the measurement problem to require a formulation of quantum mechanics that faithfully represents the physical records of situated observers and allows for a consistent narrative regarding their interactions in the context of nested measurements. He called the first condition faithfulness and the second consistency.[5]

Everett began by arguing that the standard von Neumann formulation of quantum mechanics does not provide a consistent narrative for interactions between physically situated observers. His discussion of the measurement problem in the long thesis begins with von Neumann's description of the dynamical laws:

– **Process 1**: when a *measurement* of O is made of a system S, the state ψ_S randomly evolves to eigenstate χ_S of O with probability $|\langle\psi_S|\chi_S\rangle|^2$.

– **Process 2**: when *no measurement* is made of system S, it evolves in a deterministic, linear way that depends only on its energy properties: $|\psi(t_1)\rangle_S = \hat{U}(t_0, t_1)|\psi(t_0)\rangle_S$.[6]

For Everett (1956, p. 73), the measurement problem concerns the mutual consistency of these two processes. As he put it, "[t]he question of the consistency of the scheme arises if one contemplates regarding the observer and his object-system as a single (composite) physical system. Indeed, the situation becomes quite paradoxical if we allow for the existence of more than one observer". The problem, he argued, is that von Neumann's dynamics is logically inconsistent if we allow for more than one situated observer.

Everett's (1956, p. 74) argument turns on an "amusing, but extremely hypothetical drama". His hypothetical drama was a version of the friend story that Eugene Wigner (1961) would tell later. Everett's version went as follows:

> Isolated somewhere out in space is a room containing an observer, A, who is about to perform a measurement upon a system S. After performing his measurement he will record the result in his notebook. We assume that he knows the state function of S (perhaps as a result

5 See Barrett (2011a, 2023) for discussions of Everett's empirical faithfulness condition. In brief, the condition required that we be able to find, in our model of a physically situated observer, sequences of measurement records that agree in an appropriate sense with the observer's actual records. There will be more to say about how this was to work as we go.

6 See von Neumann (1955, pp. 351–352 and 417–418) for his description of the two laws.

> of previous measurement), and that it is not an eigenstate of the measurement he is about to perform. *A*, being an orthodox quantum theorist, then believes that the outcome of his measurement is undetermined and that the process is correctly described by Process 1. ... In the meantime, however, there is another observer, *B*, outside the room, who is in possession of the state function of the entire room, including *S*, the measuring apparatus, and *A*, just prior to the measurement. *B* is only interested in what will be found in the notebook one week hence, so he computes the state function of the room for one week in the future according to Process 2 (Everett 1956, p. 74).[7]

The problem is one of reconciling *A*'s fundamentally random and nonlinear prediction with *B*'s deterministic and linear prediction of *A*'s evolution as a physically situated observer.

We can see why Everett took von Neumann's scheme to be inconsistent by considering the evolution of the quantum-mechanical state of the composite system occupying the room. Suppose that the physically situated observer *A*, their measuring device *M*, and their object system *S* begin in the state:

$$|\text{“ready”}\rangle_A \, |\text{“ready”}\rangle_M \, (\alpha \, |\uparrow_x\rangle_S + \beta \, |\downarrow_x\rangle_S).$$

Here, *M* is ready to measure the x-spin of *S*, and *A* is ready to observe and record the result. Assuming ideal measurement interactions that produce perfect correlations, *A* is calculated via Process 1 such that they will end up in either:

$$|\text{“}\uparrow_x\text{”}\rangle_A \, |\text{“}\uparrow_x\text{”}\rangle_M \, |\uparrow_x\rangle_S$$

with probability $|\alpha|^2$ or

$$|\text{“}\downarrow_x\text{”}\rangle_A \, |\text{“}\downarrow_x\text{”}\rangle_M \, |\downarrow_x\rangle_S$$

7 Everett reported writing this in 1955, and it was widely distributed the following year. Wigner published his friend story in 1961. While Everett took a course in quantum mechanics from Wigner at Princeton, there is good reason to believe that this is Everett's thought experiment. Part of the evidence is that Everett and Wigner drew very different conclusions from the story, and given the argument of the long thesis, it is likely that Everett would have discussed, and disagreed with, Wigner's conclusion, had he known it. Everett also gave a plausible account of how he started to think about situated observation in quantum mechanics in a 1977 discussion with his friend Charles Misner. A transcription of that discussion, along with other relevant source material, can be found in Barrett and Byrne (2012, pp. 299–310).

with probability $|\beta|^2$. Meanwhile, B calculates by Process 2 that A will result in the entangled superposition:

$$\Diamond \alpha \,|``\uparrow_x"\rangle_A \,|``\uparrow_x"\rangle_M \,|\uparrow_x\rangle_S + \beta \,|``\downarrow_x"\rangle_A \,|``\downarrow_x"\rangle_M \,|\downarrow_x\rangle_S.$$

Inasmuch as A and B calculate mutually incompatible states, the hypothetical drama yields a straightforward logical contradiction. Everett concluded that "[i]t is now clear that the interpretation of quantum mechanics with which we began is untenable if we are to consider a universe containing more than one observer. We must therefore seek a suitable modification of this scheme, or an entirely different system of interpretation" (1956, p. 76).[8]

As a first step in addressing the problem, Everett took B's computation of the global state of the composite system to be correct. He held that situated observers such as A are "subject to the same type of description and laws as any other systems" (Everett 1956, p. 152). B is hence fully justified in using Process 2 to calculate A's future state.

Everett also held that we might, in principle, show that the resultant state was the entangled state calculated by B by means of an interference measurement. Indeed, that it is always in principle possible to make such a measurement is why Everett held every element of the superposition to be real:

> It is [...] improper to attribute any less validity or "reality" to any element of a superposition than any other element, due to this ever-present possibility of obtaining interference effects between the elements. All elements of a superposition must be regarded as simultaneously existing (Everett 1956, p. 150).

Detecting interference between branches might be accomplished by measuring an observable O that has the state given by expression $\Diamond$ above as an eigenstate with eigenvalue $+1$ and an orthogonal state as an eigenstate with eigenvalue -1. For his part, Wigner (1961) believed that such a measurement would show that A's interaction with M had caused a collapse of the physical state by Process 1. In that case, there would be a positive probability of obtaining the result -1 on an O

8 While there is much that separates Hugh Everett's and John Bell's views regarding how we might best address the quantum measurement problem, it is significant that they both took it to be one of providing a consistent account of measurement records of physically situated observers. As Bell explained in a 1989 interview with Renée Weber: "For me it is absolutely unquestioned in my analysis of things that the real world is out there, and that I am an incident in it" (Goldstein 2022).

measurement. However, since Everett held that all interactions accord with Process 2, he believed that a measurement of O would certainly yield the result $+1$.[9]

Everett put the room containing observer A "somewhere out in space" so that nothing outside the room would become correlated with the x-spin of S and ruin B's calculation, but even so, he knew that it would be extraordinarily difficult for B to carry out any such computation in practice. This is why he characterized the drama as "extremely hypothetical". He also knew that interactions with the environment and the degrees of freedom involved in such a measurement would make it virtually impossible to perform an O-type measurement that would detect interference effects between the two branches of the state. However, Everett nevertheless required that a satisfactory formulation of quantum mechanics provide a consistent narrative for the evolution of the physical state of the composite system occupying the room. This was his litmus test for the consistency of any proposed quantum theory.

The upshot is that for a formulation of quantum mechanics to address the measurement problem for Everett, it must (1) account for the records of physically situated observers and (2) allow for consistent narratives for the interactions between such observers as described in his hypothetical drama. We will consider how he took this to work and then reflect on what more would be required to obtain a theory that also predicts the right forward-looking probabilities.

4.3. Relative records and subjective experience

Everett described his solution to the quantum measurement problem as follows:

> To assume the universal validity of the quantum description, by the complete abandonment of Process 1. The general validity of pure wave mechanics, *without any statistical assertions*, is assumed for *all* physical systems, including observers and measuring "apparata". Observation processes are to be described completely by the state function of the composite system which includes the observer and his object-system, and which at all times obeys the wave equation (Process 2) (Everett 1956, p. 77).

Among its many advantages, this approach "has the virtue of logical simplicity, and it is complete in the sense that it is applicable to the entire universe. [...] Since

9 See Albert (1986, 1992, Chapter 8) and Albert and Barrett (1995) for discussions of such interference experiments in the context of Everettian quantum mechanics and no-collapse theories more generally.

the universal validity of the state function description is asserted, one can regard the state functions themselves as the fundamental entities, and one can even consider the state function of the whole universe" (Everett 1956, p. 77).

The core of Everett's proposed theory was the standard von Neumann formulation of quantum mechanics but without the collapse postulate, a theory he called *pure wave mechanics*. His goal was to "deduce the probabilistic assertions of Process 1 as *subjective* appearances" to physically situated observers described by pure wave mechanics "thus placing the theory in correspondence with experience". In the resulting description, "the formal theory is objectively continuous and causal, while subjectively discontinuous and probabilistic" (Everett 1956, p. 77).

To accomplish this, Everett supplemented pure wave mechanics with: (1) a *physical model* of situated observers and their records; (2) a *distinction* between absolute and relative states; and (3) a *typicality measure* over relative states. He then identified subjective appearances with the relative physical records of situated observers. He called the result the relative-state formulation of quantum mechanics.[10]

In Everett's model, a situated observer performs a measurement by correlating the state of their physical record with the property of the object system they with to observe. The linear dynamics typically produces an entangled postmeasurement state, which is akin to the state ◊ that B computes for A. Everett associated the situated observer's subjective experiences with their relative records, their records as described by each branch of the full state when written in a basis that characterizes them as having measurement records in each branch. In the absolute state ◊, A has two relative records. In the first branch, they record x-spin up (relative to S being x-spin up), and in the second branch, they record x-spin down (relative to S being x-spin down).

Everett described state ◊ as two observers with different experiences from one perspective and as the same physical observer with two different experiences from

10 Pure wave mechanics is a deterministic theory that says nothing at all about situated observers, subjective appearances, absolute and relative records, typicality or probabilities of any type. As a result, Everett needed to add both definitions and auxiliary assumptions to get a theory that had any empirical content at all. He described this as showing that pure wave mechanics, "together with the necessary correlation machinery for its interpretation, forms a logically self-consistent description of a universe in which several observers are at work" (Everett 1956, p. 77). The empirical content of the theory is in terms of the typical relative records of a situated observer. In this sense, only the full relative-state formulation of pure wave mechanics has empirical content.

another. He sought to explain the two perspectives in a footnote in the long version of his PhD thesis (see quotation below).

However, there is reason to believe that not even Everett was entirely satisfied with this explanation. When he edited the long thesis for inclusion in the DeWitt and Graham (1973) anthology, he indicated in a note to his copy of the manuscript that this footnote should be omitted.[11]

> At this point, we encounter a language difficulty. Whereas before the observation we had a single observer state afterwards there were a number of different states for the observer, all occurring in a superposition. Each of these separate states is a state for an observer, so that we can speak of the different observers described by the different states. On the other hand, the same physical system is involved, and from this viewpoint it is the *same* observer, which is in different states for different elements of the superposition (i.e. has had different experiences in the separate elements of the superposition in the separate elements of the superposition) (Everett 1956, p. 121).

Everett's repeated association of subjective experiences with relative states suggests another way to put the matter. In a situation such as that described by state (1), we might say that there are *two* physically situated *relative observers* after the measurement, each associated with the same physical system but with different relative states and hence different experiences. This way of putting it has the virtue of associating incompatible experiences with different physically situated observers. That said, it remains unclear how best to understand such talk.

What is needed is a clear account of how we think of the two relative observers as physically situated with incompatible physical records. To this end, it is tempting to address Everett's language difficulty by postulating the existence of a different physical *world* corresponding to each possible postmeasurement relative observer and record.[12]

Importantly, Everett himself never mentioned worlds in any of his published works. One reason may be that inasmuch as he insisted on the ever-present

---

11 See Barrett and Byrne (2012, pp. 355 and 363) for a discussion and a copy of Everett's handwritten note.

12 See DeWitt (1971) for an early splitting-worlds interpretation of Saunders et al. (2010a) for more recent discussions. See Barrett (2011b) for a discussion of how Everett understood relative observers and his reactions to other people characterizing his theory as one involving many worlds.

possibility of obtaining interference effects between the elements of a superposition, he was committed to having just one causally closed postmeasurement world.[13] That said, having a physically distinct world for each possible measurement outcome provides a natural way of accounting for each relative observer's determinate experience. It gives us somewhere to put the observer where they can have a determinate physical record of their own. This first-pass observation explains the historical attraction of many-world interpretations of Everett.

That postulating many postmeasurement worlds is so often taken to be a natural response to the problem of accounting for multiple determinate measurement records indicates that there is more than just *language* at stake here. At root, the difficulty concerns how best to understand the theory as a faithful description of physical observers and their records. In this, the issue is one of metaphysics.

We will consider a few of the metaphysical options available for addressing Everett's language difficulty, but first, we need to clarify how he sought to deduce the probabilistic assertions of Process 1 as subjective appearances. In brief, his strategy was to show that the relative records of a *typical* relative observer will exhibit the standard quantum statistics and that it will hence *appear* to them that Process 1 produced those records. We will consider the sense in which he achieved this goal, but we will also see why we should want more. Specifically, when we return to consider metaphysical options, we will do so with an eye to obtaining the right forward-looking probabilistic predictions for situated observers, something not provided by Everett's account of probability.

4.4. Everett's account of probability

Everett clearly and repeatedly insisted that there were no probabilities in pure wave mechanics. That he took this to be a central feature of his approach is reflected in the fact that he originally entitled his PhD thesis "Wave Mechanics without Probability".[14] Rather than providing standard probabilistic predictions, he sought to show that the sequence of relative records of a typical situated relative observer, as described by pure wave mechanics, will be randomly distributed with the usual quantum relative frequencies. On the assumption that their records characterize their

13 See Albert and Barrett (1995) for a discussion of this point.

14 He chose this over such alternative options as: wave mechanics without probabilistic postulates, the multiobserver form of quantum mechanics, indeterminism in a deterministic wave mechanics, and the correlation interpretation of wave mechanics. See Barrett and Byrne (2012, p. 320) for a copy of his reflections on the title.

experience, it will therefore appear to a typical relative observer that the standard quantum probabilities held for their past measurements.

To show that records for typical relative observers exhibit standard quantum statistics, Everett needed a typicality measure over branches (elements of the superposition in the determinate-record basis), something not given by pure wave mechanics itself. He introduced a typicality measure by imposing a sequence of formal constraints until the measure was uniquely determined. Specifically, he required that the measure m (1) satisfy the standard axioms for being a probability measure, (2) be a positive function of the complex-valued coefficients a_i associated with the branches of the global superposition $\sum a_i|\chi_i\rangle$ and (3) satisfy a subbranch additive condition. Together, he took these constraints to determine m to be the norm-squared-amplitude measure over branches.[15]

Consider the state ◊ that B computes for A. Here, the typicality measure on the first branch is $|\alpha|^2$, and on the second branch is $|\beta|^2$. Importantly, this measure does not concern the probability of alternative measurement outcomes. Indeed, as Everett took all branches to be equally actual, he took the probability of each branch to be one. While the typicality measure is *across branches*, his goal was to explain the subjective appearance of quantum probabilities by the sequence of records *within a typical branch*.

Everett (1956, p. 79) was careful to explain that the typicality measure over branches involved no reference to probability. He was also careful to note that while he sometimes used the language of probability theory to discuss the measure, this was only for ease of exposition.[16] In this regard, he repeatedly instructed the reader that all talk of probabilities should be translated back to talk of the typicality measure (Everett 1956, p. 127; 1957, p. 193).

The fact that Everett did not aim to deduce probabilities over branches corresponding to alternative measurement outcomes is something that distinguishes his approach from most other formulations of quantum mechanics, including those of most Everettians.[17]

---

15 See Everett (1956, pp. 123–128; 1957, pp. 188–193) for his description of his typicality measure and how it works in his account of probability.

16 This is because, while the measure does not represent quantum probabilities, it satisfies the formal definition for being a probability measure.

17 That said, as mentioned earlier, he did want to justify using the standard quantum probabilities for practical purposes. We will return to this point later.

The following is Everett's setup and deduction. Consider a measuring device M that is ready to make and record an infinite series of measurements on each of an infinite series of systems S_k in state:

$$\alpha|\uparrow_x\rangle_{S_k} + \beta|\downarrow_x\rangle_{S_k}.$$

Suppose that M interacts with each system in turn and perfectly correlates its kth memory register with the x-spin of system S_k by the linear dynamics. This will produce an increasingly complicated entangled superposition of different sequences of measurement outcomes. After one measurement, the state of M and S_1 will be:

$$\alpha|\text{“}\uparrow\text{”}\rangle_M |\uparrow\rangle_{S_1} + \beta|\text{“}\downarrow\text{”}\rangle_M |\downarrow\rangle_{S_1}.$$

After two measurements, the state of M, S_1 and S_2 will be:

$$\alpha^2|\text{“}\uparrow\uparrow\text{”}\rangle_M |\uparrow\rangle_{S_1}|\uparrow\rangle_{S_2} + \alpha\beta|\text{“}\uparrow\downarrow\text{”}\rangle_M |\uparrow\rangle_{S_1}|\downarrow\rangle_{S_2} + \beta\alpha|\text{“}\downarrow\uparrow\text{”}\rangle_M |\downarrow\rangle_{S_1}|\uparrow\rangle_{S_2} + \beta^2|\text{“}\downarrow\downarrow\text{”}\rangle_M |\downarrow\rangle_{S_1}|\downarrow_x\rangle_{S_2}.$$

Figure 4.1 illustrates the first three steps of the branching process, the amplitudes associated with each branch and the sequence of x-spin records in each branch. The typicality measure for each branch is the norm-squared of the associated amplitude.

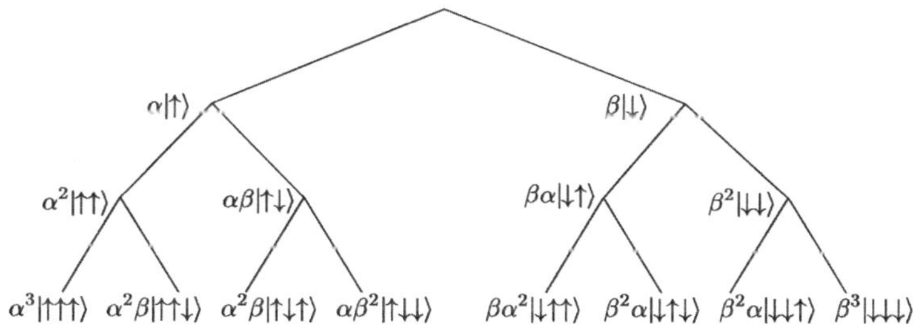

Figure 4.1. *Branches and weights*

Within this framework, Everett argued that, as the number of measurements increases, almost all branches in measure m describe sequences of measurement records that are randomly distributed with standard quantum statistics.

Indeed, we can show the following:

– **Relative frequency**: for any $\delta > 0$ and $\epsilon > 0$, there exists a k such that after k measurements, the sum of the norm-squared values of the amplitude associated with each branch where the distribution of spin-up results is within ϵ of $|\alpha|^2$ and the distribution of spin-down results is within ϵ of $|\beta|^2$ is within δ of one.

– **Randomness**: the sum of the norm-squared of the amplitude associated with each branch where the sequence of relative records satisfies any standard criterion for being random increases to one as the number of measurements increases.

The randomness result holds for any criterion of randomness that classifies at most a countable number of ω-length binary sequences as non-random.[18]

It is an immediate consequence that a typical relative observer's sequence of relative records, in the measure m sense of typical, will exhibit standard quantum statistics as the number of measurements increases. As Everett put the point:

> In the language of subjective experience, the observer which is described by a typical element... of the superposition has perceived an apparently random sequence of definite results for the observations. ...It will thus *appear* to the observer which is described by a typical element of the superposition that each initial observation on a system caused the system to "jump" into an eigenstate in a random fashion and thereafter remain there for subsequent measurements on the same system. Therefore, qualitatively, at least, the probabilistic assertions of Process 1 *appear* to be valid to the observer described by a typical element of the final superposition (Everett 1956, p. 123).

Therefore, a typical relative observer will find themselves with a sequence of measurement records that are randomly distributed with standard quantum relative frequencies. This is Everett's deduction of quantum probabilities as subjective appearances to a typical relative observer.

18 See Hartle (1968), Farhi et al. (1989) and Barrett (1999) for proofs of the first theorem and Barrett (1999) for a proof of the second when there are finitely many possible outcomes for each measurement. Barrett and Goldbring (2022) prove the analogous theorems when there are hyperfinitely many possible outcomes. Regarding the second theorem, see Barrett and Huttegger (2020) for a discussion of how we might understand quantum randomness as a variety of algorithmic randomness.

4.5. Strong adequacy and pure wave mechanics

Given the formal results of the last section, the relative-state formulation of pure wave mechanics is strongly adequate for a typical observer, in Everett's measure m sense of typical, if (1) we take it to make sense to talk of the experience of a physically situated *relative* observer; (2) we suppose that the *experience* of a relative observer is determined by their relative records; and (3) we take the prediction that a typical relative observer's records at a time after many measurements will exhibit the standard quantum statistics to be sufficient to capture the standard empirical content of quantum mechanics. However, even granting the first two assumptions, there is good reason to worry over the third.

The standard formulation of quantum mechanics predicts that observer A will record x-spin up with probability $|\alpha|^2$ and will record x-spin down with probability $|\beta|^2$. Such forward-looking probabilistic predictions are an essential part of the empirical content of quantum mechanics, but nowhere in Everett's account is there an explanation for why a particular observer should expect the standard quantum probabilities to hold for *their* future measurement results. As a result, even with the auxiliary assumptions that Everett adds to pure wave mechanics to obtain points (1) and (2) above, his theory fails to make the usual quantum predictions.

Everett's strategy was to deduce the standard quantum statistics as features of typical sequences of relative records by considering how relative records are produced when a situated observer correlates the state of their physical record with the physical property they are observing. However, that a typical sequence of relative records exhibits the standard quantum statistics does not entail that we should expect our future records to be typical. It does not even entail that we should expect our present records to be typical. To obtain expectations regarding present records, we would need an additional assumption that, together with the rest of the theory, provides expectations across alternative branches at a time rather than just statistics within typical branches, something that tells the observer that they should *expect to be typical* in the measure m sense of typical.[19] To obtain expectations regarding their future records, we would also need to say something regarding how an agent is supposed to identify themselves over time, something we will discuss in a moment.

Let us call an auxiliary assumption that allows a situated observer to infer expectations from the typicality measure a $\star$-principle. Equipped with a way of identifying observers over time, a $\star$-principle suitable for inferring forward-looking

19 See Hemmo and Shenker (2015) and Barrett (2017, 2020) for discussions of the distinction between typicalities and probabilities.

probabilities might stipulate that the probability of an observer experiencing a branch is given by the measure that m assigns to that branch conditional on the observer in fact making the measurement.[20] Note that, in contrast with Everett's approach, a principle such as this might serve as the basis for a situated observer assigning *probabilities across alternative present or future branches*.

That said, simply supplementing Everett's theory with a $\star$-principle is not enough. To make sense of forward-looking quantum probabilities, we also need a suitable way of understanding what these probabilities concern. Further, to get a strongly adequate theory, we must say something regarding the diachronic identity of observers and their records.

4.6. Splitting worlds and self-location

Suppose that we take each determinate-record branch of the postmeasurement state to be a physically distinct world that resulted from a single premeasurement world. In this case, there is one observer A before the measurement and two observers $A_{\uparrow_x}$ and $A_{\downarrow_x}$ after the measurement, the first in a world where the outcome is x-spin up and the second in a world where the outcome is x-spin down. If premeasurement A takes both future observers to be their future self, perhaps by being psychologically continuous with their premeasurement self, then they should expect the result "$\uparrow_x$" with probability 1 and the result "$\downarrow_x$" with probability 1 as well. If premeasurement A takes neither future observer to be identical to their present self, perhaps because $A_{\uparrow_x}$ and $A_{\downarrow_x}$ are symmetrically related to their present

20 There is a long tradition of adding other principles to pure wave mechanics from which we try to might derive a $\star$-principle. These auxiliary assumptions often include something like a principle of indifference. It is sometimes even suggested that we get to assume a principle of indifference for free as a basic constraint of reason. Sebens and Carroll (2018) appeal to something they call the epistemic separability principle (ESP) to get synchronic self-location probabilities. ESP is meant as a generalized principle of indifference that allows an agent to assign the standard quantum probabilities over alternative quantum worlds as self-location credences. In contrast, Wallace (2010) provides a set of 10 auxiliary assumptions that together entail the usual quantum probabilities. In this case, there is also an appeal to an indifference principle. My own sense is that there is nothing wrong with such an approach except that we do not get a principle of indifference for free. A rational inquirer need not assign their priors according to a principle of indifference. Further, in the present case, whether a particular principle of indifference in fact holds is contingent on how the world actually works. In the present case, how we set our priors will affect our resulting quantum probabilities and hence is part of the predictive apparatus of the theory itself. See Greaves (2007), Wallace (2012) and Barrett (1999, 2020) for further discussions on each side of this issue.

self but are not themselves identical and they hold that identity requires transitivity, then they should expect each result with probability 0. If they do not know how to think of their future self and their records in situations such as this, then they have no probabilities at all. In none of these cases does A obtain the right forward-looking probabilities.

While obtaining forward-looking self-location probabilities requires special care, we can obtain the right *synchronic* self-location probabilities on even a straightforward splitting-worlds account.[21] Suppose that A has performed their measurement and that the premeasurement world has split into one postmeasurement world for each possible record. Each copy of A now inhabits a world where there is a different but perfectly determinate physical record. Consider one of the two postmeasurement copies of the observer, for example, $A_{\uparrow_x}$. Suppose that they have not yet looked at the measurement record in their world. Before they do, they might ask themselves what their degree of belief should be that the physical record in the world they now inhabit is "$\uparrow_x$" and what their degree of belief should be that it is "$\downarrow_x$." The other copy $A_{\downarrow_x}$ might do the same. If the theory is supplemented with a suitable $\star$-principle, they might each assign the standard quantum probabilities as synchronic self-location probabilities. However, these are synchronic probabilities regarding what each observer will see when they observe their already present record, not forward-looking quantum probabilities for what record A will obtain before they perform her measurement.

The observer A can in this case predict before performing their measurement that each future copy of themselves will have postmeasurement synchronic self-location probabilities given by the usual quantum probabilities. This is a prediction, but its content is not the standard forward-looking quantum probabilities over possible measurement outcomes. For that, A must expect, before they perform the measurement, that the probability that *they* will record each result is given by the standard quantum probabilities. This is not at all the same as A believing that each future copy of themselves will assign the standard quantum probabilities as synchronic credences for their present record.

The former is an expectation regarding the value of their future measurement record, whereas the latter is at most an expectation regarding the epistemic states of their future selves.[22]

---

21 See Saunders et al. (2010) and Vaidman (2012, 2021) for discussions of probability in Everett and self-location probabilities more specifically. As we will see, how we understand self-location probabilities depends on the details of our metaphysics.

22 See Lewis (2007, pp. 10–12) for a reply to Vaidman (2002) and Greaves (2004) along these lines.

To predict that A will obtain record r with probability p_r, there needs to be a way of identifying A across time, and not every way will work. As we just saw, if A is in fact both $A_{\uparrow_x}$ and $A_{\downarrow_x}$, then they obtain each result with probability 1. If they are neither, then they obtain each result with probability 0. Neither are not the right forward-looking probabilities.

The point is that for a theory to predict the standard forward-looking probabilities for a situated observer, it needs a suitable $\star$-principle, and it needs to provide something appropriate for the resultant probabilities to concern, which, in turn, requires some metaphysical scaffolding. We must be able to understand postmeasurement observers, however many there are, as having determinate measurement records. Further, we need a notion of trans-temporal identity so that a premeasurement observer can ask what the theory predicts for *their* postmeasurement record.

4.7. Two metaphysical options

We will briefly consider two ways we might seek to provide metaphysical scaffolding for the forward-looking probabilities of a situated observer.[23] Each involves denying that measurements create new observers.[24] However, one clearly works better than the other.

The first approach is due to Saunders and Wallace. In the metaphysics they propose, a person is a four-dimensional entity, and in the context of an event such as A's x-spin measurement, "there are two persons present all along" (Saunders and Wallace 2008a, p. 294). Let us call these observers A and A^*. Both A and A^* are present before and after the measurement. They coincide in the same physical object

---

23 This is not meant as an exhaustive list of options. See and Albert and Loewer (1988) and Tappenden (2008) for examples of other, quite different, approaches. See the next footnote for a brief description of the Albert and Loewer metaphysics. For his part, Tappenden follows Sider's (1996) approach to identity.

24 There are other Everettian formulations of quantum mechanics that share this feature. On Albert and Loewer's (1988) many-minds theory, each mind associated with a single physical observer is present both before and after a measurement and minds never split. Forward-looking probabilities concern what a particular mind will experience after the measurement. While this approach provides perfectly ordinary forward-looking probabilities, since the minds are nonphysical, observers are not physically situated. See Barrett (1999) and Zwirn (2016) for similar approaches, closely akin to Albert and Loewer's *single-mind* theory, and Barrett (2020) and Lewis (2016) for discussions.

(at each time) before the measurement and are different physical objects with different physical records (at each time) after the measurement.[25] The two observers are said to *overlap* in the same world before the measurement and then split, occupying different worlds thereafter. Following Saunders (2010b, p. 197), we call this the overlapping-world approach.[26]

The states of the two observers before and after the x-spin measurement are considered. As indicated in Figure 4.2, A and A^* are physically identical (at each time) until the measurement occurs; subsequently, they are physically distinct (at each time), with different physical records described by the elements in the global state written on the determinate record basis.

We suppose that A is the observer who will obtain the result "$\uparrow_x$" and that A^* is the observer who will obtain the result "$\downarrow_x$".

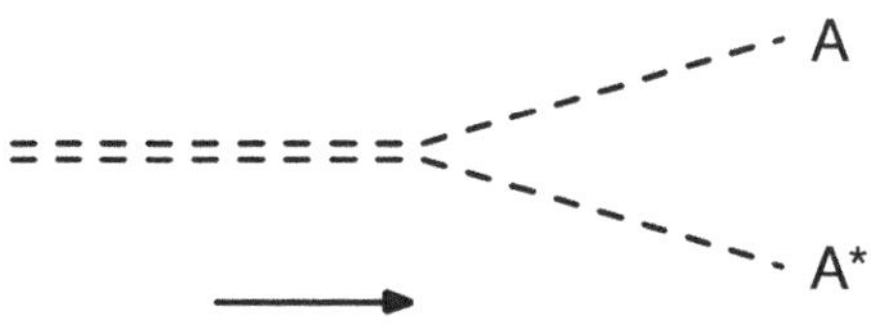

Figure 4.2. *Overlapping worlds*

To make sense of forward-looking probabilities as measures of uncertainty, there must be something concerning which each premeasurement observer is uncertain. On the overlapping-worlds account, each observer before the measurement is supposed to be uncertain about their premeasurement identity. In the present case, when a premeasurement observer contemplates making a measurement and wonders concerning what result they will obtain, they are wondering whether they are A or A^*. If they are A, then they will certainly obtain x-spin up, and if they are A^*, then they

25 This follows a proposal made by David Lewis (1976, 1983) in reply to Parfit's (1971) worries over making sense of personal identity in the context of splitting. See Lewis (2007, 2016), Tappenden (2008), Saunders and Wallace (2008a, 2008b), Saunders (2010a, 2010b) and Wilson (2011) for discussions of this approach in the context of Everett.

26 There is ample room here for terminological confusion. While overlapping worlds may sound like worlds that diverge at a time, it is customary to contrast overlapping and diverging worlds. In addition, what Saunders (2010b) and Wilson (2020) call *divergent worlds* is what Barrett (1999) called *many threads*. We will use many-threads here, but in some ways *parallel worlds* would be yet better.

will certainly obtain x-spin down. A suitable $\star$-principle, then, would tell the observer to assign $|\alpha|^2$ to the former possibility and $|\beta|^2$ to the latter. The idea is that the standard forward-looking quantum probabilities are ordinary epistemic probabilities from the premeasurement observer's perspective concerning their present identity.

While the notion of personal identity involved here is counterintuitive, that is much less of a concern if it provides a satisfactory framework for making sense of forward-looking quantum probabilities. However, Lewis (2007) has given good reason for believing that this approach does not work.[27]

Lewis's argument is as follows. At all times before the measurement, A and A^* coincide in precisely the same physical system. As a result, since temporally local events are not individuated as persons are, the premeasurement question "What result will I get?" is asked by a person-stage that is common to both A and A^*. Hence, the question involves a single act of reference where the pronoun I refers to *both* A and A^*. Therefore, the premeasurement question unpacks as "What result will A and A^* obtain?". Insofar as this question makes sense at all, it involves no uncertainty. Given how persons are identified on the present account, the theory simply predicts that A will obtain "$\uparrow_x$" and that A^* will obtain "$\downarrow_x$". Since there is no uncertainty regarding the result that A or A^* will obtain, there is nothing that an observer's premeasurement epistemic probability might represent.[28]

The problem with the overlapping-worlds account on this argument, then, is that neither observer can represent uncertainty regarding their identity before the measurement because, inasmuch as they are physically identical at each premeasurement time, neither can refer only to themselves. Whenever the premeasurement question "Which observer am I?" is asked, the right answer is always "A and A^*". Hence, if the premeasurement question "What is the probability that I am A?" is asked, the answer is 1, and the same is true for the question "What is the probability that I am A^*?".

As metaphysical arguments go, I find this one pretty compelling. If there is something wrong with it, then Saunders and Wallace's metaphysical framework must be yet more complicated than the present reconstruction suggests. If there is a more straightforward way to obtain forward-looking probabilities for physically situated observers that is at least as well motivated by the structure of Everettian quantum mechanics, then this metaphysical complexity clearly counts against their framework.

---

27 See Tappenden (2008) for a similar line of argument.

28 See Lewis (2007, pp. 6–7) for further discussion.

The many-threads approach is an example of a metaphysical framework that works in an entirely straightforward way.[29] Here, there is precisely one physical world for every possible path or *thread* through the branching structure generated by a sequence of measurement-like interactions, a metaphysics that adheres closely with the formal structure described by Everett. Many of these worlds are inhibited by perfectly ordinary physically situated observers.

In this approach, a situated observer is an ordinary three-dimensional physical object that inhabits a world and persists through time. There are two physically situated observers A and A^* at each time both before and after the measurement. Each inhabits a different world with a different, but analogous, measurement event, which results in the record corresponding to that world. When A ends up with the physical record "$\uparrow_x$", as they certainly will, they are simply discovering a previously unknown property of the non-branching world they inhabit.

Here, quantum probabilities are a situated observer's credences for inhibiting a world with specified properties. A suitable $\star$-principle for the theory would recommend that an observer adopt a prior degree of belief of inhabiting each world equal to the norm squared of the amplitude associated with the thread characterizing that world.[30]

An observer then updates by conditioning on what they learn as they explore their world. They have empirical reason to accept or reject the theory's recommendation depending on whether the assignment of priors recommended by the $\star$-principle yields predictions that agree with their degree of success when they use the recommendation for forward-looking prediction.

On this approach, observers A and A^* perform their x-spin measurements, and A obtains the record "$\uparrow_x$" and A^* obtains the record "$\downarrow_x$" because that is simply what happens in their worlds. Quantum probabilities result from each observer's initial

29 See Barrett (1999, pp. 179–184; 2005, 2020, pp. 184–187) for discussions of the many-threads approach. It is closely related to many-histories formulations. See, for example, Gell-Mann and Hartle's (1990) reconstruction of Everett. Perhaps because it does not appeal in any way to the customary Everettian intuition of splitting, Lewis (2007) did not include this strategy in his survey of ways of making sense of forward-looking probabilities. Importantly, Saunders (2010b) also saw virtues in this approach in contrast with overlapping worlds. See Wilson (2011, 2013, 2020) for a closely analogous approach.

30 This is akin to the distribution postulate in Bohmian mechanics. If the theory is right and we violate its recommendation, we likely make the wrong empirical predictions. See Barrett and Goldbring (2022) for a mathematical framework that is well suited to this approach.

uncertainty concerning which non-splitting world they inhabit. Each observer starts with the standard quantum priors for each world being their own and then conditions on the events they in fact observe. Their resulting credences serve as forward-looking epistemic probabilities regarding what they will see later in their world.

Figure 4.3. *Many threads*

In the present case, before the experiment, both A and A^* assign probability $|\alpha|^2$ to being situated in a world where they record x-spin up and $|\beta|^2$ to being situated in a world where they record x-spin down. After they condition on their results, A will assign probability 1 to inhabiting an x-spin up world, and A^* will assign probability 1 to inhabiting an x-spin down world. Then, they use these updated credences to predict what will happen on subsequent measurements.

Since a particular observer A only needs to explain their own physical records, there is nothing to prevent them from supposing instead that their world is the only world. In this case, quantum probabilities would reflect their credences regarding each possible world being actual given their $\star$-principle. This would be akin to a hidden-variable theory such as Bohmian mechanics, with the $\star$-principle playing the role of the distribution postulate.[31] That said, Everett clearly did not have a picture like this in mind, as he explicitly distinguished his approach from Bohmian mechanics (Everett 1956, pp. 76 and 153).

4.8. Interacting worlds

In accordance with how Everett used the notion of relative states, we might individuate worlds on the many-threads approach by supposing that measurement interactions produce perfect correlations and then using relative states in the determinate-record basis to specify the full set of possible threads. Alternatively, we might individuate worlds on the basis of decoherence considerations.[32] In either

31 In this case, for reasons we will discuss in the next section, the determinate classical properties of A's world at a time would be insufficient to determine how it evolves.

32 See the discussions of many-threads and decohering histories in Barrett (1999).

case, a suitable ⋆-principle would yield the standard quantum probabilities as self-location credences for an observer finding themselves in each particular world.

However we individuate postmeasurement worlds, Everett required that they remain causally connected. A formulation of quantum mechanics was only satisfactory if it allowed us to provide a consistent narrative for his hypothetical drama. Since he required that process 2 characterize all physical interactions, if an O-type measurement (of the sort we considered earlier) is performed when A is in the global state $\lozenge$, the result would indicate that state $\lozenge$ obtains and that there are hence two relative copies of A, each with a different relative record. Consequently, we need to say how observers associated with different worlds might be physically distinct in a sense that explains their determinate subjective experiences and at the same time causally connected in a sense that explains such in-principle interference effects. This is the heart of the puzzle that Everett sought to address in his language-difficulty footnote.[33]

One way to account for such nested measurements via the many-threads approach is to simply accept that alternative worlds are physically distinct but causally connected. In this view, the value of a record at a time is determined by the local physical state of the record in a particular world, but how each relative record is produced is determined by interactions between worlds mediated by the evolution of the absolute global state.[34] Here, worlds are always physically distinct and never split. Each A observer has a determinate but different postmeasurement physical record, and they will obtain the same result again if they immediately repeat their measurement in their world. However, as Everett insisted, observer B is also right regarding the evolution of A's absolute state. As a result, from their calculation, B correctly predicts what would happen in the event of a Wigner's-friend-type measurement on the composite system containing A, their measuring device M and their object system S in every world.[35]

4.9. Discussion

Everett showed that a typical relative observer (in his specified sense of typical) will result in a sequence of measurement records that exhibit standard quantum

---

33 See Albert and Barrett (1995) for a discussion of this problem.

34 Of course, we need not think of this as *causal* influence between worlds. The point is just that a postmeasurement record sometimes fails to be determined by the premeasurement state of the world in which the measurement is performed.

35 See Albert (1992, Chapter 8) for a discussion of the observables that observers A and B can measure. See Sebens (2015) for another approach to interacting worlds.

statistics. While this provides a sense in which the relative state formulation of pure wave mechanics can be taken to be strongly adequate, we might want more.

As we noted initially, Everett (1957, p. 196) himself wanted a formulation of quantum mechanics that justified the use of standard probabilistic predictions for practical purposes. However, if we understand this as wanting a theory that makes the standard forward-looking predictions, he failed to provide it. That said, a many-threads formulation of Everettian quantum mechanics does precisely this.

On the many-thread approach, there is one never-splitting world for each path through the branching structure described by Everett. Each observer is situated in a world with perfectly ordinary physical records. To obtain the standard quantum probabilities, we supplement the relative-state formulation with a ⋆-principle that recommends that each situated observer adopt a prior degree of belief of inhabiting each physical world, equal to the norm squared of the amplitude associated with that world. Finally, we have discussed some of the ways we might fill in the remaining details to yield a particular many-threads theory.

The present approach is as motivated by the formal structure of Everett's theory as the overlapping-worlds approach. They both rely on how we individuate branches to ensure non-splitting observers. The difference is that the premeasurement observers who go their separate ways when a measurement is performed to occupy physically distinct worlds on the Saunders–Wallace approach always occupy physically distinct worlds on the many-threads approach. As a result, saying what a premeasurement observer's forward-looking quantum probabilities concern is entirely unproblematic in the latter case. They are simply synchronic epistemic probabilities regarding which world the observer already inhabits.[36]

4.10. References

Albert, D.Z. (1986). How to take a photograph of another Everett world. *Annals of the New York Academy of Sciences*, 480, 498–502. doi: 10.1111/j.1749-6632.1986.tb12452.x.

Albert, D.Z. (1992). *Quantum Mechanics and Experience.* Harvard University Press, Cambridge, MA.

Albert, D.Z. (2000). *Time and Chance.* Harvard University Press, Cambridge, MA.

Albert, D.Z. and Barrett, J.A. (1995). On what it takes to be a world. *Topoi*, 14(1), 35–37.

36 I would like to thank Peter Lewis and Paul Tappenden for helpful comments on an earlier draft of this paper.

Albert, D.Z. and Loewer, B. (1988). Interpreting the many worlds interpretation. *Synthese*, 77(2), 195–213.

Barrett, J.A. (1994). The suggestive properties of quantum mechanics without the collapse postulate. *Erkenntnis*, 41, 233–252.

Barrett, J.A. (1999). *The Quantum Mechanics of Minds and Worlds*. Oxford University Press, Oxford.

Barrett, J.A. (2005). Relativistic quantum mechanics through frame-dependent constructions. *Philosophy of Science*, 72, 802–813.

Barrett, J.A. (2011a). On the faithful interpretation of pure wave mechanics. *British Journal for the Philosophy of Science*, 62(4), 693–709.

Barrett, J.A. (2011b). Everett's pure wave mechanics and the notion of worlds. *European Journal for Philosophy of Science*, 1(2), 277–302.

Barrett, J.A. (2015). Pure wave mechanics and the very idea of empirical adequacy. *Synthese*, 192(10), 3071–3104.

Barrett, J.A. (2017). Typical worlds. *Studies in History and Philosophy of Science Part B*, 58, 31–40.

Barrett, J.A. (2020). *The Conceptual Foundations of Quantum Mechanics*. Oxford University Press, Oxford.

Barrett, J.A. (2021). Situated observation in Bohmian mechanics. *Studies in History and Philosophy of Science*, 88, 345–357.

Barrett, J.A. (2023). Everettian quantum mechanics. Thesis, Stanford Encyclopedia of Philosophy [Online]. Available at: https://plato.stanford.edu/entries/qm-everett/ [Accessed 30 July 2024].

Barrett, J.A. and Byrne, P. (eds) (2012). *The Everett Interpretation of Quantum Mechanics: Collected Works 1955–1980 with Commentary*. Princeton University Press, Princeton.

Barrett, J.A. and Goldbring, I. (2022). Everettian mechanics with hyperfinitely many worlds. *Erkenntnis*, 89(4), 1–20.

Barrett, J.A. and Huttegger, S.M. (2020). Quantum randomness and underdetermination. *Philosophy of Science*, 87, 391–408. doi: 10.1086/708712.

Bell, J.S. (1981). Quantum mechanics for cosmologists. In *Speakable and Unspeakable in Quantum Mechanics*. Cambridge University Press, Cambridge.

DeWitt, B.S. (1971). The many-universes interpretation of quantum mechanics. In *Foundations of Quantum Mechanics*. Academic Press, New York.

DeWitt, B.S. and Graham, N. (eds) (1973). *The Many-Worlds Interpretation of Quantum Mechanics*. Princeton University Press, Princeton.

Everett, H., III. (1956). The theory of the universal wave function. Thesis, Princeton University.

Everett, H., III (1957). "Relative State" formulation of quantum mechanics. *Reviews of Modern Physics*, 29, 454–462.

Farhi, E., Goldstone, J., Gutmann, S. (1989). How probability arises in quantum mechanics. *Annals of Physics*, 192(2), 368–382.

Frigg, R. and Werndl, C. (2024). Philosophy of statistical mechanics. In *The Stanford Encyclopedia of Philosophy*, Zalta, E.N. and Nodelman, U. (eds). Stanford University Press, Stanford [Online]. Available at https://plato.stanford.edu/archives/spr2024/entries/statphys-statmech/.

Gell-Mann, M. and Hartle, J.B. (1990). Quantum mechanics in the light of quantum cosmology. In *Complexity, Entropy, and the Physics of Information*, Zurek, W.H. (ed.). Addison-Wesley, Redwood City, CA.

Goldstein, S. (2022). Excerpts of 1989 interview of John Bell by Renée Weber [Online]. Available at: https://sites.math.rutgers.edu/oldstein/bell-weber.html [Accessed 22 August 2022].

Greaves, H. (2004). Understanding Deutsch's probability in a deterministic multiverse. *Studies in History and Philosophy of Science Part B*, 35, 423–456.

Greaves, H. (2007). Probability in the Everett interpretation. *Philosophy Compass*, 2, 109–128. doi: 10.1111/j.1747-9991.2006.00054.x.

Hartle, J.B. (1968). Quantum mechanics of individual systems. *American Journal of Physics*, 36, 704–712.

Hemmo, M. and Shenker, O. (2015). Probability and typicality in deterministic physics. *Erkenntnis*, 80(3), 575–586.

Ismael, J.T. (2007). *The Situated Self*. Oxford University Press, New York.

Ismael, J.T. (2016). *How Physics Makes Us Free*. Oxford University Press, New York.

Lewis, D. (1976). Survival and identity. In *The Identities of Persons*, Rorty, A.K. (ed.). University of California Press, Berkeley.

Lewis, P.J. (2007). Uncertainty and probability for branching selves. *Studies in History and Philosophy of Science Part B*, 38, 1–14.

Lewis, P.J. (2016). *Quantum Ontology: A Guide to the Metaphysics of Quantum Mechanics*. Oxford University Press, New York.

von Neumann, J. (1955). *Mathematical Foundations of Quantum Mechanics*. Princeton University Press, Princeton.

Parfit, D. (1971). Personal identity. *Philosophical Review*, 80, 3–27.

Saunders, S. (2010a). Many worlds? An introduction. In *Many Worlds? Everett, Quantum Theory, and Reality*, Saunders, S., Barrett, J., Kent, A., Wallace, D. (eds). Oxford University Press, Oxford.

Saunders, S. (2010b). Chance in the Everett interpretation. In *Many Worlds? Everett, Quantum Theory, and Reality*, Saunders, S., Barrett, J., Kent, A., Wallace, D. (eds). Oxford University Press, Oxford.

Saunders, S. and Wallace, D. (2008a). Branching and uncertainty. *British Journal for the Philosophy of Science*, 59(3), 293–305.

Saunders, S. and Wallace, D. (2008b). Reply. *British Journal for the Philosophy of Science*, 59(3), 315–317.

Saunders, S., Barrett, J., Kent, A., Wallace, D. (eds) (2010). *Many Worlds? Everett, Quantum Theory, and Reality*. Oxford University Press, Oxford.

Sebens, C.T. (2015). Quantum mechanics as classical physics. *Philosophy of Science*, 82(2), 266–291. doi: 10.1086/680190.

Sebens, C.T. and Carroll, S.M. (2018). Self-locating uncertainty and the origin of probability in Everettian quantum mechanics. *British Journal for the Philosophy of Science*, 69, 25–74.

Sider, T. (1996). All the world's a stage. *Australasian Journal of Philosophy*, 74, 433–453.

Tappenden, P. (2008). Saunders and Wallace on Everett and Lewis. *British Journal for the Philosophy of Science*, 59(3), 307–314.

Vaidman, L. (2002). Many-worlds interpretation of quantum mechanics. In *The Stanford Encyclopedia of Philosophy*, Zalta, E. (ed.), Stanford University Press, Stanford [Online]. Available at: plato.stanford.edu/archives/sum2002/entries/qm-manyworlds/.

Vaidman, L. (2012). Probability in the many-worlds interpretation of quantum mechanics. In *Probability in Physics*, Ben-Menahem, Y. and Hemmo, M. (eds). Springer, Berlin, Heidelberg.

Vaidman, L. (2021) Many-worlds interpretation of quantum mechanics. In *Stanford Encyclopedia of Philosophy*. Stanford University Press, Stanford.

Wallace, D. (2010). How to prove the born rule. In *Many Worlds? Everett, Quantum Theory, and Reality*, Saunders, S., Barrett, J., Kent, A., Wallace, D. (eds). Oxford University Press, Oxford.

Wallace, D. (2012). *The Emergent Multiverse: Quantum Theory according to the Everett Interpretation*. Oxford University Press, Oxford.

Wigner, E. (1961). Remarks on the mind-body problem. In *The Scientist Speculates*, Good, I.J. (ed.). Basic Books, New York.

Wilson, A. (2011). Macroscopic ontology in Everettian quantum mechanics. *Philosophical Quarterly*, 61, 363–382.

Wilson, A. (2013). Objective probability in Everettian quantum mechanics. *The British Journal for the Philosophy of Science*, 64, 709–737.

Wilson, A. (2020). *The Nature of Contingency: Quantum Physics as Modal Realism*. Oxford University Press, Oxford.

Zwirn, H. (2016). The measurement problem: Decoherence and convivial solipsism. *Foundations of Physics*, 46, 635–667.

5

Quantum Theory from a Pragmatist Perspective

Richard HEALEY
Emeritus of Philosophy, University of Arizona, USA

5.1. A pragmatist perspective

Pragmatism is not a doctrine but an approach to conceptual problems. It regards concepts as intellectual tools that we create and use for particular purposes, improving them in light of experience. From its American source in the work of Charles Peirce and William James, the widening river of pragmatist thought has split into many channels, reuniting as well as further dividing. The pragmatist perspective on quantum theory sketched out here and further developed in other work (Healey 2026) draws on more recent neopragmatist resources, particularly that provided by Brandom (1994, 2000) and Price (2013, 2025). I begin by describing my general pragmatist perspective before applying it to quantum theory.

Humans, including scientists, are evolved organisms. We differ from other organisms in many ways, including our capacity to learn and use language and other symbol systems to communicate with each other. Once equipped with such a symbol system, each of us can also use it to enrich their ability to communicate with their future self in writing or other recording media, to deliberate and plan future actions, and more generally to think about thoughts that it makes newly available. This last use of language is notable for two reasons. For a pragmatist, even our ability to entertain most private thoughts depends on the public linguistic community of which we are members, and the core function of these thoughts is to facilitate future action.

Quantum Physics and Cosmology,
coordinated by Hervé ZWIRN.

I see an important contrast here with a philosophical tradition that goes back at least to Descartes and perhaps continues through Husserl to contemporary phenomenologists. As I understand it, this tradition begins with the individual person's mind, with its thoughts and experiences, and seeks to recover an "external world" of physical objects and other persons either by inference or through some other process of "constitution". From a pragmatist perspective, this view is backward, as an individual human is above all an evolved organism: a part of the natural world.

My pragmatist perspective is thoroughly naturalistic. However, it differs from other forms of naturalism in the way that Price (2011, 2013) has clarified by distinguishing what he called subject naturalism from object naturalism.

Object naturalism has an ontological aspect and an epistemological aspect. Ontologically, it is the view that in some important sense, all that exists *is* the world studied by science. Epistemologically, it is the view that all genuine knowledge is scientific knowledge. For a subject naturalist, on the other hand, philosophy needs to begin with what science tells us about ourselves. In Price's view, subject naturalism does not imply object naturalism. Indeed, it undermines it by rejecting a key assumption needed to validate that inference that he calls "*Representationalism*":

> The function of statements is to represent worldly states of affairs, and true statements succeed in doing so.

From my pragmatist perspective, the function of many statements is not to represent worldly states of affairs, and such statements may be true without doing so. Mathematical truths, moral truths and truths about meaning have different functions, none of which is to represent anything in the world studied by natural science. A statement concerning Fermat's "last theorem" is a mathematical truth that represents relationships between positive integers. These exist, but not in the world studied by scientists, who nevertheless use them to express empirical truths.

Pragmatists have always regarded the view that truth consists of correspondence as, at best, an unhelpful metaphor with no non-vacuous literal content. Some (including Peirce and James) have tried to add such content in the form of a theory of truth. From my pragmatist perspective, this was a mistake, as Frank Ramsey and others have realized. Rather than attempting to define, analyze or theorize truth, a pragmatist should first ask why we have or need such a concept in the first place. From my pragmatist perspective, truth, reference and representation should all be given a deflationary treatment by offering not an analysis but an account of their functions in our practices.

One function is to enable us to easily say things that would be hard or even impossible to say without deploying these concepts. I can, for example, say that most of what Donald Trump says is not true, although even Donald Trump himself does not know just what he says. However, Price (2003) argued for a further normative role for the notion of truth in spurring reasoned debate among people within a linguistic community. I believe that the notion of relative truth can be given a similar deflationary treatment and that this has an important application to quantum theory, where it can play a similar normative role within a physically isolated branch of a linguistic community.

Bertrand Russell once said in a lecture that probability is the most important concept in modern science, especially as no one has the slightest notion of what it means (Hajek 2023). The confusion clearly emerges from the diversity of attempts to make sense of quantum theory, in which probability plays a central role. From a pragmatist perspective, probability plays essentially the same role in quantum theory as it does elsewhere: it gives good advice to anyone in a physical situation with limited epistemic access to certain events on the appropriate degree of confidence to hold regarding alternative possible outcomes of those events.

Charles Peirce, the founder of pragmatism, began by criticizing Descartes' method of skeptical doubt as well as his view of knowing what we mean when our ideas on some matters are clear and distinct (Peirce 1878). In his view, a scientific inquiry should be fuelled only by a *genuine* doubt on a matter, and he proposed the application of his pragmatic maxim as a further necessary step to be taken by anyone seeking to make their ideas clear. Pragmatists have continued Peirce's attempt to probe the grounds of meaning in our language in various ways.

My pragmatist perspective includes the idea that linguistic meaning arises from the use of statements and specifically from their use in inferences of one kind or another. Sellars coined the term 'material inference' to identify the relevant kinds of inference and to stress that meaning-conferring inferences are not logically valid. Brandom (1994, 2000) then developed his own inferentialist pragmatism, a resource on which I will draw on later, in offering a pragmatic perspective on quantum theory. Brandom himself takes his inferentialist pragmatism to conflict with philosophical naturalism because of its commitment to a basic notion of a social norm. However, for a subject naturalist, there need be no such conflict, as Jaroslav Peregrin (2024) has recently argued.

Subject naturalism diverges from object naturalism through its rejection of representationalism. Even when some terms in scientific theory represent something in the natural world, this may not be its function. From a pragmatist perspective, the goals of natural science may be successfully pursued by developing a theory whose function is something other than faithfully representing the natural world. These goals include the explanation, prediction and control of natural phenomena. A theory may further these goals without itself offering a representation of the world in which those phenomena occur.

This may be hard to see if we continue to think of a theory as a set of statements, perhaps including axioms or laws and their deductive consequences. However, physical theory is better thought of as providing a family of mathematical models that may be applied to a variety of physical systems to explain, predict and possibly control their behavior. One way to apply a model is to take it to offer a representation of those systems and their behavior. However, this is not the only way a model of a theory can be applied. An alternative strategy for applying a theory is to use its models directly to guide a physically situated agent on what degree of belief to hold regarding alternative possible outcomes of events to which the agent lacks direct epistemic access. A user of the theory may then make use of this advice to make predictions (by putting themselves in that physical situation), to control phenomena (by suitably engineering that situation) or to provide explanations (of phenomena observed by one in that situation). From my pragmatist perspective, that is how models of quantum theory are applied. From this perspective, the novel nonrepresentational function of its models is what makes quantum theory revolutionary in physics.

5.2. How quantum theory implements the novel strategy

As Wallace (2020) noted, quantum theory is best thought of as a framework in which a variety of specific theories have been, and are being formulated. These include nonrelativistic quantum mechanics, relativistic quantum mechanics, quantum field theories (relativistic and nonrelativistic, Lagrangian, algebraic, axiomatic, quantum electrodynamics and standard model gauge theories), string theory, loop quantum gravity, etc. Some theroies have been extraordinarily successful, yielding extremely accurate predictions and powerful explanations, while others have failed. Others have not yet led to novel predictions or have not yet reached any settled formulation.

While these theories all have what Wittgenstein called a family resemblance, they do not make the same assumption about the structure of space-time, and many are often taken to differ in their ontologies (fields or particles). Nevertheless, they do share some abstract structural features: all of their models contain some mathematical objects (such as a wave-function) whose application generates probabilities and other mathematical objects (linear operators) associated with measurable magnitudes (as the operator $-i\hbar\partial/\partial x$ is associated with the x-component of momentum in the Schrödinger-picture position representation of nonrelativistic quantum mechanics).

From a pragmatist perspective, the main function of a quantum state is to generate probabilities in an application of the Born rule. A quantum model is applied via the Born rule. In a simple quantum model, this takes the form (for single probabilities):

$$\text{prob}_\rho(M_s \varepsilon \Delta) = \text{Tr}(\rho \mathbf{P}_M[\Delta])$$

Here, ρ is a density operator representing a quantum state assigned to some system s, and the statement "$M_s\varepsilon\Delta$" locates the value of magnitude M (on s) in the Borel set Δ of real numbers. $\mathbf{P}_M[\Delta]$ is the relevant projection operator from the spectral family defined by the unique self-adjoint operator $\hat{M}$ corresponding to magnitude M.

Like other objective probabilities, those generated in a legitimate application of the Born rule offer reliable advice to any agent able to understand and follow it. This is advice as to the degree of belief we should have in each of a set of alternative possible events if we were in a specific physical situation, blocking better epistemic access to what actually happens. The spatiotemporal location of a localized agent blocks epistemic access to its future: in a relativistic space-time, it can access information transmitted only from its backward light cone. However, a situated agent may face additional physical barriers to informational access, as does Wigner (1962), while he remains outside the isolated laboratory, where his friend has agreed to perform a quantum measurement inside.

An objective probability distribution offers advice tailored to a physical situation that a localized agent might occupy: an agent need not occupy that situation to benefit from this advice, nor choose to heed it even if they do occupy it. This makes Born probabilities (like other objective probabilities) relative not to any actual agent or their opinions (and so not in that sense subjective). Born probabilities are relative not to an agent but to a *physical agent-situation*. They serve the epistemic and practical needs of any agent who happens to be in that situation: by aligning degrees of belief about possible events to their Born probabilities, that agent can make better decisions about uncertain consequences of their actions. However, an agent not in that situation (even a situation one supposes to be located in a world containing no agents) can still use Born probabilities to explain phenomena they can neither control nor observe. Applications of quantum theory extend beyond the enhancement of an agent's ability to make prudentially rational decisions about how to act.

Because its primary function is to yield Born probabilities, a quantum state is also objective but relative to a physical *agent-situation*. A system is not *in* a quantum state: a quantum state may be correctly assigned to a system *relative to* a physical agent-situation. A system may be correctly assigned more than one quantum state at once, each relative to a relevantly different agent-situation. We can even describe circumstances in which it is correct to assign a system different *pure* quantum states, each relative to a different agent-situation (Wang et al. 2022). These different state assignments yield different Born probabilities for the same possible events, but each assignment is correct relative to the corresponding agent-situation. A statement may truly represent a quantum state or Born probability. However, this is not its function, and neither the state nor the probability is or represents what Bell (2004) called a beable of quantum theory: a purported element of physical reality.

According to most textbooks, quantum theory predicts the Born probabilities of our observations of alternative outcomes when we measure a magnitude. However, those textbooks do not say what a measurement is, when it occurs, or whether there must be someone observing its outcome. The second advisory function of a quantum state is to license the application of the Born rule to yield probabilities of those mutually exclusive magnitude claims $M_s \varepsilon \Delta$ with enough content to be evaluable as true or false. According to *inferentialist pragmatism* (Brandom 1994, 2000), how much content a claim has is a function of the reliability of inferences to and from that claim. We should apply the Born rule to a magnitude claim only if many such inferences would be very reliable.

The quantum state in a model of environmental decoherence does not describe or represent the physical process to which it is applied. Its function is rather to help anyone apply the model to gauge inferential reliability. Each magnitude claim that $M_s = m_j$ may be assigned a Born probability if the reduced quantum state of S in environment E is robustly diagonal in a "pointer basis" of eigenstates $|m_i>$ of $\widehat{M}$, the self-adjoint operator associated with M. The Born rule is applicable only to a magnitude claim $M_s = m_j$ on a system whose interaction with its environment closely meets that condition: I then call the physical context a *decoherence environment* for M.

When quantum theory is applied to model naturally occurring physical interactions, decoherence environments occur naturally, with no agents making quantum measurements. Decoherence in such models is extremely rapid, robust and practically irreversible. The Born rule is then applicable, whether or not an agent has arranged the conditions in which physical interactions produce a measurement outcome.

Decoherence environments are also relative. What counts as a decoherence environment is specified by the decoherence of a system's quantum state in a model. Since the quantum state is related to the agent-situation, what counts as a decoherence environment is also related to the agent-situation. This means that whether a measurement can be said to have an outcome is relative to the agent-situation. The correlations between the outcomes of space-like separated operations to measure different spin components on each of two ions assigned an entangled EPR-Bohm state $1/\sqrt{2}(|\uparrow\downarrow> - |\downarrow\uparrow>)$ provide an example (see Figure 5.1).

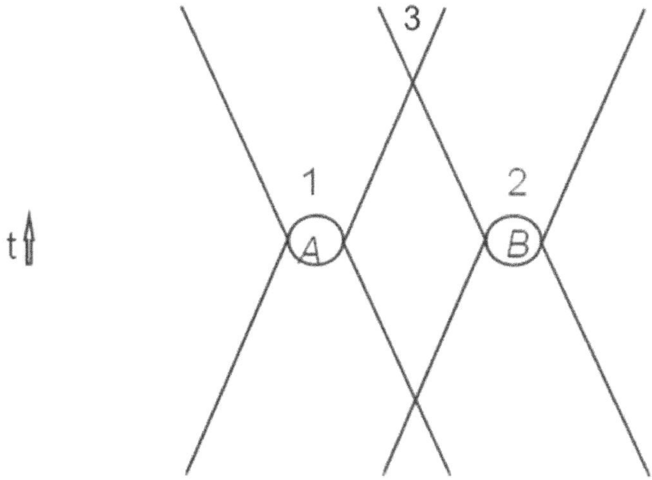

Figure 5.1. *Relative outcomes in the EPR-Bohm state*

This example shows how a measurement resulting from Alice's operation in region *A* may have an outcome relative to one agent-situation (1) but not relative to another (2). However, note that the primary relativity here is not to an agent-situation but to a decoherence environment. Suppose that we alter exactly which space-time location is the agent-situation relative to which the quantum state of each ion is assigned within the future light cone of a region marked in the figure by the numeral "1" and "2", respectively (where the two future light cones of the regions marked by the letters "*A*" and "*B*" do not overlap). There is no significant alteration in the relevant decoherence environments, nor in whether Alice's or Bob's measurement (induced by his operation in region *B*) has an outcome relative to each altered agent-situation. The outcome of a quantum measurement is relative to a decoherence environment and relative to an agent-situation only to the extent to which that environment is itself relative to the agent-situation.

Suppose that Alice interacts with her ion in region *A*, measures *x*-spin up and reassigns the state of her ion to $1/\surd 2(|\uparrow> + |\downarrow>)$ to successfully predict that her subsequent remeasurement will also yield spin up (as in Wang et al. 2022). By assigning state $1/\surd 2(|\uparrow> - |\downarrow>)$ to Bob's ion, Alice correctly predicts probability ½ for each possible outcome of the *z*-spin measurement on his ion implemented by his interaction in region *B*. However, since no quantum state describes an intrinsic property of the system to which it is assigned, Alice cannot infer that Bob's ion had *x*-spin down at any point in or on the light cone of region *B*. A model of environmental decoherence of Alice's ion that extends region *A* and includes her own agent-situation in the future light cone of *A* permits her to take "*x*-spin up" to be a meaningful claim about her ion, and relative to that decoherence environment, this truly reflects the outcome of her *x*-spin measurement.

Similarly, Bob measures *z*-spin up and reassigns the state of his ion to $|\uparrow>$ to successfully predict that his subsequent remeasurement will also yield spin up. A model of the environmental decoherence of Bob's ion that extends region *B* and includes his own agent-situation in the future light cone of *B* permits him to take "*z*-spin up" to be a meaningful claim about his ion, and relative to that decoherence environment, this truly reflects the outcome of his *z*-spin measurement.

However, no model of the environmental decoherence of Bob's ion extends region *B* as well as Alice's agent-situation outside the future light cone of *B*, so Bob's measurement has no outcome relative to such an agent-situation. Moreover, no model of the environmental decoherence of Alice's ions extends region *A* as well as Bob's agent-situation outside the future light cone of *A*, so Alice's measurement has no outcome relative to such an agent-situation. On the other hand, there is a model of environmental decoherence extending both regions *A* and *B* as well as any agent-situation in the overlap of the future light cones of regions *A* and *B*. Thus, both Alice's and Bob's measurements may be said to have outcomes relative to such an agent-situation.

5.3. No measurement problem

Many attempts to understand quantum theory have focused on what has come to be known as the quantum measurement problem. The quantum measurement problem involves reconciling quantum theory with the fact that measurements have outcomes of the sort of the theory was created to predict and (as far as possible) explain. Attitudes toward this problem vary widely. Optimists deny that there is any problem, whereas pessimists maintain that there is a serious problem that can be solved only by modifying quantum theory.

An intermediate opinion is that, while quantum theory does face a prima facie measurement problem, this problem is readily resolved by applying the unmodified theory to interactions that efficiently delocalize the phase of a system into its environment during a quantum measurement. Although widespread, this opinion remains controversial. Attempts to support it by argument have met objections from both optimists and pessimists intended to show why simply applying quantum theory to establish the rapid and effectively irreversible diagonalization of the reduced density operator of a system does not solve any problem posed by measurement in quantum theory.

Adopting a pragmatist perspective permits satisfactory resolution of the quantum measurement problem. The problem is most acute if we assume that a system's quantum state gives a complete description of that system. The linearity of quantum theory thus implies that measurements typically fail to have outcomes, in manifest contradiction to experimental reports. This problem disappears once we recognize that it is not the function of the quantum state to describe or represent the behavior of a quantum system. From a pragmatist perspective, the residual problem of when, and to what, to apply the Born rule is then resolved by judicious appeal to decoherence.

The resolution is exemplified by the situation depicted in Figure 5.1. Alice's operation in region A produces an outcome of her x-spin measurement only later (relative to an x-spin decoherence environment in the future light cone of region A and to an agent-situation there relative to which this is an x-spin decoherence environment). Alice's measurement has no outcome relative to an x-spin decoherence environment or agent-situation in region A. Moreover, as the agent-situation that Bob occupies changes, Alice's measurement has no outcome relative to Bob's measurement until he occupies an agent-situation in region 3. Parallel points apply to the outcome of Bob's z-spin measurement after interchanging "1" with "2", "x" with "z", and "A" with "B" and "Alice" with "Bob".

Both Alice's and Bob's measurements have outcomes related only to a joint decoherence environment for x-spin on Alice's ion and z-spin on Bob's ion and thus relative to an agent situation in region 3. In a realization of this scenario, experimental results are recorded, thereby contributing to statistics matching the joint probabilities yielded by application of the Born rule to the Bohm-EPR state. Each record of an individual outcome would still be related to some agent-situation in region 3. There is no absolute fact about the outcome of a quantum measurement, which raises questions about the objectivity of measurement outcomes and whether the statistics of our records of quantum measurement outcomes give us good reasons to accept quantum theory as a contribution to objective scientific knowledge. I will postpone discussion of these general questions about the objectivity of relative facts until section 5.7.

5.4. No nonlocal action

Some maintain that experimentally confirmed violations of Bell inequalities show that the world is nonlocal; that an event in space-time region R_1 that may result from a free human action influences what happens in a space-like separated region R_2 (Goldstein et al. 2011; Maudlin 2014). From a pragmatist perspective, quantum theory accounts for observed violations of Bell inequalities while denying any such nonlocal influence.

Bell inequalities constrain a set of probabilities for the outcomes of various quantum measurements, including joint measurements on each of a pair of quantum systems initiated by operations performed at distant locations. An experiment is taken to violate these when statistics of actual measurement outcomes significantly fail to match probabilities constrained in this way (while closely conforming to probabilities predicted by quantum theory). A Bell test therefore compares statistical data with rival probabilistic models of that data, refuting those constrained by Bell inequalities while supporting those predicted by quantum theory.

Quantum theory explains a violation of Bell inequalities because probabilities following from application of the Born rule to the correct quantum state imply those of a data model that best fits observed statistics, where each individual outcome is an effect of locally initiated causes that have no influence on any space-like separated outcome. The relativity of the measurement outcome to the agent-situation is crucial here because it implies that there is no outcome of a locally initiated quantum measurement relative to a space-like separated agent-situation, and so nothing there to influence. Bob's operations in region *B* of Figure 5.1 have no effect on Alice's measurement outcome, and nor do Alice's operations in region *A* have any effect on Bob's measurement outcome relative to any agent-situation outside region 3.

There is another, and perhaps better, way to think of a pattern of correlations in violation of a Bell inequality. Each joint probability in this pattern for the outcomes of a pair of measurements initiated in distant (including space-like separated) regions *A* and *B* is also a single probability for the outcome of a measurement of a single magnitude pertaining to the system pair and represented by an operator in its four-dimensional tensor product spin Hilbert space. This is an outcome relative only to an agent-situation in the overlap of the future light cones of regions *A* and *B*. Quantum theory explains the probability of all such outcomes by deducing it from an application of the Born rule to the spin quantum state assigned to the pair, given that each outcome is a joint effect of causes in *A* and *B*.

In his discussion of patterns of correlation violating Bell inequalities, Bell (2004, pp. 239–241) himself introduced a condition he called Local Causality and used perfect correlations for joint measurement outcomes that quantum theory predicts in an EPR-Bohm quantum state of photons to argue that quantum theory does not meet this condition.

> A theory will be said to be locally causal if the probabilities attached to values of local beables in a space-time region 1 are unaltered by specification of values of local beables in a space-like separated region 2, when what happens in the backward light cone of 1 is already sufficiently specified, for example, by a full specification of local beables in a space-time region 3 [a thick "slice" that fully closes the backward light cone of region 1 wholly outside the backward light cone of 2] (Ibid., p. 240).

Since the Local Causality condition is stated in terms of what Bell calls local beables, it cannot be applied without first specifying what the local beables are in the Gedanken-experimental situation he describes. This scenario includes two ideal polarizers placed facing each other, each oriented to pass vertically polarized photons but block horizontally polarized photons, with a (perfectly efficient) counter placed behind each to detect any photons that pass that polarizer. Pairs of photons are prepared between the polarizers, assigned what came to be known as the Bell state $\Phi^{+} = 1/\sqrt{2}(HH + VV)$.

Quantum theory predicts a perfect correlation between successes and failures in detecting photons on the two sides of the experiment: the event of a counter detecting or failing to detect a photon is described by a value of a local beable indicating the outcome of a measurement of the vertical polarization of that photon. Bell takes these events to occur in space-like separated regions 1 and 2.

A further difficulty arises in applying Bell's Local Causality condition to quantum theory. This theory attaches a probability (or chance) to values of such local beables in a region 1 or 2 only relative to an agent-situation, but its value is different relative to different agent-situations. This is illustrated in Figure 5.2 when the angle between the polarizer axes a and b is set to 0°. V_A represents the detection of a photon in region 1, given the detection of a photon in region 2.

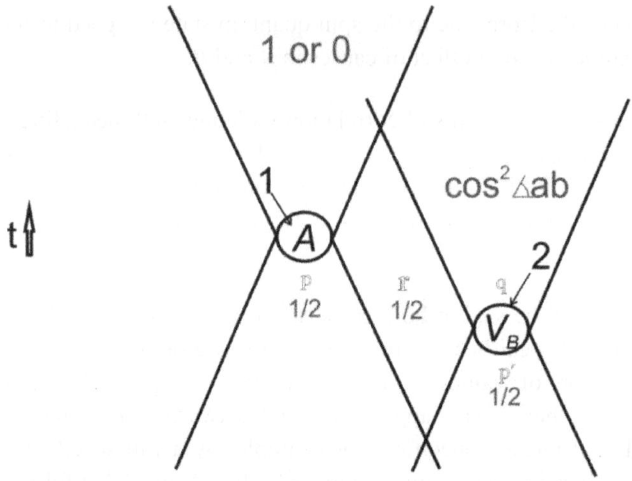

Figure 5.2. *"The" chance of V_A. For a color version of this figure, see www.iste.co.uk/zwirn/quantum.zip*

When applying quantum theory to this scenario, Bell mentions:

> The theory requires a perfect correlation of 'yeses' or 'nos' on the two sides. Therefore, specification of the result on one side permits a 100% confident prediction of the previously totally uncertain result on the other side (Ibid., p. 241).

This is true of the chance relative to an agent-situation in the future light cone of region 2 (such as that near *q*) but false of the chance relative to an agent-situation in the past light cone of region 1 (such as that near *p*), even if those agent-situations occur at the same time in the laboratory frame.

The key relation among probabilities Bell derives from his condition of Local Causality is this, where the value +1 or −1 of *A*, *B* indicates the answer "yes" or "no", respectively, to the question: "is a photon detected by a counter on that side of the experiment?".

$$Pr(A, B|a, b, c, \lambda) = Pr(A|a, c, \lambda). Pr(B|b, c, \lambda) \text{ (Factorizability)}$$

Here, *c* represents the values of any variables describing the experimental set-up, and λ represents the values of any hypothetical variables supposedly completing the quantum description of the scenario (of which there are none, from the present perspective). These values are assumed to give a complete specification of at least a region of space-time earlier than regions 1 and 2 and extending across their past light cones where they no longer overlap.

Bell took correlations satisfying Factorizability to be locally explicable because the values of A and B are probabilistically independent of one another and of the remote polarizer settings but are probabilistically dependent only on the local polarizer settings a and b, respectively, and on past causes c (if not also λ). The experimental demonstration that probabilities correctly predicted by quantum theory violate Factorizability led him to maintain that the observed correlations are locally inexplicable. While true given his understanding of "locally inexplicable", this does not imply that they are explicable by reference to space-like influences, say from one wing to another.

Experiments demonstrating the failure of Factorizability do not show that there are any such influences, and the application of quantum theory describes none. From a pragmatist perspective, quantum theory explains why Factorizability fails by deriving the probabilities it involves from the Born rule and presenting each (relative) outcome contributing to experimental statistics that they model as a probabilistic effect of causes in its past light cone. The quantum world is not nonlocal.

5.5. Application of quantum field models

Relativistic quantum field theories have been applied very successfully, both in the standard model of high-energy physics and elsewhere. However, there is still no agreement on what these theories are about (on the ontology of quantum field theory). Particles and fields are the two usual suspects, or candidates for this honor. However, philosophers have presented similar reasons for disqualifying both (Fraser 2008; Baker 2009). From a pragmatist perspective, this issue may be resolved by acknowledging that it is not the function of a quantum field-theoretic model (or any other quantum model) to describe or represent the nature or behavior of any systems to which it is applied.

A quantum field-theoretic model describes neither fields nor particles. However, when correctly applied, the model yields Born probabilities for alternative possible events described by magnitude statements, and those statements may be about quasiclassical fields or quasiclassical particles. Just as in nonrelativistic quantum mechanics, a system's quantum state offers advice on when it is legitimate to apply the Born rule, and to which magnitude statements, through application of a model of quantum decoherence.

In some circumstances, it is legitimate to apply the Born rule to a quantum field system to yield a probability distribution over a set of mutually incompatible but collectively exhaustive significant magnitude statements about one or more quasiclassical fields: in other circumstances, it is legitimate to apply the Born rule to a quantum field system to yield a probability distribution over a set of mutually incompatible but collectively exhaustive significant magnitude statements about one or more quasiclassical particles. The prefix "quasi" indicates that the field or particle does not possess the full complement of properties ascribed to it by a classical theory, even though it has enough of them to warrant application of the associated term.

Models of the quantum field theory of light (quantized vacuum electromagnetism) have been successfully applied to lasers. The quantum state of a laser may be represented as a vector in Fock space, the symmetrized direct sum of n-dimensional complex Hilbert spaces for all integral n. An m-dimensional subspace is often associated with the presence of m photons of a certain energy in the field, with superpositions of states in subspaces of different dimensions associated with an indefinite photon number.

From the present perspective, relative to one kind of decoherence environment in a low-powered laser, laser light contains m photons, each with a certain energy, which is significant enough to be assigned a probability by the Born rule. Relative to a different kind of decoherence environment in a higher powered laser, a statement that the laser light has a frequency in a certain narrow range is significant enough to be assigned a probability by the Born rule (this uses a coherent state decomposition of its Fock space quantum state: "adding a photon of energy hv" to a coherent state of frequency v leaves that state unchanged).

An interaction that induces the first type of decoherence environment would permit measurement of the number of photons in the laser light, whereas an interaction that induces the second type of decoherence environment would permit a measurement of the frequency of that light. In neither case would it be correct to say that its outcome revealed the number of (quasi)-particles or the frequency of the (quasi)-field. In each case, the outcome would be related to the particular decoherence environment in which it was obtained. A quantum field is a mathematical object in a quantum-field theoretic model that represents nothing present in any system to which the model is applied. In that sense, a quantum field theory has no ontology.

5.6. Wigner's friend

Wigner (1962) described a thought experiment and used it to support the view that "it was not possible to formulate the laws of quantum mechanics in a fully consistent way without reference to the consciousness". While failing to convince many to accept that view, the scenario he described in his thought-experiment continues to play an important role in discussions of the conceptual foundations of quantum theory.

The scenario presents what is commonly referred to as the paradox of Wigner's friend. Wigner imagines that his friend (a competent physicist) has agreed to perform a particular quantum measurement on a system inside her laboratory after it has been completely physically isolated from all external influences, whereas Wigner himself remains outside the laboratory. Wigner and his friend agree beforehand that they will prepare the system so that it is correctly assigned a quantum state that is a nontrivial superposition of eigenstates of a particular observable; say the state $1/\sqrt{2}(|\uparrow> - |\downarrow>)$ of a spin ½ particle, whose *z*-spin she is to measure.

His friend performs their measurement and records the outcome: *z*-spin up or *z*-spin down. The spin measurement was chosen such that they correctly reassigned the particle to the corresponding eigenstate, $|\uparrow>$ or $|\downarrow>$. Meanwhile, Wigner correctly assigns some quantum state Ψ to the entire laboratory and all of its contents and unitarily evolves that state during the entire period in which his friend was to perform their measurement. After his friend's measurement, neither the eigenstate $|\uparrow>$ nor $|\downarrow>$ is compatible with the evolved state $U\Psi$, so Wigner and his friend disagree about the quantum state of the particle.

Assuming that a quantum eigenstate of a system represents it as having the corresponding eigenvalue, they also disagree about whether a measurement has been performed inside the laboratory. If Wigner now enters the laboratory, his quantum state $U\Psi$ predicts that the outcome of his own measurement of the particle's spin will be either *z*-spin up or *z*-spin down, each with probability ½. However, it does not predict that his outcome will match what his friend took to be the outcome of their own measurement. For Wigner, there was no measurement outcome until the outcome that he himself obtained after entering the laboratory, which he would find to contain multiple records of that outcome, including his friend's reported memories.

In summary, Wigner and his friend disagree about whether there is an outcome of the friend's measurement, the resulting quantum state of her lab and its contents and the probability that Wigner will observe a particular outcome upon entering the laboratory.

In later publications, Wigner came to doubt that his own appeal to consciousness resolves this "paradox". However, an alternative strategy for resolving it seems available to one that rejects the inference from a system's quantum eigenstate to its possession of the corresponding eigenvalue. In this view, his friend's measurement has an outcome that is recorded in their lab, which Wigner comes to know only after entering their laboratory. The outcome of their measurement is an absolute fact. However, unitary reversal of the quantum state of the entire lab would erase all traces of this outcome.

In an earlier publication (Healey 2013), I adopted this strategy from a pragmatist perspective that acknowledged the limited relativity of quantum state assignments to agent-situations. I took these limitations to follow from a result proven in a previous paper (Brun et al. 2002) that purports to show how far such assignments may differ. I later realized that while the proof is valid, it assumes that a pragmatist should reject that all permissible assignments must be consistent with the existence of some unique pure state assignment. The fact that different pure states may be assigned relative to different agent-situations led me to realize that measurement outcomes as well as quantum state assignments must be considered relative to an agent-situation. The outcome of a quantum measurement is not absolute but a relative fact.

5.7. Extending Wigner's friend

Recent arguments based on extensions of Wigner's original scenario have put more pressure on the idea that the outcome of a quantum measurement is an absolute fact. The scenario of Wigner's friend may be extended by adding friends in isolated laboratories, each with their own external "superobserver" with exquisite quantum control over their friend's laboratory. The following is a schematic account of an extended Wigner's friend scenario (EWFS).

Each friend is an observer in a (different) physically isolated laboratory who is able to measure a magnitude on a quantum system and note the outcome. There are other "superobserver" agents: each agent has exquisite quantum control over everything in their friend's entire laboratory (including the friend). *Either* a superobserver can (unitarily) "undo" their friend's quantum measurement and then measure a different magnitude *or* perform a specific precise measurement on their friend's entire lab that also erases all of the friend's records of their own measurement outcome. Here, I focus on an EWFS that takes the first option.

Bong et al. (2020) derived probabilistic local friendliness inequalities for absolute outcomes in one EWFS in violation of QM predictions, and Haddara and Cavalcanti (2023) showed that outcomes that QM deems possible cannot all be absolute in a related EWFS based on the Hardy state. Both make a locality assumption.

– *Local agency*: any intervention (a free choice of variables) by an observer is uncorrelated with any set of relevant events outside its future light cone.

These arguments do not go through if the outcome of a quantum measurement is not absolute but relative to the observer, although there are other ways of blocking the conclusion. However, what is "an observer"? And when is "an observation" made?

The following scenario demonstrates the Local Friendliness inequalities. Inside their laboratories, the friends, Charlie and Debbie, measure a pair of particles assigned an entangled state, producing their outcomes labeled as *c* and *d*, respectively.

Outside of their laboratory, each "superobserver", Alice or Bob, makes an independent choice of measurement labeled as *x* and *y*, respectively, on the entire content of their laboratory, with outcomes labeled as *a* and *b*, respectively: Alice on Charlie and Bob on Debbie. The choices and measurements are both space-like separated.

Assume that each friend and superobserver observes an outcome every time a measurement is repeated on a different system such as that (for each "run"). Quantum theory predicts probabilistic correlations between pairs of these outcomes (in the EWFS of Haddara and Cavalcanti (2023), some of these correlations are perfect). If each outcome is an absolute event that is relative to no one and nothing else, then their relative frequencies in many runs must have a joint statistical distribution.

However, the probabilistic correlations between pairs of these outcomes predicted by quantum theory are incompatible with the existence of a joint probability distribution for all outcomes at once. Hence, if the quantum predictions are correct, then at least some of these measurement outcomes must be relative, not absolute (given the other plausible assumptions).

From my pragmatist perspective, a quantum measurement has an outcome only relative to a suitable decoherence environment: that outcome is only a relative fact. It follows that Charlie's and Debbie's measurements have no absolute outcomes. However, each has an outcome relative to the decoherence environment in their respective lab prior to the action of its external superobserver. This implies that there is no joint probability distribution for all outcomes at once. However, many questions must be addressed, including the following.

Is there an absolute fact about Alice's or Bob's measurement outcome? Is there even an absolute fact about a measurement outcome relative to a suitable decoherence environment? If there are no absolute facts about quantum measurement outcomes, do the statistics of those outcomes give us any objective reason to accept quantum theory? My answers to these questions are presented below.

Briefly, a relative fact is what is stated by a statement that is true relative to a context of assessment. However, this answer requires elaboration of the notions of relative truth and context of assessment. Fortunately, MacFarlane (2014) has provided such elaboration in his book *Assessment Sensitivity*, although his motives for doing so were quite different to mine, and his applications of relative truth there have nothing to do with quantum mechanics.

The notion of relative truth may be explained by contrast with the plain notion of truth. Among analytic philosophers, pragmatism is often characterized by a conception of truth associated with the writings of William James, who is commonly interpreted as holding that truth is whatever works in practice or is useful to believe, a view that Russell and Moore effectively criticized. However, Charles Peirce, the founder of pragmatism, and other 20th-century pragmatist thinkers hold very different views of truth. The one thing that pragmatists agree on is the negative thesis that it is unhelpful to say that truth consists of correspondence to the facts.

As I noted in the first section of this chapter, from the present pragmatist perspective, truth is not a substantive general property shared by statements or sentences of a certain kind that require philosophical analysis. In that sense, this pragmatist view of truth is deflationary. Instead, what is required is an account of why we have a concept of truth; of its function in a linguistic community such as ours.

One function is to permit us to say things it would be difficult or impossible to say without it. Donald Trump says a lot, and probably no one (including Trump himself) knows exactly what he says, and so cannot say of every one of his statements *S* "Trump says that *S*, and *S*", or "Trump says that *S*, but not *S*", where, for example, *S* is the statement "The 2016 US presidential election was rigged". However, after fact checking some statements by Trump, many people could use the term 'true' to say "most of what Donald Trump said was not true".

As noted above, Price (2003, 2025) argued that the concept of truth has an additional normative function within a linguistic community, the function of spurring reasoned debate between community members who disagree. Without the concept of truth, each party can simply continue to assert their own view of the matter while acknowledging that the other holds a contrary view. However, if they take themselves to be arguing about the truth of the matter, they have an added incentive to continue to offer reasons as to whose opinion is correct (if either) or so should be adopted by both. In this way, the concept of truth serves the social function of fostering agreement within the community, thereby facilitating collective action on community projects that may benefit all but that no individual community member could pursue alone.

In a deflationary view, plain, nonrelative, notions of truth and fact may be characterized by two principles.

– *Truth:* a statement that *P* is true if and only if *P*.

– *Fact*: a statement states a fact if and only if it is true.

Corresponding relative notions may be characterized by these alternative principles:

– *Relative truth*: a statement that *P* is true-relative to- *c* if and only if *P*-relative to- *c*.

– *Relative fact*: a statement states a fact-relative to- *c* if and only if it is true-relative to- *c*.

Here, *c* is a context at which a statement is assessed, not a context in which it is made. MacFarlane (2014) distinguishes these contexts when developing his own account of relative truth. A variety of truth-relativism is associated with a class of statements and contexts for which no plain notion of truth and fact is applicable, but only notions of relative truth and relative fact.

It follows from the plain concepts of truth and fact that a statement that *P* states a fact if and only if it is true. That is, if and only if *P*. In accordance with these concepts, a fact is a fact without regard to context. This irrelevance of context makes it natural (if overblown) to call a plain fact absolute or *transcendently objective*: it is absolute insofar as it is not relative to anything like a context or viewpoint, and transcendently objective as it is not limited by such things.

A perspectival statement is a statement assessable for truth only at a context. A relative fact is what is stated by a perspectival statement that is true as assessed at a context. A statement that is true relative to *c* may not be true relative to a different context c^*: it may even be false relative to c^*, and so strongly relative. A context c provides a semantic viewpoint on those perspectival statements assessable for truth at c. A perspectival statement assessed as true at context c states a perspectival fact, a fact in the perspective offered by c: the set of such perspectival facts constitutes the perspective P_c of c.

A relative fact in perspective P_c may be a relative fact in the perspective P_{c^*} of *every* context c^* at which it is assessed. That would not make this a plain fact because it is not simply true: the plain notion of truth is not applicable to a perspectival statement. It would rather make it a statement of (what may be called) an *immanently objective fact.*

After introducing the general notion of an immanently objective fact, we can now apply it to the specific case of reports of quantum measurement outcomes. In this case, a context of assessment for one or more magnitude claims is a decoherence context in which those claims are sufficiently meaningful to be assessed for truth or falsity. Such decoherence contexts occur when one or more magnitudes are measured. This application of the notion of an immanently objective fact is important because statistical summaries of the outcomes of such quantum measurements are the data scientists treat as the evidence against which they test predictions of the Born rule. When these statistics are judged to match the predicted Born probabilities in a wide range of cases while never significantly deviating from them, that is convincing evidence warranting acceptance of quantum theory.

Figure 5.2 illustrates the idealized scenario of one Bell test, in which statistics for joint outcomes of polarization measurements are collected for various values of the angle between axes a and b, as those settings are chosen randomly and independently at space-like separation (as in Aspect's famous experiments and their more recent improvements). These statistics violate Bell inequalities and match the Born probabilities for the entangled state assigned to the particle pairs. The statistics are collected in the overlap of the future light cones of regions 1 and 2. For any agent-situation in that overlap region, there is a decoherence environment relative to which both measurements of polarization in a recorded coincidence have outcomes. These provide an item contributing to the statistical data provided by the experiment. While there is no absolute fact about that item of data, there is the same relative fact in each of these decoherence environments. Therefore, the statistical distribution of the data collected is immanently objective.

The same analysis applies to Alice's and Bob's outcomes in an EWFS, such as those discussed in Bong et al. (2020) and Haddara and Cavalcanti (2023). This answers the second and third questions posed just now. Those outcomes constitute immanently objective relative facts, even though there is no (absolute) fact about any of Alice's or Bob's measurement outcomes. However, this raises further concerns. Since neither these nor any other outcomes of our quantum measurements constitute an objective fact, do the statistics of all of these outcomes give us any reason to regard them as providing us with any evidence for quantum theory, let alone the overwhelming evidence they are generally taken to give us?

More generally, if there were not immanently objective (relative) facts, then there would be nothing scientists would be justified in acknowledging as *data* capable of confirming or refuting scientific knowledge claims. Science depends on immanent objectivity but not transcendent objectivity. Because scientists and other observers in our world come to share essentially the same agent-situation, with a common decoherence environment, all of our quantum measurement outcomes are immanently objective (relative) facts. However, the perspectival character of these facts implies no relativism of scientific knowledge based on quantum theory because our changing physical situations always privilege a single, shared perspective, in which reports of the outcomes of our quantum measurements are immanently objective (relative) facts. These findings provide us with all the evidence we need to warrant the acceptance of quantum theory. Of course, because science is fallible, we should bear in mind that subsequent evidence may warrant a different attitude toward any specific quantum theory or even toward the whole framework of quantum theories.

In a recent paper, Riedel (2025) posed a challenge to the view that the outcome of a quantum measurement is a merely relative fact. He argues that any view that accepts both of two plausible conditions must extend this relativity even to these *relative* outcomes in such a way that the resulting relativity iterates an arbitrary finite number of times. These conditions are a locality requirement much weaker than that needed to derive a Bell inequality and a tracking requirement that operationalizes the idea that, when a scientist asks an experimenter what their measurement outcome was, they can take their sincere reply simply to report the actual outcome of their experiment.

In some of his writings, Carlo Rovelli already seemed ready to embrace such iterated relativism within his own relational quantum mechanics (RQM), although the principle of cross-perspective links expressing a tracking condition within RQM was made explicit recently in the joint paper (Adlam and Rovelli 2023). However, from my pragmatist perspective, there is no such iterated relativism even when we accept the needed modifications of both Riedel's weak locality condition and his tracking condition. The need for such modifications becomes clear when we appreciate the differences between what different relativistic views take quantum measurement outcomes to be relative to.

5.8. Four kinds of quantum relativism

This pragmatist perspective is one of many ways of viewing quantum theory. Notably, a century after Heisenberg's seminal paper, there was still no consensus on how to understand the theory that grew from the seeds it planted. Popularizations often say that quantum theory describes the microscopic world of atomic and subatomic particles, but its applications are not restricted in this way. Moreover, a variety of views deny that quantum theory simply describes physical systems, taking any application of the theory to be in some sense relative to something other than the system to which it is applied. This is how quantum theory appears from the present pragmatist perspective. It is also how it appears from at least three other perspectives: Rovelli's (1996, 2022) RQM, QBism (Healey 2023) and Zwirn's convivial solipsism. It may be worth concluding by saying why I think the view from a pragmatist perspective is clearer and free of the distortions introduced by viewing quantum theory from any of these alternative relativist perspectives.

In presenting and defending RQM, Carlo Rovellli has lauded its naturalist credentials. In his view, quantum theory says what a physical system is like, always relative to a second physical system with which it has interacted. Other physical systems need be in no way special: in particular, the second system need not be a human observer who has used this interaction to perform a quantum measurement on the first system and become aware of the outcome. There is no appeal to thought or experience here or elsewhere. If the second system is in fact a human observer, a second human may learn of the outcome of their measurement by asking them. However, that, too, is simply a further physical interaction between the two human systems: the second human is effectively a physical apparatus whose pointer reading coincides with that of the first human. There is no need to describe this transfer of information as a communication of thoughts between two conscious agents. Information is physical, here and elsewhere.

RQM has been subjected to a variety of criticisms (Muciño et al. 2021; Pienaar 2021; Healey 2022) that at least point to residual unclarities in the view. It remains to be seen exactly what kind of interaction counts as a measurement, what exactly it is a measurement of, and just when it can be said to have led to an outcome. In its original form, the relativity inherent in RQM apparently iterates so that any comparison between what a system is like relative to a second system and what it is like relative to a third system must itself be made relative to some further system. This led to the objection that RQM left us with no overarching perspective on the quantum world but merely a collection of unrelated perspectives from what Pienaar termed "island universes". A paper by Adlam and Rovelli (2023) sought to bring order to this apparent chaos by proposing a principle called cross-perspective links. To my mind, this improved RQM, bringing it closer to my pragmatist perspective (Healey 2022).

The naturalism Rovelli claims as a virtue of RQM is a particularly robust form of object naturalism. From a subject naturalist perspective, it requires too much. By requiring that comparisons between perspectives are conceptually possible only when they are physically implemented in an interaction. Moreover, quantum information science addresses phenomena such as quantum teleportation, in which information is transmitted by the transfer of quantum states that are not themselves physical, from my pragmatist perspective.

QBism (Fuchs et al. 2014; Fuchs and Schack 2015; DeBrota and Stacey 2019) is another form of quantum relativism. For a QBist, both quantum states and Born probabilities are relative to the agent applying quantum theory. This goes along with De Finetti's personalist view of all probability assignments, according to which the world contains no probabilities except those in the minds of (or revealed by the behavior of) agents making decisions under uncertainty. Indeed, quantum theory as a whole is an empirically motivated addition to decision theory, and the Born rule does not provide a statistical interpretation of the quantum state but may be stated solely as a prudential norm governing the degrees of belief of an agent applying quantum theory to make wise decisions.

This lies in the face of many physicists' commitment to some inchoate form of frequency interpretation of quantum probabilities, according to which their objective status is manifested by tests in which statistics of alternative outcomes of repeated measurements of one or more quantum observables closely match their predicted Born probabilities. In addition, it drains the question of whether quantum theory shows that the world is ultimately indeterministic of any metaphysical significance. However, philosophers have convincingly argued that any attempt clearly to state and effectively argue for a frequency interpretation of probability is doomed to failure (Hajek 2023).

From my pragmatist perspective, personalist and frequency interpretations both offer mistaken views of probability. A better view emerges if we ask why we have a concept of probability at all, bearing in mind that we have arrived at this concept only relatively recently. Hacking (1975) tracks the emergence of our concept from the 17th-century discussions between Pascal and Fermat. The measure-theoretic formalization of probability mathematics dates back to the work of Kolmogorov in the 1930s.

We often find ourselves in a situation where we face some generic barrier to knowledge that is important to us in deciding what to do or simply what to believe. The most obvious barrier is provided by time and there are many things about the future that we presently have no way of knowing: who will win the next election (or tomorrow's horse race); will it rain next week; will a tossed coin land on heads or tails? In contrast, there is much we can learn from past events that is worth considering when forming opinions about possible future outcomes. Probability provides a way of agglomerating and packaging such accessible information.

This is what a modern weather forecast seeks to do. Firstly by collecting a large amount of meteorological data covering a wide geographical area near the target site. Secondly, by running large computer simulations based on physical models of the atmosphere throughout that area, each with a different minor variation from the original input data (to allow for the chaotic nature of the phenomenon). Finally, by using the resulting ensemble of data to issue a probabilistic prediction of rain in the target area the next week. Anyone may then use this probability to form their own corresponding degree of belief that it will rain there next week.

The function of this, and other, objective probabilities, is to guide the informed opinion of any suitably situated epistemic agent concerning matters to which they have no epistemic access from that situation. If the epistemic barrier is simply temporary, it is sometimes wiser to wait until the situation improves. However, that is not always possible, as when we must decide now whether to make plans to go camping next week.

Sometimes the epistemic barrier is not temporal. Wigner cannot access his friend's laboratory without disturbing its physical isolation: no one can know the exact phase-space location of all of the molecules in my coffee, not to mention its cup and the air in my room. However, Wigner can use the Born probabilities yielded by his (relative) quantum state to form an opinion of what he will find when he enters his friend's laboratory, and I may be able to use some form of statistical mechanics to reassure myself that the coffee in my cup will be cold an hour from now.

What makes a probability objective is not what in the world it represents but that it establishes a social norm as best guide to setting our degrees of belief concerning events to which our physical situation bars epistemic access for some generic reason. Born probabilities are objective in just that sense: their relativity, as well as that of the quantum state that yields them, arises from the fact that what an agent has epistemic access to often depends on features of their physical situation. This is why both quantum states and Born probabilities are related to the agent-situation from a pragmatist perspective. This is also why I reject QBist subjectivism about probability, quantum states, Hamiltonians and other important elements of quantum theory.

However, I have learned several things about quantum theory from QBists that we can now see from the present pragmatist perspective. The most basic negative lesson is that a quantum state neither is nor represents what Bell (2004) called a beable of quantum theory, an element of physical reality according to that theory. Its main function is rather to generate probabilities for each mutually exclusive and collectively exhaustive set of possible events. However, I disagree with the QBists about what these events are. I take them to be physical, associated with measurement only because they result from interactions of a type that lends itself to the performance of a quantum measurement by any agent capable of using it for that purpose. Unlike QBists, I take a quantum state to be an objective feature of a world, not a subjective feature of an agent, and to exist even in worlds containing no agents.

I agree with QBists that the discontinuous change in a quantum state during measurement is not a physical process but rather the necessary update of that state by one applying quantum theory to incorporate information made available as an agent's situation changes. However, that change would be in the first place physical, and only secondarily epistemic, and the new knowledge would be acquired by ordinary physical means, not by some primitive (mental) act of measurement. Indeed, I regard recent QBist attempts to align their talk of measurements with phenomenologists' emphasis on individual lived experience as a step in the wrong direction.

I will try to conclude my comparison with QBism on a more positive note. I agree that, correctly understood, quantum theory neither faces a formidable measurement problem nor implies that the world is nonlocal (in the sense that nearby events exert an immediate influence on distant (spacelike separated) events). In addition, I share what I took to be the QBist view that a user of quantum theory is an agent only in the sense assumed in decision theory. Here, an agent need not be an individual conscious human but may be a collective entity such as a firm or an embodied alien or artificial intelligence, provided that this was able to use quantum theory in pursuit of its own ends and, specifically, to maximize the utility revealed by its choice behavior.

However, the current fascination with phenomenology has made me appreciate that QBists are now opting for a narrower conception of an agent: one with a unified consciousness capable of experiencing the outcome of a quantum measurement (Bitbol et al. 2024). This strikes me as a step in the wrong direction.

The fourth kind of quantum relativism I discuss is Zwirn's (2016, 2020, 2024) convivial solipsism. I see this as a variant of a many-minds view (Albert and Loewer 1988) that itself arose as a variation of Everett's (1957) relative state formulation of quantum theory. In common with those and other views, convivial solipsism begins by assuming that it makes sense to speak of a quantum state of the universe – indeed a pure quantum state – and that this always evolves unitarily in the Schrödinger picture. From a pragmatist perspective, as for the QBist, there is no such state, although an agent may model the large-scale degrees of freedom of the universe by assigning a quantum state that fails to model the small-scale degrees of freedom whose inclusion would be required if we were to try to extend the model to a situated agent such as the one applying the model.

The convivial solipsist next expresses this massively entangled quantum state on a preferred basis, assumed to be defined somehow by decoherence interactions among the universe's subsystems. Unitary evolution of the universal quantum state is taken to define an evolving branch structure within it. The next step is to assign at least some of these branches a representational function. Whether to assign this function also to the entire quantum state as describing the evolution of the entire universe is here left an open question.

What a branch represents is not what contemporary Everettians sometimes call a world within the multiverse. Rather, a branch is taken to represent only the perceptual experience of any observer associated with that branch. The branches split as the universal quantum state evolves unitarily, and the experiences of observers associated with the stem prior to the split diverge in the "twigs" into which it branches.

A key assumption of convivial solipsism is that the experience of any observer is confined to but "hangs on to" a unique branch throughout this splitting process. A quantum measurement induces (or is at least associated with) a branch split, corresponding to the observers' experiencing different outcomes relative to the different resulting "twigs" in the superposed quantum state following the unitary evolution associated with the measurement process. Each observer then assigns a quantum state to the measured system that corresponds to the outcome they experienced, so there is an effective "collapse" of the relative state consistent with global unitarity.

Convivial solipsism offers a picture of the quantum world as a whole and how it evolves. Hervé calls this empirical reality: it remains unclear how seriously this picture should be taken within convivial solipsism. Perhaps in addition to the universe described by its evolving quantum state, this includes all of the experiences of all branch observers. However, no observer can see this picture because each observer's experience is always confined to a single branch, within which it constitutes that observer's phenomenal reality.

No observer has experiential or epistemic access to an observer hanging on to a different branch: that is, the sense in which the view is solipsist. However, it is convivial insofar as in any attempt by an observer hanging on to one branch to communicate or otherwise engage with an observer hanging on to a different branch, what the first observer will subsequently experience will cohere internally as a sequence of experiences as of what the second observer has experienced, despite failing to correspond to what the second observer has in fact experienced.

I present three objections to convivial solipsism. From a pragmatic point of view, its approach to quantum theory is too representationalist. It assigns both a universal quantum state and branch-relative quantum states a representational function that they lack. It does not take these to represent anything in the physical world but rather the perceptual experiences of observers, who think of themselves as flesh and blood denizens of it. It speaks of systems being in quantum states rather than being assigned quantum states, and it applies an analog of the eigenstate–eigenvalue link connecting those states to properties of a system in it; albeit mental rather than physical properties.

My second objection is that the central notion of experience cannot bear the weight that is put on it in convivial solipsism. From a pragmatist perspective, perceptual experience remains an obscure notion. A pragmatist such as Dewey had a rich notion of experience that involved active engagement with the world, the sense in which an experienced teacher is not merely a person who has had perceptual experiences while teaching but also one who has successfully engaged in the practice of teaching over an extended period of time. All of the perceptual experiences of which we know, are present only in evolved organisms, but even here, we argue about their extent and physical basis. A pragmatist should recoil from the suggestion that we can accept as primitive within a fundamental physical theory purely passive, Cartesian notion of experience, especially one that stands alone with no physical basis. Hervé occasionally speculates about finding such a basis in a "hanging on mechanism" in human (and other) brains, but this strikes me as a wild-goose chase motivated by reversion to the wrong kind of (object) naturalism.

My final objection is against the solipsism inherent in convivial solipsism. In my view, RQM was significantly improved by the explicit adoption of cross-perspective links. This was an improvement because it acknowledged the vital importance of securing evidence for quantum theory (or any other scientific theory) in the form of publicly recognized and publicly available data. Because each observer's experiences remain essentially private in convivial solipsism, there can be no such data. There, my experiences as of a physical world including measurement outcomes radically mislead me as to its properties and even its existence, and I lack all access to the experiences of an observer hanging on to a different branch, although I can fool myself into thinking otherwise. For a convivial solipsist, we have no objective evidence for quantum theory.

5.9. References

Adlam, E. and Rovelli, C. (2023). Information is physical: Cross-perspective links in relational quantum mechanics. *Philosophy of Physics*, 1(1) [Online]. Available at: https://philosophyofphysics.lse.ac.uk/articles/10.31389/pop.8.

Albert, D. and Loewer, B. (1988). Interpreting the many-worlds interpretation. *Synthese*, 77(2), 195–213.

Baker, D. (2009). Against field interpretations of quantum field theory. *British Journal for the Philosophy of Science*, 60(3), 585–609.

Bell, J. (2004). *Speakable and Unspeakable in Quantum Mechanics*, revised edition. Cambridge University Press, Cambridge.

Bitbol, M., Zwirn, H., Fuchs, C. (2024). Round Table: QBism and Convivial Solipsism. Collège de Physique et de Philosophie, Paris [Online]. Available at: https://www.youtube.com/watch?v=7e6nUVe5qyc.

Bong, K.-W., Utreras-Alarcón, A., Ghafari, F., Liang, Y.-C., Tischler, N., Cavalcanti, E.G., Pryde, G.J., Wiseman, H.M. (2020). A strong no-go theorem on the Wigner's friend paradox. *Nature Physics*, 16, 1199–1205.

Brandom, R. (1994). *Making it Explicit.* Harvard University Press, Cambridge, MA.

Brandom, R. (2000). *Articulating Reasons.* Harvard University Press, Cambridge, MA.

Brun, T., Finkelstein, N., Mermin, N.D. (2002). How much state assignments can differ. *Physical Review A*, 65, 032315.

DeBrota, J. and Stacey, B. (2019). FAQBism. *arXiv:1810.13401.* doi: 10.48550/arXiv.1810.13401.

Everett, H. (1957). Relative state formulation of quantum mechanics. *Reviews of Modern Physics*, 29(3) 454–62.

Fraser, D. (2008). The fate of 'particles' in quantum field theories with interactions. *Studies in History and Philosophy of Modern Physics*, 39(4), 841–859.

Fuchs, C. and Schack, R. (2015). QBism and the Greeks: Why a quantum state does not represent an element of physical reality. *Physica Scripta*, 90(1), 015104.

Fuchs, C., Mermin, N.D., Schack, R. (2014). An introduction to QBism with an application to the locality of quantum mechanics. *American Journal of Physics*, 82(8), 749–754.

Goldstein, S., Norsen, T., Tausk, D., Zanghi, N. (2011). Bell's Theorem. *Scholarpedia*, 6(10), 8378 [Online]. Available at: http://www.scholarpedia.org/article/Bell%27s_theorem.

Hacking, I. (1975). *The Emergence of Probability*. Cambridge University Press, Cambridge.

Haddara, M. and Cavalcanti, E. (2023). A possibilistic no-go theorem on the Wigner's friend paradox. *New Journal of Physics*, 25, 093028.

Hajek, A. (2023). Interpretations of probability. In *The Stanford Encyclopedia of Philosophy (Winter 2023 Edition)*, Zalta, E. and Nodelman, U. (eds). Stanford University, Stanford [Online]. Available at: https://plato.stanford.edu/archives/win2023/entries/probability-interpret/.

Healey, R. (2013). Observation and quantum objectivity. *Philosophy of Science*, 80(3), 434–453.

Healey, R. (2022). Securing the objectivity of relative facts in the quantum domain. *Foundations of Physics*, 52(88), 1–20.

Healey, R. (2023). Quantum-Bayesian and pragmatist views of quantum theory. In *The Stanford Encyclopedia of Philosophy (Winter 2023 Edition)*, Zalta, E. and Nodelman, U. (eds). Stanford University, Stanford [Online]. Available at: https://plato.stanford.edu/archives/win2023/entries/quantum-bayesian/.

Healey, R. (2026). *Pragmatism Works: Essays on Quantum Theory, Science and Metaphysics*. Oxford University Press, Oxford.

MacFarlane, J. (2014). *Assessment Sensitivity*. Clarendon Press, Oxford.

Maudlin, T. (2014). What Bell did. *Journal of Physics A*, 47(42), 424010.

Muciño, R., Okun, E., Sudarsky, D. (2021). Reply to Rovelli's response to our "Assessing Relational Quantum Mechanics". *arXiv:2107.05817*. doi: 10.48550/arXiv.2107.05817.

Peirce, C. (1878). How to make our ideas clear. *Popular Science Monthly*, 12(January), 286–302 [Online]. Available at: https://courses.media.mit.edu/2004spring/mas966/Peirce%201878%20Make%20Ideas%20Clear.pdf.

Peregrin, J. (2024). *Normative Species: How Naturalized Inferentialism Explains Us*. Routledge, New York.

Pienaar, J. (2021). A quintet of quandaries: Five no-go theorems for relational quantum mechanics. *Foundations of Physics*, 51(97), 1–27.

Price, H. (2003). Truth as convenient friction. *Journal of Philosophy*, 100(4), 167–190.

Price, H. (2011). *Naturalism Without Mirrors.* Oxford University Press, Oxford.

Price, H. (2013). *Expressivism, Pragmatism and Representationalism.* Cambridge University Press, Cambridge.

Price, H. (2025). *Facts and the Function of Truth, Extended Edition.* Oxford University Press, Oxford.

Riedel, T. (2025). Is quantum relativism untameable? Revenge Wigner arguments for relative facts. *British Journal for the Philosophy of Science*, 76(3) [Online]. doi: 10.1086/732830.

Rovelli, C. (1996). Relational quantum mechanics. *International Journal of Theoretical Physics*, 35, 1637–1678.

Rovelli, C. (2022). The relational interpretation. In *Oxford Handbook of the Interpretations of Quantum Mechanics*, Freire Jr, O. and Bacciagaluppi, G. (eds). Oxford University Press, Oxford.

Wallace, D. (2020). On the plurality of quantum theories: Quantum theory as a framework, and its implications for the quantum measurement problem. In *Scientific Realism and the Quantum*, French, S. and Saatsi, J. (eds). Oxford University Press, Oxford.

Wang, P., Zhang, J., Luan, C.-Y., Um, M., Wang, Y., Qiao, M., Xie, T., Zhang, J.-N., Cabello, A., Kim, K. (2022). Significant loophole-free test of Kochen-Specker contextuality using two species of atomic ions. *Science Advances*, 8(6), eabk1660.

Wigner, E.P. (1962). Remarks on the mind-body question. In *The Scientist Speculates*, Good, I.J. (ed.). Heinemann, London.

Zwirn, H. (2016). The measurement problem: Decoherence and convivial solipsism. *Foundations of Physics*, 46, 635–667. doi: 10.1007/s10701-016-9999-5.

Zwirn, H. (2020). Nonlocality versus modified realism. *Foundations of Physics*, 50(1), 1–26.

Zwirn, H. (2024). Convivial solipsism as a maximally perspectival interpretation. *Foundations of Physics*, 54(39), 1–16.

6

The Role of the Observer: Convivial Solipsism

Hervé ZWIRN[1,2]

[1]*Institut d'Histoire et de Philosophie des Sciences et des Techniques, Paris, France*
[2]*École Normale Supérieure de Paris-Saclay, Centre Borelli, France*

As presented in the introductory chapter of this book, the measurement problem has given rise to numerous attempted, divergent solutions. In my opinion, if we set aside solutions that introduce a modification of formalism, then the interpretations that have been proposed, some of which are presented in the introductory chapter, all suffer from flaws that make them unacceptable.

Without being exhaustive, which is not possible, I first present my objections to some of these interpretations and then provide the reasons that led me to develop convivial solipsism (ConSol) and the positive arguments in favor.[1]

6.1. Current interpretations

6.1.1. *Philosophical positions*

The solutions that different authors have developed to solve the problem of measurement depend largely on the philosophical position they adopt. In fact,

1 A detailed presentation of Convivial Solipsism and its consequences can be found in the following series of articles: Zwirn (2000, 2016, 2017, 2020a, 2020b, 2021, 2022, 2023, 2024a, 2024b).

Quantum Physics and Cosmology,
coordinated by Hervé ZWIRN.

depending on the philosophical position adopted, a solution may seem acceptable or absurd. Each of the interpretations we analyze can be linked to one of these positions to varying degrees.

6.1.2. *Realism*

Realism is the thesis that there is a reality external to any observer and that it exists independently and would exist in the same way if there were no observer. More precisely, this thesis alone is called metaphysical realism. This metaphysical realism can be declined into various versions, depending on whether we assume that this reality is accessible to us or that it is beyond our grasp (like Kant's "thing-in-itself", of which we can have no knowledge). Realism, which postulates that this reality is accessible to us because it corresponds to what we perceive directly of the world, is sometimes called naive realism or man-in-the-street realism. It consists of thinking that the world we see and live in is indeed, as it appears to us, that the objects we manipulate exist outside of us and that there is an identity between what we perceive and what exists. If we add the assumption that this reality can and must be described by science, this position takes the name of scientific realism.

Admittedly, this is a natural point of view and one that we spontaneously adopt in our everyday lives. Only philosophers have challenged it since antiquity. The vast majority of physicists adopt this position, which results from the fact that the description of reality given by science can be qualified as objective in the sense that it is a description that is valid independently of the presence of observers, who are there to observe the possible results of experiments carried out on the world. Science aims to describe the world as it is, in absolute terms, and this description is valid in the same way for any observer. This was the position of Einstein, Schrödinger, de Broglie, Bell, Bohm and many others among the great physicists. Even today, this is the type of description that many physicists would like to see, even if it is unlikely that it will ever be possible to produce a theory that satisfies it.

6.1.3. *Idealism*

Idealism, the direct opposite of realism, professes the view that perceptions take precedence and that the reality to which we ascribe existence is merely a construction of our mind to order the perceptions we have. Berkeley, one of the leading philosophers of idealism, famously said, "esse est percepi aut percipere" (to be is to be perceived or to perceive). This view culminates in Husserl's phenomenology, for which (in simplified terms) everything starts from what he calls first-person experience, meaning the direct perceptions of each individual.

Although the two are related, idealism should not be confused with dualism. The latter professes the existence of two types of entities, matter and spirit, neither of which is reducible to the other.

6.1.4. *Instrumentalism and pragmatism*

Instrumentalism, which involves the bangs of realism and idealism, takes no position on the existence of an independent reality. Instead, it considers that the purpose of science is primarily to help us make predictions about the results of experiments we can carry out. Instrumentalism does not deny the possibility of an independent reality but sees science as unable to accurately describe that reality, if it exists. The sole role of science is to be an instrument for predicting the results of experiments.

Pragmatism, the cousin of instrumentalism (associated with the names Charles Sanders Peirce, William James and John Dewey), emphasizes the effectiveness of science. According to pragmatism, what is true is what works in practice.

6.1.5. *Changes in formalism*

Let us briefly review the interpretations that require a change in the mathematical formalism of quantum mechanics. The two best known theories are the de Broglie Bohm theory (dBB theory) and the theory developed by Ghirari, Rimini and Weber (GRW theory). They are briefly described in the introductory chapter. These theories have the advantage of solving the measurement problem. They are also realistic and produce an objective description of reality.

However, there is far from a consensus among physicists. This is due to a number of conceptual problems, including the difficulty of giving them a relativistic extension.

6.1.6. *The Copenhagen school*

The interpretation known as the "Copenhagen School" is briefly presented in the introductory chapter. Owing primarily to Bohr and Heisenberg, it remains the preferred interpretation of all physicists not involved in reflections on the foundations of quantum physics. Simplifying it to retain its main features consists of considering that a measurement is made when a quantum system interacts with a macroscopic measuring apparatus and that a classical language is indispensable for communicating experimental results. Bohr introduced the notion of complementarity, which refers to the fact that a quantum system can appear in different and apparently incompatible forms, such as a wave or a particle, depending on the experimental setup used. In Young's famous hole experiment, for example, electrons may appear as particles arriving one by one on the screen if only one hole is open or as waves producing interference if both holes are open. This means that the properties measured do not belong to the system itself but rather to the whole system and measuring device. The difficulty introduced by the fact that it is not possible to define a boundary precisely between microscopic and macroscopic (which has often been opposed to this interpretation) is in fact resolved by a better understanding of Bohr's subtle thinking.

A device must be considered not subject to quantum mechanics only when it is used to make a measurement. On the other hand, the same device can be considered a quantum object if we want to measure it via another device. This pragmatist view of things gives the observer an essential role, although this is not made explicit by the proponents of this interpretation, who want it to be objective. Although Bohr was cautious on the subject and Heisenberg instead explained that quantum mechanics does not directly describe the world but rather our knowledge of it, this interpretation is often associated with a standard realism that makes the reduction of the wave packet a physical event affecting the state of the system. This leads to a number of difficulties, which we will discuss below.

6.1.7. *The dualist interpretation of von Neumann, Wigner, London and Bauer*

In this interpretation, the observer's consciousness is placed at the center of the measurement operation. At the end of any measurement, an observer becomes aware of a precise result, and it is at this point that we can be truly certain that a measurement has been made. Wigner (1961) devised an experiment (now famous and widely developed, including recent extensions, known as Wigner's friend's experiment), in which he argued that the only reasonable solution in the situation described by his experiment was to consider that a measurement was made only when a conscious observer noted a result. He was thus led to exclude consciousness from the field of quantum mechanics and thereby postulate a kind of Cartesian dualism in which consciousness, although non-physical, nevertheless has a physical action on the system by projecting the superposed state toward a proper state of the measured observable. Similar considerations have been made by von Neumann, London and Bauer.[2]

2 See section 6.2 on convivial solipsism for some quotes along these lines.

This interpretation has never been taken seriously since dualism is a philosophical position widely rejected by most physicists. However, it is based on a fairly obvious observation: the primacy of perception. We shall see that ConSol takes up this idea but avoids the trap of Cartesian dualism by postulating no action of consciousness (or perception) on matter.

6.1.8. *Pragmatism*

We mentioned pragmatism in the chapter entitled "A Brief Introduction to Quantum Mechanics and a Review of Interpretations" (Chapter 1). Pragmatism is a cousin of instrumentalism (which sees science as a tool for predicting the results of experiments, not a description of reality) and comes in several versions.

Instrumentalism takes no position on the existence or non-existence of an independent reality and remains neutral on the subject. I have to confess that I have little sympathy for instrumentalism, which, although coherent in a philosophical position, seems to me to renounce the main motivation of scientific discourse too easily.

Pragmatism is more moderate in this respect but resigns itself to a more or less pronounced abandonment of the same type in regard to delving into the deeper explanation of certain phenomena. It satisfies itself with the fact that theory produces predictions in line with experiments but does not seek to understand why. For example, in regard to the problem of measurement, the pragmatism professed by Richard Healey (see Chapter 5) explicitly admits that it does not attempt to solve the problem but rather is content to state that the appearance of a single result in a measurement is an experimental fact. This position is logically coherent, but in my view, it falls far short of the goal of understanding the deeper reasons why the world is the way it is (or appears to us the way it appears to us); however, if not impossible, this goal may be achieved in its entirety.

6.1.9. *Relational interpretation*

As I mentioned in Chapter 1, this interpretation maintains that the state of a system has no meaning in absolute terms but only in relation to another system. I mention it here only for the sake of completeness, but it suffers from a major lack of coherence by conflating correlations and measurement.

6.1.10. *QBism*

I will repeat here part of the description I provided in Chapter 1. QBism, originally developed by Fuchs and Schack (2015) and Fuchs and Stacey (2019), does not consider the wave function to represent the actual physical state of the system but rather to be a tool enabling the observer (who is called an agent in this framework) to predict (or bet) the results they will obtain in a future experiment. For QBists, the word "experience" should be understood here as "personal experience", meaning the experience that the world has induced in each agent over the course of their life. This is partly an instrumentalist (or at the very least pragmatist) position, since for QBists, quantum mechanics is merely a tool enabling agents to calculate their probabilistic expectations for their future experience on the basis of knowledge of the results of their past experience. It is not intended to describe the external world (if indeed there is one, and the QBists' position on this is not entirely clear). QBists adopt the subjective interpretation that the probabilities of quantum formalism represent the degree of belief of an agent and, therefore, are specific to that agent. The entities of formalism (wave functions, probabilities) are relative to a particular agent, who uses formalism to make predictions about the experiments they will perform. For example, the quantum state of a system is not an objective property of that system. It is merely a tool for assigning a subjective probability to the agent's future experience. A measurement (in the usual sense) is only a special case of what QBism calls experience, in other words, any action performed by an agent on their external world. A measurement does not reveal a preexisting state of affairs but creates a result for the agent. The way QBists solve the measurement problem is very simple: they assume that the direct internal perception of our private experience is the only phenomenon accessible to an agent that they do not model with quantum mechanics and that the agent's perception is the result of the experience. There is therefore no longer any ambiguity about the use of the reduction postulate, which says nothing about the "real state" of the system being measured but which is nothing more than the updating of the agent's state assignment on the basis of the agent's experience. In particular, there is no measurement when there is no agent.

QBism allows us to free ourselves from the concept of non-locality (which I will discuss later), in a way that is quite similar to that of ConSol.

Although close in many respects to ConSol, QBism seems to be unsatisfactory for reasons I have analyzed in detail in Zwirn (2022). The main criticism is the vagueness of the underlying ontology. For example, what meaning should be given to the term "agent"? Is it a conscious observer, or could it be something else? What is the meaning of the fact that an agent creates a result when making a measurement? Is this result in the agent's mind or in the outside world? When an agent creates a measurement result through an experiment, is this result specific to the agent in question, or does it become information shared with other agents? The answers to these questions are often contradictory and differ according to the writings of the authors of this interpretation, which undermines the clarity of the world image depicted by QBism.

6.1.11. *Everett's interpretation*

Hugh Everett's aim was to eliminate the reduction postulate, which he found "hideous". He also wanted quantum mechanics to be able to describe the Universe as a whole – something not possible with the standard interpretation that requires a measurement made by a device or observer outside the system being measured. He therefore proposed simply dispensing with this postulate. In his interpretation (sometimes referred to as the many-worlds interpretation), there is no collapse, and the universal wave function (including observers) evolves in a unique way under the Schrödinger equation.

In Chapter 1, I provided a brief description of this interpretation, to which I refer to the reader.

As I shall reiterate below, this infinite (and even uncountably infinite) multiplication of worlds was deemed unacceptable by Bernard d'Espagnat. The latter led to the development of ConSol, which, like Everett, rejects the postulate of wave packet reduction as a physical process.

Beyond this rejection, which might be considered a matter of personal taste, Everett's interpretation poses serious problems for the concept of probability and even for the definition of multiplying worlds. I refer the reader to the chapters by Lev Vaidman (Chapter 3) and Jeffrey Barrett (Chapter 4) for an analysis of these difficulties.

6.2. Convivial Solipsism

I developed ConSol to address several motivations.[3] The first, of course, is to solve the problem of measurement in an unmodified formalism. The second is to

3 I have developed Convivial Solipsism through a series of articles (Zwirn 2000, 2016, 2017, 2020a, 2021, 2022, 2023, 2024a, 2024b). A presentation for the general public can also be consulted (Zwirn 2020b).

take into account what seems to me to be the most obvious: the fact that we have to rely on what we perceive, because ultimately this is the only thing we can be reasonably sure of. The greatest physicists have stressed that consciousness and perception are at the most fundamental level. As Planck (1931) pointed out:

> I regard consciousness as fundamental. I regard matter as derivative from consciousness. We cannot get behind consciousness. Everything that we talk about, everything that we regard as existing, postulates consciousness.

Von Neumann (1955 [1932]) went even further:

> Experience only asserts something like: an observer has made a certain (subjective) perception, but never such as: a certain physical quantity has a certain value." "It is inherently entirely correct that the measurement or the related process of the subjective perception is a new entity relative to the physical environment and is not reducible to the latter." "But in any case, no matter how far we calculate, [....], at some time we must say: and this is perceived by the observer. That is, we must always divide the world into two parts, the one being the observed system, the other the observer.

Wigner (1983 [1976]) noted the following:

> When is the measurement completed? We will see that [...] the measurement is completed only when we have observed its outcome.

Although I disagree with the dualism to which this has led them,[4] I believe that these remarks contain a profound truth that is often ignored or even rejected by physicists who adhere unconditionally to realism. Their obsession with science that recaptures the objectivity of classical physics sometimes leads them to adhere to interpretations whose coherence falls far short of the mark.

For my part, I have been guided by the desire to place the need for logical coherence above all else. On the basis of this requirement, I think that the consequences of a logically indisputable construction should be accepted, even if they are strange and offend common sense.

---

4 Von Neumann was cautious on the subject. Wigner teamed up with London and Bauer to go so far as to say that consciousness was responsible for the physical reduction of the wave packet, although he later backtracked on this idea; London and Bauer's ideas are in fact more subtle than that.

ConSol grew out of a remark by d'Espagnat (1971), who was dissatisfied with Everett's interpretation because of the infinite proliferation of worlds it implies. Everett's abandonment of the wave-packet reduction postulate is a tempting way of trying to solve the problem of measurement – a problem that lies precisely in the impossibility of finding a precise rule prescribing when to apply it. However, in this case, we still need to explain why an observer perceives only one result when making a measurement. Everett postulates that the observer splits into as many observers as possible results, which d'Espagnat rejects. ConSol solves this problem while avoiding this infinite multiplication of observers and worlds. It does this by means of what I have called "the hanging-on mechanism" and makes the observer play a central role, in whose absence there is no measurement.

6.2.1. *The hanging-on mechanism*

The measurement problem arises, as we have seen, from the fact that when an observable is measured on a system whose state is a superposition of eigenvectors of the observable in question, the application of Schrödinger's equation results in an entangled state between the measuring device and the system.[5] The result is not a single value but rather a global state (device plus system) involving all possible outcomes. If we add an observer looking at the device to the description of this situation, then we must consider that the observer themself is entangled with the system and the device and therefore participates in this mixture of results. However, when an observer looks at the measuring device, they see a single result. As von Neumann and Wigner noted above, the only time we can be sure of having obtained a unique result is when an observer can declare having seen it.

We therefore differentiate between the physical state of the observer's brain, which can be superposed, and the perception experienced by the observer, which is always a definite result. Of course, this does not mean that we are adopting a type of Cartesian dualism. In Everett's interpretation, such dualism is not adopted either, where the observer is aware of only one branch, while they are physically in a superposed state that is never reduced. What is assumed is that our brains are not equipped to make us aware of all of the subtleties of a superposed state and that the "effect" such a state has on us is limited to what we call a definite result. Importantly, there is absolutely no effect of consciousness on matter, as was assumed in Wigner, London and Bauer's conception.

5 In what follows, it is assumed that quantum mechanics has universal validity and applies equally to microscopic systems, macroscopic measuring devices and even human observers who consult the devices.

In Everett's interpretation, this entanglement with all possible results is seen as describing the overall state of the Universe, which is never reduced but in which the observer splits into as many different observers as there are measurement results, each seeing a different result. This is not the case with ConSol, which distinguishes between the physical state of the observer, who is indeed in an entangled state with the system and apparatus, and the observer's perception of this situation.

Let us assume, for the sake of simplicity at this stage, that the global entangled state involving all possible outcomes does indeed describe the physical state of the world as it is. This is the assumption made by Everett, who in fact adopts a realistic framework. The hanging-on mechanism then stipulates that the observer in such a state (i.e. confronted with a superposition of possible outcomes) does not have the perceptual capacity to "see" this state in all its richness but that for reasons linked to the observer's perceptual limitations (for example, due to filters in the brain), they can only perceive a partial version of it, corresponding to a cross-section according to one of the possible outcomes. Let us use an analogy to better understand the phenomenon: color blind people do not have the perceptive capacity to discern certain colors, or more precisely, they cannot distinguish between red and green. As a result, when confronted with colors that they cannot see in all their richness, they will see a projection of the nuances they are able to apprehend. They will not see a red apple amid the green leaves of an apple tree. In the famous Ishihara test, they do not perceive the numbers contained in a circle of colored dots. Another example is that of a viewer watching a color film with yellow filter glasses. The viewer would also not see the richness of all of the colors in the film but would see a "yellowed" version. These images are, of course, limited, but they can help us understand what perception filters (located in our brains) can do to prevent us from grasping a situation in its entirety. Numerous examples of this type can also be found in images that can be viewed in two different ways.[6] The brain switches from one possibility to the other but is unable to see both aspects at the same time. The hanging-on mechanism is the application of this idea to superposed states. Our perception is not developed enough to see a superposition of states. It therefore interprets it as just one of these states. Once a given result is selected, the observer's perception will remain attached to the component of the superposed state corresponding to the perceived result (called a branch). In subsequent measurements, it will only be able to attach itself to branches that are connected to this one. In this sense, the term "hanging-on" should be understood. Once this attachment has been made,

6 For example, the image of the young–old woman (easily found on the Internet) or that of the dancer, a silhouette that can be seen turning on itself either clockwise or anticlockwise (see, for example, the video at www.youtube.comlwatch?v=9CEr2GfGilw).

everything then happens as if the other terms of the superposed state no longer exist for the observer. This fact is essential to explain why, if the observer makes a second measurement of the same observable on the same system (in the absence of any other perturbation), the same result will be obtained, as predicted by quantum mechanics.

It then becomes possible to understand what a measurement is: it is the perception by an observer of one of the possible results of a superposed state. A measurement is therefore not a physical action that modifies the state of the system. This remains unchanged. It is simply the fact that a given observer perceives this state in a certain way that is unique to them. The observer and their perception are at the heart of measurement, so it is easy to understand why it makes no sense to speak of measurement in the absence of an observer.

Here, we return to the major role played by the observer's perception during a measurement, which von Neumann and Wigner emphasized. However, we can immediately see that ConSol avoids the stumbling block their thesis encountered. Here, there is no physical action of consciousness on the system, resulting in a physical reduction of the state of the system to a state of its own. The state of the system does not change and remains superposed. Seeing a single result is merely a kind of illusion created by the limits of our perception and concerns only a given observer. In fact, we shall see that nothing forces another observer to see the same result…

Another aspect of the hanging-on mechanism is that the choice of the result (or branch) to which the observer's perception hooks is made probabilistically, following Born's rule. If a position is measured and the superposed state is, for example, the state $(a|x\rangle + b|y\rangle)$ with $|a|^2 + |b|^2 = 1$, the probability of seeing the system at position x will be $|a|^2$ and that of seeing it at position y will be $|b|^2$.

We can therefore see that, with the hanging-on to a given branch and the probability of choosing this branch in accordance with Born's rule, we find the usual characteristics of quantum mechanics. The hanging-on is equivalent to the projection of the state of the system onto the eigenvector associated with the eigenvalue obtained as a result. The probability of obtaining a result, which is given by Born's rule, corresponds here to the probability of hanging-on to a given branch. ConSol therefore conforms in every respect with traditional quantum mechanics, but the difference lies in the way we understand how it works.

In the case of standard quantum mechanics, measurement (for which we are unable to give a precise definition) produces a physical result on the system by changing its state into a new state, which is that of the eigenvector associated with the result obtained. There are several difficulties with this way of looking at things. The first, as we have already mentioned, is that we do not know what a measurement is, and as a result, we do not know when it is legitimate to apply this measurement process with wave packet reduction, instead of the Schrödinger equation. The second is that there is no explanation for the reduction mechanism itself. Why would a measurement (even if we knew how to define it) have the effect of transforming the initial superposed state into a reduced state? What physical process would cause this? Quantum mechanics provides no answer to this question.

In ConSol, on the other hand, things are much clearer. Measurement is not a physical process and does not change the physical state of the system. It perceives the superposed state of the system under observation from a certain angle. A measurement is therefore entirely relative to the observer making it. There is no longer any ambiguity as to the choice to be made between Schrödinger's equation or wave-packet reduction. The evolution of the system's state obeys only the Schrödinger equation. The fact that the observer perceives a unique result is due to the hanging-on mechanism and concerns only the observer, not the system. Now, this choice is made at random, and the observer has no control over it.

6.2.2. *Relativity yet conviviality*

The description we have just given has an important and surprising consequence: since the choice of the component to which an observer's perception hooks onto is random, there is no reason why two observers observing a system in the same superposed state should hook onto the same branch. The result obtained is entirely relative to a single observer. An objection immediately arises in this case; two observers could disagree on the result of the same measurement, which is never observed in practice (except in the case of error on the part of one of the observers or of lying, but we exclude these cases as irrelevant to our purpose). Relativity often leads to disagreement. This is the case, for example, in Einstein's theory of relativity, in which two observers located in different reference frames can disagree about their observations. Disagreement between observers located in different reference frames has been experimentally confirmed for relativity, and as counterintuitive as it may be, it is now a proven experimental fact. However, nothing of this kind exists in quantum mechanics, where two physicists participating in the same experiment in a laboratory agree most of the time on the result they have obtained. Such a consequence would lead us to reject ConSol. Fortunately, nothing of the sort ever happens, although understanding why this is so leads us to push the logical consequences of our hypotheses to the limit and to paint a very different picture of reality from traditional realism.

Notably, we have postulated that quantum mechanics is a universal theory and that it applies to everything from microscopic systems to macroscopic devices, including the observers themselves. As a result, for a given observer (let us call her Alice according to the first names now established by usage), another observer, Bob, must be considered a physical system subject to quantum mechanics and described in the same way as the system studied or the measuring apparatus used. Suppose, then, that Alice makes a measurement of an observable A on a system that is in a superposed state of eigenvectors of A. This measurement involves the use of a device P, and we have seen[7] that the interaction between the system and the device results in an entangled overall state (system plus device) of the following type: $|\Psi_{SP}\rangle = \sum c_i|v_i\rangle \otimes |P_i\rangle$ where the $|v_i\rangle$ are the eigenvectors of A associated with the possible outcomes and the $|P_i\rangle$ are the device states corresponding to the outcomes obtained. The hanging-on mechanism is then the observer perceiving only one of the components $|v_i\rangle \otimes |P_i\rangle$; this choice is made randomly with a probability that is $|c_i|^2$.

If we now assume that Bob is also making a measurement of the same observable on the same system, from Alice's point of view, this is nothing more than the fact that Bob, as a physical system, interacts with the system and the device; again, for Alice, the state of the system encompassing Bob (system plus device plus Bob) is given by $|\Psi_{SPB}\rangle = \sum c_i|v_i\rangle \otimes |P_i\rangle \otimes |B_i\rangle$, where $|B_i\rangle$ is Bob's state corresponding to having seen result *i*.

Let us then ask what happens when, having made her measurement and found the result i_0, Alice asks Bob what result he has found. To answer this question, we must first remember that having obtained a result, meaning having hooked onto one of the components of the global state, the observer remains hooked onto this component for all subsequent measurements. Everything happens as if the other components had disappeared from the observer. Therefore, Alice is attached to the component $|v_{i0}\rangle \otimes |P_{i0}\rangle \otimes |B_{i0}\rangle$. Furthermore, any communication or exchange between two observers involves physical processes. Therefore, for Alice, asking Bob about the result he has obtained is equivalent to making a measurement of Bob's observable whose eigenvectors are precisely the $|B_i\rangle$. When Bob is in the state $|B_{i0}\rangle$, he has seen the result i_0 and will say so if questioned about it. Therefore, if Alice questions Bob while she is hooking on the $|v_{i0}\rangle \otimes |P_{i0}\rangle \otimes |B_{i0}\rangle$ component, we can see that she can obtain the same i_0 result from Bob as she obtained herself. It follows that there can be no disagreement between the result she has obtained and what Bob will tell her about that same result. It is in this sense that ConSol is convivial. There can be no conflict between observers.

---

7 Please refer to Chapter 1, in which the problem of measurement is described.

Therefore, where does Solipsism come from? The fact that Alice hears Bob gives her the same result as she obtained does not mean that Bob has, as far as he is concerned, obtained that result. This is perhaps the most difficult aspect of this interpretation to understand. The description that was given for Alice, namely, that after the interaction between the system and the device, when Alice perceives the entangled state $|\Psi_{SP}\rangle = \sum c_i|v_i\rangle \otimes |P_i\rangle$, she will randomly select one of the components of this state and that this will determine her personal perception, is also valid from Bob's point of view. He will therefore also randomly select one of the components of this intricate state. However, it does not have to be the same as Alice's state. Bob may well have selected a different result and be hooked on a different branch than Alice does. However, how can he then have seen a different result from Alice's and yet tell Alice when she questions him that he saw the same result? This is where we need to understand that this is possible because of the application of the notion of superposed states, including to observers.

From Alice's point of view, Bob must be seen as a physical system similar to any other system. He can be compared to a measuring device. Let us analyze what it means for a measuring device P to be in a superposed state. Let us imagine that the results i are indicated by the device by a certain position of a needle on a pointer graduated in a circle (like the hours on a clock face): i_0, i_1,..., i_n. The i_0 result is obtained when the needle is on the i_0 graduation of the pointer. Assuming that the device is in a superposed state $\sum c_i|Pi\rangle$ means that the pointer occupies no precise position but is in a superposition of the positions of each of the graduations[8]. Of course, no one has ever seen such a situation, and the reason is that when faced with such a superposition, the hanging-on mechanism means that only one position is chosen randomly by our perception. The fact remains, however, that this selection only concerns our perception and that the physical state of the device remains unchanged and superposed. Hard as it may be to imagine, this analysis must be extended to Bob. Bob is in a superposed state $\sum c_i|B_i\rangle$. Just as the pointer occupies no definite position, Bob is in a superposed state where he has seen all possible outcomes and therefore in a superposed state where he gives all corresponding answers. Of course, it is difficult to imagine "Bob in a superposition of possible answers". However, imagining a pointer in a superposition of positions is not easier. When Alice questions Bob, in accordance with the hanging-on mechanism, she selects only one of the possible answers, and as we have seen, she can select only the one that corresponds to her own result. However, there is absolutely no link between Alice's and Bob's selection mechanism. Bob perception may therefore be

8 This way of describing things is not entirely accurate, since in an entangled state, none of the subsystems has a state of its own. It is, however, a convenient way of talking about the part of the overall system that interests us.

different from that of Alice, given that these two perceptions concern a personal way of making the cut that enables us to perceive the entangled state of the system (device plus system), which is in no way affected by these perceptions.

Importantly, a complete distinction must be made between the system in a superposed state (which, for the sake of pedagogical presentation, we present for the moment as if it existed independently in itself and were described "in the absolute" by this entangled state), Alice's perception of it and Bob's perception of it. In a way, Bob has his own perception of the overall system (system plus device), and Alice has her own perception of the overall system (system plus device plus Bob). Alice is the only one to have made a measurement on both the system and Bob, whereas Bob has only made a measurement on the system. As we can see, it becomes obligatory when making a statement to specify who the transmitter is (or from what point of view it is made). One observer has no way of directly accessing the perceptions of another. Alice cannot know what Bob has "really perceived". At most, she can ask him what he has perceived, and she will then make a measurement of Bob, but this will not give her any indication of how Bob himself hooked into a branch of the entangled state.

The results are therefore totally relative to each observer, and it is in this sense that this interpretation introduces a kind of solipsism: the world in which an observer lives is their own, and there is no reason why it should be identical to that of another observer. However, this solipsism is convivial because when one observer questions another, they are bound to agree with that observer (on matters of common measurement results, of course, not on general subjects such as politics, which would be off topic here).

6.2.3. *Empirical reality and phenomenal reality*

The above considerations introduce some strange consequences into our way of seeing the world. What kind of reality, then, is the consequence? Clearly, we have to abandon traditional realism, for which we all share the same world that appears identical to each of us. The relativism we mentioned leads to the fact that what is true for one observer may be false for another. This type of situation, where what is true depends on our personal position, is called perspectivalism (Leifer 2018; Zwirn 2024a).

ConSol is based on two levels of reality. The first, called empirical reality, is defined by mathematical formalism with the description given by superposed states and entanglement. We can now clarify that, contrary to the somewhat naive way in which we have presented it up to now, this empirical reality should not in fact be considered an external reality that exists "outside the observer". Rather, it is a field of potentiality that attaches itself to the observer. The relativity we have already mentioned is even more extreme, since the characterization of this empirical reality by state vectors or interactions described by the use of certain Hamiltonians via the Schrödinger equation is also relative to the observer. This empirical reality seems to share certain features in common with what d'Espagnat (2003) called the veiled reality. However, we need not push this comparison too far, as d'Espagnat had a much more "real" conception of the veiled reality than I wish to confer on the empirical reality of ConSol. For d'Espagnat, the veiled reality, although not knowable in its totality, was nevertheless independent of any observer and shared by observers. The empirical reality of ConSol, on the other hand, is relative to each individual observer and cannot be shared.

Using this empirical reality as a starting point, the observer creates personal results during measurement operations through the hanging-on mechanism, which corresponds in some way to the actualization produced by the observer of the potentialities of their empirical reality. Each conscious observer constructs their own world, which we call their phenomenal reality, by selecting a possible result of their empirical reality modeled by the state vector they use.

An observer's phenomenal reality is what the observer commonly calls "reality". This is the world in which they live. From this point of view, every observer's phenomenal reality behaves exactly like the ordinary reality of traditional realism. The only difference is that, unlike the reality of traditional realism, each observer's phenomenal reality is unique to themselves, and they have no way of sharing it with other observers, even though it is impossible for them to realize this, and they have the illusion (by communicating with other observers) that they are living in the same reality as they are.

6.2.4. *The Einstein–Podolsky–Rosen paradox*

The first problem solved by ConSol is that of measurement, since it was developed for this purpose. However, it also avoids the awkward concept of non-locality – which is now generally accepted despite the inconsistencies it gives rise to and which physicists accept with astonishing ease.

The concept of non-locality stems from the analysis of an argument put forward by Einstein et al. (1935), initially aimed at demonstrating that quantum mechanics cannot be a complete theory. These authors relied on the measurement of two incompatible properties (such as position and momentum) on two particles that were entangled but separated by a great distance. Quantum mechanics predicts that it is impossible to know the value of two incompatible quantities simultaneously, such as velocity and momentum, or spin in two different directions. Einstein et al. (1935) considered two particles entangled such that the total impulse is zero and the positions remain symmetrical with respect to the initial starting point. They rely on the following criterion of reality:

> If the value of a physical quantity of a system can be predicted with probability 1, without disturbing the state of the system in any way, then there is an element of reality corresponding to this physical quantity.

They then reasoned that if we measure the position of one particle, we can infer the position of the other. However, if we choose to measure the impulse of one particle, we can infer the impulse of the other. Since the choice to measure the position or impulse of one particle has no effect on the other, which may be very far away at the moment of measurement, they deduced that the other particle necessarily has a well-defined position and impulse, which is not allowed by quantum mechanics, and which is therefore an incomplete theory. We do not analyze Niels Bohr's (rather obscure) response to this line of reasoning, but we cite David Bohm's slight modification of the experiment to transform the two particles of zero total impulse into two particles of spin ½ whose total spin is zero. The corresponding entangled state is called the singlet state and is written as follows:

$$|\psi\rangle = \frac{1}{\sqrt{2}}[\,|+\rangle^{U}\,|-\rangle^{V} - \,|-\rangle^{U}\,|+\rangle^{V}]$$

In this equation, $|+\rangle^{U}$ represents the state of the first particle called U when its spin along Oz is + (the same for other notations). The consequence of this is that if we measure the spin of particle U and find "+", we will instantly know that a measurement of the spin of the other particle will yield "–" and vice versa. This would not be a problem if we could assume that as soon as the particles separate, they each possess a definite spin opposite that of the other. However, in the singlet state described above, quantum mechanics does not allow us to consider that the spin of the particles is determined until a measurement has been made on at least one particle. Moreover, the spin obtained is determined randomly, but prior to measurement, it cannot be assumed that it already has a value. It therefore seems that making a measurement on U will have an effect on the spin of V, which will be determined instantaneously – however, far V is from U. We are therefore faced with a situation in which one of two things is true: either the spin is determined before the measurement, and quantum mechanics is therefore incomplete, since it cannot model the fact that the two particles already have a defined spin when they separate, and this before any measurement, or the spin is not determined before the measurement, and consequently, an instantaneous effect of the measurement of one particle on the other at a distance. This instantaneous action strongly displeased Einstein, who dubbed it "spooky action at a distance"[9] – understandably, since, as the father of the theory of special relativity, which had shown that nothing can go faster than the speed of light, this phenomenon seemed unacceptable to him.

The debate between Einstein and Bohr (who, for his part, was not bothered by this remote action) remained philosophical until John Bell (1964) succeeded in writing down equations (or rather an inequation) that were experimentally testable to determine whether nature was governed by hidden variables (completing quantum mechanics) or whether it used these instantaneous remote actions. The experiment in question was carried out by several physicists, and Alain Aspect (who won the Nobel Prize in 2022 for it) conducted the experiment in 1982, which was convincing enough to win over physicists. Bell's inequality is violated as predicted by quantum mechanics, and not as a hidden-variable theory would have it. The instantaneous effect at a distance, known as non-locality, must therefore be retained. Bohr has thus apparently won out over Einstein.

6.2.5. *Is there really such a thing as non-locality?*

The experiment clearly indicates that Bell's inequality is violated, as predicted by quantum mechanics. This rules out the possibility that the correlation observed between the result of a measurement on the first particle and that on the second can be explained by the fact that, as soon as they separate, each carries with it the information needed to guarantee this correlation (this is what we call local hidden variables). This leads most physicists to consider that this instantaneous action at a distance (or non-locality, as should be called) is now an established fact. Therefore, how do they address this apparent violation of the theory of relativity? Quite simply, by pointing out (rightly) that it is impossible to use this non-locality to transmit information at a distance instantaneously. It is true that the result of the measurement made on the first particle is random and that the experimenter cannot decide what it will be. If this were not the case, it would be easy to imagine an instantaneous information transmission process: to send a "+" instantaneously to the other end of the galaxy, all we would have to do is impose a "–" result on the measurement made on the first particle. In this way, messages of arbitrary length can

9 Ghostly action from a distance.

be encoded. However, this is not the case, since the result of the measurement is random. The simple fact that it is not possible to communicate instantaneously (let alone transmit matter or energy) is enough to reassure some physicists, who accept this non-locality without further precaution.

An example to underline the strangeness of this non-locality is: suppose that you (in Paris) are tossing a coin with a friend (who was on one of the first trips to Mars).[10] You have synchronized your watches and each of you toss a coin at regular intervals (e.g. every 10 s from noon). Now imagine that you compare your list of results and realize that, at each toss, your coin and your friend's coin landed on the same side. Admittedly, you would not be able to communicate a voluntary message, but you must admit that you would find this extremely bizarre and would be desperate to understand how it could be.

Furthermore, we have presented non-locality as the correlation between a first measurement on particle U, which has the effect of determining the spin of particle V according to that obtained on particle U. The image we then have is that of an instantaneous effect of the first measurement on U on the measurement on V. The measurement on U is seen as the instantaneous cause of the result of the measurement on V.

This is strange enough, but we can take it a step further. The theory of relativity teaches us that when two events are separated by what is known as a space-like interval (they occur at times and distances such that light would not have time to travel the interval between them),[11] it is possible to find frames of reference in which one occurs before the other and frames of reference in which the opposite is true. In other words, neither occurs absolutely before the other. Suppose that the measurements of U and V are separated by a space-like interval. It becomes impossible to know which measurement came first.

In some frames of reference, we can say that the measurement on U, which occurred first, caused the result of the measurement on V. In other frames of reference, we can say that the measurement on V occurred first and caused the result of the measurement on U. The image we give of the causality of one measurement on the other, already strange because of the instantaneity of its action, is no longer acceptable, since neither of the two measurements can be considered the cause of the other. Non-locality advocates rarely point out this inconsistency, preferring to ignore

10 Depending on the time of year, it takes light just under 4 min to travel from Earth to Mars.

11 For example, two events, one occurring at noon on Earth and the other at noon and 1 min on the Sun. It takes light 8 min and 20 s to cover the Earth–Sun distance. No ray of light can link these two events, which are therefore separated by a space-like interval.

it. However, it is disturbing. This correlation is indeed instantaneous, but it cannot be considered to have a cause, since neither the common cause that would come from local hidden variables (the most natural explanation) nor the instantaneous cause linked to the effect of one of the two measurements (a less acceptable but nonetheless understandable explanation) are possible. This non-locality is decidedly mysterious.

This way of presenting non-locality (and accommodating its incoherence without further embarrassment) is linked to a realistic vision of quantum mechanics. Indeed, it is implicit in the position of many physicists, who consider that the wave function describes the real physical state of the system and that the collapse of the wave function is a physical process that has a real impact on the state of the system.

Within this framework, the only way to understand experiments on entangled particles is indeed to consider that there is some kind of instantaneous effect at a distance, but which, as we have seen, cannot be attributed to either of the two measurements in particular. This view (which often falls within the interpretation of the Copenhagen School) is in fact unacceptable because it overlooks the problem of measurement (which remains unresolved) and accepts a concept of non-locality whose coherence leaves much to be desired without further questioning! However, this is the position shared by the majority of today's physicists, who refuse to reflect on the foundations of quantum physics because they consider that the questions raised are metaphysical and do not concern the work of a "real" physicist.

However, certain interpretations make it possible to escape this non-locality and show that it is merely an illusion. These interpretations do not consider that a measurement results in a reduction in the wave packet that has a physical effect on the system. This is the case of ConSol, in which there is no non-locality. The analysis of spin measurements (separated by a space-like interval) on two particles U and V is different.

First, in ConSol, a measurement is nothing more than the fact that a given observer selects one of the components of the superposed state and perceives the value attached to this component. The state of the system is not affected by this measurement. No physical effect is produced. Therefore, how does this analyze the experiment with entangled particles in a state of zero total spin? Let us refer again to Alice as the first observer making the measurement on particle U, and suppose that Bob is making the measurement on the distant particle V. Let us look at this from Alice's point of view. Alice will find a result (say "+") when measuring the spin of U. According to the hanging-on mechanism, she will therefore hang-on to the corresponding component of the intricate state and therefore hang-on to the $|+\rangle^U\,|-\rangle^V$ component. Now, let us assume that Bob has made the measurement on particle V. For Alice, Bob's measurement is not a measurement; it is just Bob's entanglement with the system as a whole. Therefore, from Alice's point of view, after Bob's measurement, the global state of the two particles and Bob becomes the following:

$$|\psi\rangle = \frac{1}{\sqrt{2}}[\,|+\rangle^U\,|-\rangle^V\,|-\rangle^B - |-\rangle^U\,|+\rangle^V\,|+\rangle^B]$$

where $|+\rangle^B$ is the state of Bob having obtained the result "+" (id. for "–"). In this situation, nothing special happens on V when Alice makes her measurement on U. She simply hung onto the branch $|+\rangle^U\,|-\rangle^V\,|-\rangle^B$. Only later will she be able to see the result of Bob's measurement on V (and this will necessarily be within a timeframe compatible with normal travel at sublight speed). Now, given the branch she is hooked onto, Bob's measurement (the question she will ask him) can only give the result "–". This correlation, however, is not an instantaneous action at a distance. This results directly from the fact that Alice hooked onto the corresponding component during her first measurement on U. Moreover, it should be remembered that hearing Bob's reply that he has seen "–" will not mean that Bob has, from his own point of view, seen "–", since each observer's perceptions are personal to that observer and can in no way be shared.[12] The disappearance of non-locality is quite similar to what also happens in QBism.

ConSol thus avoids the inconsistencies in the concept of non-locality, which are present in the usual interpretations. Admittedly, this entails a vision of reality that is quite far removed from what common sense teaches us. However, is it not better to have a conception that is coherent, even if it is surprising, than one that seems commonplace but is not logically coherent?

6.3. Conclusion

While ConSol presents an extremely strange worldview, its logic is based on assumptions that many physicists accept without drawing the necessary conclusions.

---

12 This is an important difference from the standard description of the situation: to hear Bob say a posteriori that he perceived the result "–" is to retroactively infer the fact that the result on V was really and for all observers "–" from the immediate instant when Alice made her measurement. This inference, which is the very cause of non-locality, is no longer possible in convivial solipsism.

These assumptions are as follows:

– to accept the unmodified standard formalism of quantum mechanics and not to modify or supplement it with additional equations or hidden variables;

– to consider that quantum mechanics is a universal theory and that it applies to all scales, from microscopic to macroscopic, including the observers themselves;

– to recognize that it is impossible, in the unmodified formalism, to provide a clear and unambiguous definition of the concept of measurement if we limit ourselves to considering physical systems that we measure and macroscopic measuring devices while excluding from the analysis any conscious observer;

– to refuse to accept any bending of logic.

The conclusions reached by ConSol are in line with recent results obtained in the so-called extended Wigner's friend experiments. These findings show that, under certain assumptions, events (i.e. measurement results) are not considered absolute (valid for any observer) but rather relative (valid for a single observer)[13]. This is exactly what ConSol achieves. These (thought) experiments are based on stronger assumptions than the ones we have just mentioned, such as the one called local friendliness, or the fact that it is possible for a "superobserver" to "undo" a measurement made by another observer. Of course, we can refuse to accept some of these hypotheses (in particular, the one concerning the possibility of "undoing a measurement"), but the strength of these arguments is that the conclusion they lead to is a theorem mathematically demonstrated if we accept the hypotheses. ConSol does not have the force of a mathematical theorem – it seems difficult to demonstrate that, under the assumption we have cited, it is the only possible interpretation. On the other hand, it is possible to provide several ontological and metaphysical versions (a discussion we have not been able to enter into in this chapter). However, it is interesting to note that, starting from a different point of view, the conclusions are identical to those of these experiments. Moreover, it is much more explicit and precise about the vision to be adopted than these extended experiments, which merely refute the fact that events are valid for all observers.

The main reason why this is an interpretation to consider is that it is the only one that satisfies logical rigor in the context of the assumptions cited above. At the beginning of this chapter, we explained why none of the interpretations proposed to date meet all of the requirements of these assumptions. Some are unsatisfactory in terms of the objective pursued (pragmatism), some accommodate inconsistency

13 We cannot describe these experiments here, but interested readers with sufficient theoretical background can consult (Schmid et al. 2024) for a review article presenting the main results.

(RQM), others leave important points unclear (Copenhagen School, QBism), and others fail to explain certain fundamental features (Everett theory).

It is, however, possible to reject this interpretation if the formalism is modified, as is the case with dBB theory and GRW theory (or other theories of the same type). In this case, the measurement problem is unquestionably solved without any logical error, and we can re-establish an objective physics that goes without any observer. In a way, this is the dream of certain physicists who are reluctant to engage in philosophy (and a fortiori metaphysics). They would like to return to the comfortable framework of classical physics, even if they are forced to accept certain oddities of quantum mechanics (such as the existence of apparent instantaneous actions at a distance or the impossibility of assuming that all non-compatible observables of a system simultaneously have well-defined values for a given system) are inevitable in any theory.

6.4. References

Bell, J.S. (1964). On the Einstein–Podolski–Rosen paradox. *Physics*, 1(3), 195–290.

Einstein, A., Podolsky, B., Rosen, N. (1935). Can quantum-mechanical description of physical reality be considered complete? *Physical Review*, 47, 777–780.

d'Espagnat, B. (1971). *Conceptual Foundations of Quantum Mechanics*. Benjamin, New York.

d'Espagnat, B. (2003). *Veiled Reality*. Taylor & Francis, Oxford.

Fuchs, C.A. and Schack, R. (2015). QBism and the Greeks: Why a quantum state does not represent an element of physical reality. *Physica Scripta*, 90(1), 015104.

Fuchs, C.A. and Stacey, B.C. (2019). QBism: Quantum theory as a hero's handbook. In *Proceedings of the International School of Physics "Enrico Fermi"*. Italian Physical Society, Bologna.

Leifer, M.S. (2018). What are Copenhagenish interpretations and should they be perspectival? Online Talk, Boston [Online]. Available at: https://www.youtube.com/watch?v=C-C_K-gK6q4.

Planck, M. (1931). Interviews with great scientists. *The Observer* [Online]. Available at: https://www.newspapers.com/article/the-observer-max-planck-observer-12531/25590070/?locale=en-GB.

Schmid, D., Ying, Y., Leifer, M. (2024). A review and analysis of six extended Wigner's friend arguments. *arXiv*. doi: 10.48550/arXiv.2308.16220.

Von Neumann, J. (1955 [1932]). *Mathematische Grundlagen der Quanten-mechanik (Mathematical Foundations of Quantum Mechanics)*. Springer, Berlin.

Wigner, E.P. (1961). Remarks on the mind-body question. In *The Scientist Speculates*, Good, I.J. (ed.). Heinemann, London.

Wigner, E.P. (1983 [1976]) Interpretation of quantum mechanics. In *Quantum Theory and Measurement*, Wheeler J.A. and Zurek, W. (eds). Princeton University Press, Princeton.

Zwirn, H. (2000). *Les limites de la connaissance.* Odile Jacob, Paris.

Zwirn, H. (2016). The measurement problem: Decoherence and convivial solipsism. *Foundations of Physics*, 46, 635–667.

Zwirn, H. (2017). Delayed choice: Complementarity, entanglement and measurement. *Physics Essays*, 30, 3.

Zwirn, H. (2020a). Non locality versus modified realism. *Foundations of Physics*, 50, 1–26. doi: 10.1007/s10701-019-00314-7.

Zwirn, H. (2020b). L'observateur, un défi pour la physique quantique. *Pour la Science*, 509.

Zwirn, H. (2021). Is the past determined? *Foundations of Physics*, 51, 57.

Zwirn, H. (2022). Is QBism a possible solution to the conceptual problems of quantum mechanics? In *The Oxford Handbook of the History of Quantum Interpretations.* Freire, Jr., O., Bacciagaluppi, G., Darrigol, O., Hartz, T., Joas, C., Kojevnikov, A., Pessoa, Jr., O. (eds). Oxford University Press, Oxford.

Zwirn, H. (2023). Everett's interpretation and convivial solipsism. *Quantum Reports*, 5, 267–281. doi: 10.3390/quantum5010018.

Zwirn, H. (2024a). Convivial solipsism as a maximally perspectival interpretation. *Foundations of Physics*, 54, 39. doi: 10.1007/s10701-024-00773-7.

Zwirn, H. (2024b). Is intersubjectivity proven? A reply to Khrennikov and to QBists. *International Journal of Theoretical Physics*, 63, 269. doi: 10.1007/s10773-024-05806-5.

7

Copenhagenish Interpretations of Quantum Mechanics

David SCHMID[1], Yìlè YĪNG[1,2] and Matthew S. LEIFER[3]

[1]Perimeter Institute for Theoretical Physics, Waterloo, Canada

[2]Department of Physics and Astronomy, University of Waterloo, Canada

[3]Institute for Quantum Studies, Schmid College of Science and Technology, Chapman University, Orange, USA

In this chapter, we define a class of *Copenhagenish interpretations* encompassing modern interpretations that follow the Copenhagen spirit. These interpretations are characterized by four postulates: observers observe, universality, anti-ψ-ontology and completeness. We explain why such interpretations are not equivalent to the textbook (or orthodox) interpretation, to the view that we should shut up and calculate, or to strict operationalism. We then discuss what lessons are implied for Copenhagenish interpretations by the measurement problem, the Wigner's friend thought experiment and the simple variants of the Wigner's friend thought experiment that we term Wigner's enemy, stalkee and penpal.

In particular, we discuss how Copenhagenish interpretations give multiple distinct descriptions of each experiment, where these descriptions are each individually true, yet cannot be combined into any single description. To make such interpretations consistent, we require epistemological constraints forbidding certain perspectives from being combined. We discuss these constraints, their motivations and some of the challenges they introduce.

Quantum Physics and Cosmology,
coordinated by Hervé ZWIRN.

7.1. Introduction

More than a century after the inception of quantum mechanics, the Copenhagen interpretation remains popular among physicists. However, the very meaning of the term "Copenhagen interpretation" is debated among scholars of the foundations and interpretations of quantum mechanics. This is because the views of the Copenhagen founders differed in significant ways, are self-contradictory even in different writings by the same author, and can thus be difficult to interpret (especially the in Bohr's writings)[1].

Several more coherent modern interpretations of quantum mechanics draw inspiration from the Copenhagen interpretation. Examples include Bub and Pitowsky's "information" interpretation (Bub 2005; Bub and Pitowsky 2010 Bub 2016), Healey's quantum pragmatism (Healey 2012, 2017, 2018, 2020), Rovelli's relational quantum mechanics (Rovelli 1996, 2018), the QBism of Fuchs et al. (2014) and Fuchs and Stacey (2019), Zwirn's "Convivial Solipsism" (Zwirn 2016, 2024) and the views of various others, such as Peres (1978, 1995) and Brukner and Zeilinger (2003, 2005). These interpretations are more consistent and have been worked out in more detail than the original Copenhagen view. We call them Copenhagen*ish* interpretations[2].

To date, there has been no precise definition of a Copenhagenish interpretation, which makes it difficult to understand the common principles underlying them. Indeed, Copenhagenishts are often more keen to emphasize what *distinguishes* their view from other Copenhagenish interpretations, so that they can be seen to be contributing something new to the debate. Meanwhile, critics may find it convenient to dismiss all variants of Copenhagenishism by pointing at the impenetrability and incoherence of certain specific versions thereof, particularly those due to Bohr. In this work, however, we aim to present Copenhagenish interpretations in as charitable and robust a way as possible, to identify sensible core principles that underpin them, and to attempt to make these principles consistent. This can also be useful for distinguishing which criticisms of Copenhagenish ideas attack the core premises from those that aim to discriminate among specific variants, and can help us understand why certain features of specific variants may be desirable.

1 Howard (2004) argues that the idea of a single "Copenhagen Interpretation" is largely an invention of Heisenberg in the 1950s. Before that time, people just spoke of ideas "in the Copenhagen spirit".

2 Another common name is *neo-Copenhagen* interpretations, but it is difficult to find a terminology that all proponents can agree upon. Fuchs objects to "neo-Copenhagen" on the grounds that "Copenhagen" is most closely associated to the views of Bohr, whereas QBism draws more strongly from Pauli (Fuchs 2011, 2017). Thus, the *ish* in Copenhagenish is meant to indicate a vague family resemblance rather than a direct line of influence.

In this work, we propose a precise definition of what constitutes a Copenhagenish interpretation based on four postulates:

1) observers observe;
2) universality;
3) anti-ψ-ontology;
4) completeness.

Roughly, these postulates state the following. Observers observe states that sufficiently complex agents (such as humans) observe definite outcomes when making quantum measurements. Universality states that quantum theory (including the unitary description of dynamics, the Born rule and the measurement postulates) can be applied to arbitrary subsystems of the universe. Anti-ψ-ontology states that the wavefunction is not an intrinsic property of an individual quantum system. Finally, completeness states that there is no deeper description of reality beyond the quantum formalism.

We believe that this characterization should be acceptable to most proponents of these interpretations. Although we do not attempt a detailed historical analysis[3], our view is that the original Copenhagen interpretation fits into this class, as do the examples of Copenhagenish interpretations given earlier, as well as the views of most physicists who say that they support "the Copenhagen interpretation".

Copenhagenish interpretations are often described as antirealist, positivist, empiricist, instrumentalist, operationalist, etc. (in this chapter, we use these words interchangeably), usually by those who intend these terms as insults.

Although we do not believe these terms are strictly accurate, it is true that Copenhagenish interpretations attribute few, if any, intrinsic ontological properties to systems being described according to quantum mechanics. This allows them to easily evade the measurement problem (Maudlin 1995) and no-go theorems such as Bell's theorem (Bell 1964; Wood and Spekkens 2015) or noncontextuality no-go theorems (Mazurek et al. 2016).

3 We will not engage in the common practice of quoting Bohr at length to try to argue that his interpretation is Copenhagenish in our sense. That is a job for historians. Indeed, several of the issues currently regarded as central to the foundations of quantum theory, such as the nature of the quantum state, were not formulated precisely at the time of Bohr's writing, so you will not find precise statements on them. It is a fool's errand to try to pin down his views on such topics. Much more interesting is to argue the merits of contemporary Copenhagenish interpretations that are in a similar spirit, but have been worked out to a greater degree of clarity and coherence.

Indeed, many no-go theorems about quantum theory start with an assumption like "assume every system has some objective physical properties, λ, that exist independently of the observer". This renders the no-go theorems irrelevant to Copenhangenish interpretations, as Copenhagenishts either deny that such λ's exist or deny that there is any meaningful way of describing them. However, one class of arguments that do not require assumptions of this sort are the Wigner's friend thought experiment (Wigner 1995) and modern extensions of it (Schmid et al. 2023; Brukner 2015; Frauchiger and Renner 2018; Bong et al. 2020; Cavalcanti and Wiseman 2021).

In his famous thought experiment, Wigner (1995) considered what happens when an observer – his "friend" – makes a quantum measurement and, some time later, asks the friend what they observed. This can be analyzed from two perspectives. Before Wigner interacts with them, he can model the system and friend as two quantum systems interacting unitarily, so that they end up in an entangled superposition state. For Wigner, the collapse to a definite measurement outcome supposedly only happens later, when he interacts with his friend. However, from the friend's perspective, the collapse must have occurred earlier when they actually made the measurement and presumably saw the outcome. Wigner argued that these two perspectives are contradictory. This is because he analyzed the experiment according to the *orthodox* interpretation (also called the *textbook* or *Dirac-von Neumann* interpretation) of quantum mechanics. In this interpretation, the two descriptions prescribe different objective physical properties to the system and so are contradictory. Not being a solipsist, Wigner believed that his friend does indeed experience a definite outcome, so he argued that only the friend's description is correct and that the quantum state must objectively collapse by the time a measurement outcome enters the awareness of a conscious agent[4].

However, as we explain in more detail later, the orthodox interpretation is not Copenhagenish. Copenhagenish interpretations assign far fewer intrinsic objective properties to quantum systems than the orthodox interpretation, and as a result the two descriptions arising in Wigner's argument can coexist without contradiction.

Consequently, Wigner's friend arguments (unlike many other no-go theorems in quantum foundations) can more meaningfully be used to obtain insights into the Copenhagenish picture of the world. In this work, we give a Copenhagenish analysis of a variety of simple Wigner's friend arguments, including the original Wigner's

4 Wigner used this as an argument against the materialist view of consciousness, and in favor of the view that consciousness has an effect on matter. He was not particularly concerned with the problems of measurement in quantum mechanics in the paper that introduced his friend (Wigner 1995), although modern treatments tend to view the Wigner's friend experiment as a refinement of Schrödinger's cat or the measurement problem.

friend thought experiment and three variants that we term Wigner's enemy, Wigner's stalkee and Wigner's pen pal. These analyses serve as a basis for the more difficult task of providing a detailed Copenhagenish analysis of more complicated thought experiments, such as extended Wigner's friend scenarios (Schmid et al. 2023; Brukner 2015; Frauchiger and Renner 2018; Bong et al. 2020; Cavalcanti and Wiseman 2021) (a task we leave to future work).

A key lesson illustrated by these simple thought experiments is that Copenhagenish interpretations necessarily provide multiple complementary descriptions of an experiment, where each description is valid, but where we cannot combine the different descriptions into a single description. Any given agent may adopt any given description, but no agent may combine two incompatible descriptions. For consistency, this implies that Copenhagenish interpretations must impose epistemological constraints on how we combine certain types of information. Our work elucidates the need for these constraints as well as the difficulties they introduce.

7.2. Copenhagenish interpretations

Broadly speaking, a *Copenhagenish* interpretation is one that bears a family resemblance to the ideas of the Copenhagen founders. In particular, Copenhagenishts want to view quantum theory as a fundamental physical theory, potentially applicable to any system in the universe, and in no need of "completion" with additional postulates or variables, while at the same time denying that the quantum formalism supplies a literal description of the intrinsic properties of a quantum system. It may seem that the measurement problem (Maudlin 1995), Schrödinger's cat (Schrödinger 1935), Wigner's friend (Wigner 1995) and related thought experiments already pose problems for such interpretations, but we argue that they do not. These thought experiments show that there are situations in which different observers will assign different quantum states to the same quantum system, but this need not be a problem if the quantum state is not an intrinsic property of an individual quantum system, but rather represents something more like the observer's knowledge of a quantum system. Just as different observers may assign different probability distributions to the same classical system if they have access to different information, the same can be true of quantum state assignments. With this in mind, we now explain the four defining postulates of Copenhagenish interpretations.

7.2.1. *Four defining postulates*

The four defining postulates are:

1) observers observe;

2) universality;
3) anti-ψ-ontology;
4) completeness.

We explain each of these in detail in the following sections.

7.2.1.1. *Observers observe*

Observers observe says that there exist many entities in the universe, or at least on planet Earth, that experience definite outcomes when they make quantum measurements. I know, from personal experience, that I am one such entity[5]. Assuming that solipsism is incorrect, I infer that other similarly constituted entities, such as students in a quantum mechanics lab, also count as observers who have definite experiences when they make quantum measurements.

Note that this postulate does not state that human observers, consciousness, or anything like that is *required* for something to count as an observation. We could, for example, accept an account of when definite outcomes occur based on environmental decoherence. Cats, robots, sufficiently complicated experimental equipment, or even (in some views such as Rovelli (1996, 2018)) any physical system could well be capable of registering a definite outcome in a quantum experiment. The exact account of the nature of observation will, of course, depend on the precise Copenhagenish interpretation under consideration. The point is simply that, whatever else counts as an observation, we know from experience that we ourselves perceive different outcomes and, assuming solipsism is false, we infer that other human observers do as well. Human observation is thus a *sufficient* condition for a definite experience, but it need not be necessary.

Although observers observe says that observers experience definite outcomes, it does not specify whether the outcomes exist objectively for everyone, or just for the observer who makes the measurement. Copenhagenish interpretations fall into two categories depending on their stance on this question.

1) Objective: when you make a measurement and observe the result, then the result that you have obtained is an objective fact about the universe. Other observers may not have access to this objective fact, and the subsequent evolution of the universe may make you forget this fact and/or make it fundamentally impossible for other observes to acquire information about this fact, but it is an objective fact nonetheless.

5 The authors are all theorists, so it has been a long time, but most of us have taken laboratory classes as undergraduates.

2) Relativist[6]: when you take a measurement and observe the result, the result you register is an objective fact *for you*. There is no fact of the matter about the result *for me* unless and until I repeat the measurement myself on the same system or interact with you, for example, you tell me the result you obtained.

Note that the objective version of observers observe is essentially the same as the absoluteness of observed events (AOE) assumption that has been used in several extended Wigner's friend arguments (Schmid et al. 2023; Brukner 2015; Bong et al. 2020). Of the Copenhagenish interpretations mentioned above, Rovelli's relational quantum mechanics and QBism are clearly relativist, and so reject this assumption.

The objective view is in line with straightforward empiricism, in which our observations (the only things to which we have reliable direct access) are the foundation on which science is built. Relativist Copenhagenish interpretations threaten to undermine this view, because they render scientific observations subjective. Although we reject the idea that any Copenhagenish interpretation is wholesale empiricism, objective Copenhagenish interpretations are empiricist in this one basic aspect, and this has been part of the Copenhagen view from the beginning. Heisenberg arrived at matrix mechanics by focusing on the relations between observable quantities, such as atomic emission spectra, instead of trying to construct a mechanical explanation based on potentially unobservable entities. The objectivity of these observed phenomena is not under question in objective Copenhagenish interpretations; they are the very bedrock on which the interpretation is built. This led Heisenberg and Bohr to incorporate this aspect of empiricism into their view.

---

6 In earlier talks on this topic, we used the term "perspectival" rather than "relativist". We did this because, in philosophy of science, "relativism" is often associated with extreme views such as social constructivism about scientific theories. To avoid raising that specter, we used "perspectivism". However, perspectivism is most commonly understood in epistemic terms: the apparent objectivity of knowledge is an illusion and there is no perspective-free view from which we can ascertain the nature of reality in itself. *All* Copenhagenish interpretations are perspectival in this sense, so this term cannot be used to distinguish two classes of Copenhagenish interpretation. Relativism is really the correct term to use, as it is broad enough to encompass relative notions of truth as well as relative notions of knowledge. To be precise, in the terminology of Baghramian and Carter (2025), the notion of relativism we need is strong, local, alethic relativism about the results of quantum measurements. "Strong" means that what is true for one observer may be false for another, not just inaccessible or undefinable. "Local" means that we are being relativist about one specific thing, in this case the outcomes of quantum measurements in situations where observers disagree about where the Heisenberg cut must be placed, and not wholesale about everything. "Alethic" means that we are being relativist about truth. Perspectivism is also a variety of relativism, about knowledge rather than truth, but we need the stronger notion here. The relativism we need is extreme, in that it is about truth, but it is also limited in scope. It does not imply global relativism and, in particular, it does not imply cultural relativism of any sort.

> It seems sensible to discard all hope of observing hitherto unobservable quantities, such as the position and period of the electron [...] Instead it seems more reasonable to try to establish a theoretical quantum mechanics, analogous to classical mechanics, but in which only relations between observable quantities occur (Heisenberg 1925).

> [. . .] all unambiguous information concerning atomic objects is derived from the permanent marks such as a spot on a photographic plate, caused by the impact of an electron left on the bodies which define the experimental conditions (Bohr 1963).

This empiricism has also survived into modern objective Copenhagenish interpretations:

> Consistency requires that the result of observing the detectors by another instrument is the same as if the detectors themselves are considered as the ultimate instrument. This is what is meant by the claim that there is an *objective* record of the experiment. The role of physics is to study relationships between these objective records. Some people prefer to use the word "inter-subjectivity", which means that all observers agree about the outcome of any particular experiment (Peres 1995).

In contrast, relativist Copenhagenish interpretations are much more radical. Depending on the specific interpretation, a measurement and its outcome may only exist relative to: an observer or agent (Fuchs et al. 2014; Fuchs and Schack 2013); another interacting physical system (Rovelli 1996, 2018; Di Biagio and Rovelli 2021); the placement of the Heisenberg cut[7]; or which measurements will be made in the future (Renes 2021). If measurement outcomes only have meaning from one of these perspectives, then combining statements about *different* observers' perspectives into a consistent global picture will not always be possible.

Perhaps one of the main reasons why so many physicists pay lip service to the Copenhagen interpretation is because they approve of the apparent hard headed empiricism of the objective Copenhagenish view. They do not want to get derailed into long discussions about interpretation. If they only have to focus on regularities in observed phenomena, then they can just get on with their work. From this point of

7 We introduce the Heisenberg cut in section 7.2.3. In our view, it is not clear whether we can meaningfully take measurement outcomes to be defined only relative to an agent's placement of the Heisenberg cut, given that the cut is a construct for reasoning rather than a physical entity. See also the discussion in section VIII.D of Schmid et al. (2023).

view, relativist Copenhagenish interpretations are shocking because they threaten to undermine the existence of objective experimental records on which empirical reasoning is based.

In a relativist interpretation, there need be no single, global perspective that is consistent with all of the individual perspectives. Rather, the world is comprised of a set of disconnected realities. This sort of far fetched conclusion is precisely what many physicists are trying to avoid by advocating Copenhagen. But as we will see when analyzing various Wigner's friend thought experiments, objective Copenhagenish interpretations lead to constraints on epistemology that have no obvious justification (apart from being required for internal consistency of the interpretation), whereas relativist Copenhagenish interpretations make these constraints arise naturally as a consequence of the (admittedly radical) metaphysics they support.

7.2.1.2. *Universality*

Universality states that quantum theory is a fundamental physical theory, and that anything in the universe (if perhaps not everything at the same time) can in principle be described by quantum theory. In other words, there are no fundamentally "classical" or "non-quantum" systems in the universe. The key components of this universal description are the unitary dynamics, the Born rule for computing the probabilities of measurement outcomes and the state-update rules upon measurement.

Note that we do not regard the projection postulate as an axiom of quantum theory. Whether it holds depends on how the measurement is implemented. For example, it does not hold when the system is absorbed by the measuring device. Rather, we take the "measurement postulates" to mean the Born rule together with the necessary change to the quantum state upon learning the measurement outcome, as described by a *quantum instrument* (Nielsen and Chuang 2010). The projection postulate is just one example of a quantum instrument, and while projective measurements can be implemented in such a way that it holds, it is not the most general possibility.

Provided that the system of interest is sufficiently isolated, universality implies that its evolution can always be described unitarily. In particular, this applies to macroscopic systems (such as measuring devices and observers) in addition to microscopic systems (such as atoms). This is in conflict with collapse theories, which postulate a dynamics that is fundamentally nonunitary in order to effect an *objective* collapse of the quantum state. However, the Copenhagenish position on universality *also* differs from "unitary-only" interpretations (Kastner 2020) that reject the fundamental status of the measurement postulates, such as Everett/many-worlds and the de Broglie–Bohm theory. In a Copenhagenish view, unitary dynamics and the measurement postulates have equal status.

There is a certain tension in this, and, indeed, between universality and observers observe. In a measurement interaction, the joint system consisting of the system to be measured and observer[8] is an isolated system, and so their interaction should be describable unitarily. On the other hand, since observers observe, the observer must see a definite measurement outcome, so the post-measurement state would be given by the measurement postulates. On the face of it, these two descriptions seem contradictory, as they lead to two different quantum states. However, as we shall see in the next section, the anti-ψ-ontology postulate says that quantum states do not represent intrinsic properties of a system, so two different ways of updating the state, leading to two different state assignments, do not necessarily represent contradictory statements about reality. All that matters is whether they agree on the probabilities for the outcomes of measurements that are actually performed.

Modern Copenhagenish interpretations are normally quite explicit about universality. As one example, Brukner Brukner (2015) states:

> The measurement instrument and the observer can be included in the quantum mechanical description, and then observed by someone else, a "superobserver", for whom the original measurement instrument loses its previous status as a means for acquiring knowledge.

Here, Brukner is explicit that observers and measuring devices can either be described quantum mechanically or as classical systems, and that the two descriptions serve different purposes. Rovelli states that "*any* physical system can play the role of Copenhagen's 'observer'" (Rovelli 2018). Universality is implicit in this because the interaction between two microscopic systems would normally be described unitarily, so Rovelli is stating that this should not be treated differently from the interaction of a system with a measuring device. In both cases, there is unitary evolution *and* the realization of a definite outcome. Here, contradiction is avoided by making all physical properties relative to other physical systems, that is, the measurement has an outcome relative to the measuring device, but it does not have an outcome relative to an external system that has not interacted with the system or measuring device. From the perspective of this latter system, the interaction is described unitarily.

Although universality states that any part of the universe can be described according to quantum mechanics by some observer, it does not mean that there is a

8 We typically define the "observer" to include not just the human observer, but also the measurement device and as much of the environment as necessary such that its interaction with the system can be described by unitary dynamics. If there is environmental decoherence, then the part of the environment that is involved in the decoherence process is included as part of the "observer".

meaningful description in which *every* part of the universe is described according to quantum mechanics. In particular, as we shall explain, the definite outcomes experienced by observers cannot be described within the quantum formalism in a Copenhagenish interpretation. Hence, if you want to reason the outcomes of some given observer, then that observer must be described classically[9]. This is so even if you think that the observer, system and measuring device together form an isolated system, which you would otherwise describe as a composite quantum system evolving unitarily.

Despite appearances, this is consistent with observers being quantum systems and there being no limit on what can be described according to quantum mechanics. *I* can describe *you* according to quantum mechanics and *you* can describe *me* according to quantum mechanics, but, if I want to reason about my own measurement outcomes, *I* am not allowed to describe *myself* according to quantum mechanics. Therefore, you can describe me together with the systems I am measuring via unitary evolution provided you do not want to reason about my measurement results. The reason why I cannot describe myself according to quantum mechanics in such a case is not that I am not a quantum mechanical system, but rather because I have chosen to analyze my observations, and these can only be described in classical terms. And if *you* wish to reason about my outcomes, then you too must describe me classically, and for the same reasons. Thus, any valid description can be adopted by any particular observer with the relevant information, as we would expect.

In complex enough experiments with multiple observers, it can be difficult to see how the different descriptions we can give of the experiment all fit together. In order to maintain consistency, it will become clear that the different descriptions generally cannot be combined into a single perspective. We will return to these considerations repeatedly in the coming sections (perhaps most explicitly in section 7.2.3).

7.2.1.3. *Anti-ψ-ontology*

Anti-ψ-ontology is the view that the quantum state $|\psi\rangle$ is not an intrinsic property of an individual quantum system[10]. This has to be formulated carefully, because Copenhagenishts often disagree about what the quantum state *does* represent. It may represent knowledge (Peres 1995), information (Bub 2005; Bub and Pitowsky 2008), beliefs (Fuchs and Schack 2009; Fuchs and Stacey 2019) or advice (Healey 2012, 2017), about a quantum system, where "about a quantum system" can mean

9 By classically, we just mean that the Heisenberg cut is placed before the observer, so that they are not included in the quantum description and the outcomes of their observations can be treated as classical random variables. Contra Bohr, we do not mean that they must be given a description terms of the quantities of classical mechanics, for example, position and momentum.

10 In the language of ontological models, it is not an ontic state.

"regarding the outcomes of measurements made on the system", "about responses to your interventions made on the system", or other formulations of actions and responses. The main point is that it does not mean "about preexisting intrinsic properties of the system".

This is to be contrasted with the ψ-ontic (Harrigan and Spekkens 2010) view, in which the quantum state of an individual quantum system is physically real. For example, in de Broglie–Bohm theory, quantum particles have positions, but the quantum state also exists physically, as it determines how the positions of the particles evolve in each individual run of an experiment. In Everett/many-worlds, the quantum state of the universe is not only physically real, it is moreover the *only* fundamental ontology. All appearances are to be explained in terms of it and its unitary evolution[11].

Operationally, if Alice prepares a system in the quantum state $|\psi\rangle$ and hands it to Bob, who knows nothing about Alice's preparation procedure, then Bob would not be able to determine $|\psi\rangle$ by making a quantum measurement on the system. This is a prediction of quantum mechanics, on which all interpretations agree. However, suppose that instead of just being given the quantum system, Bob somehow learns the complete description of all intrinsic physical properties of the system. In a ψ-ontic interpretation, he would thereby be able to infer $|\psi\rangle$, but in an anti-ψ-ontic interpretation he would not. He would need something like an ensemble of independent and identically prepared systems in order to gain enough information to reliably estimate $|\psi\rangle$.

The term ψ-epistemic is often used for the anti-ψ-ontic view, but that is too narrow for our purposes because it singles out "knowledge" as the key term (whereas, as noted above, different Copenhagenish interpretations may prefer to focus on information, beliefs, advice, etc.). It is too narrow also because the term ψ-epistemic is defined in the framework of ontological models (Harrigan and Spekkens 2010), in which the quantum state represents knowledge about some underlying physical reality. However, in a Copenhagenish interpretation in which $|\psi\rangle$ represents knowledge, that knowledge is about the system's response to actions that may be performed upon it.

Copenhagenish interpretations differ on whether there is an objectively correct quantum state to assign to a system, given the available information about the system. For example, in QBism, quantum state assignments are always subjective. They depend on the degrees of *belief* of the agent using quantum theory, and two agents may assign different quantum states to the system even in the case where they both

11 This is a ψ-complete interpretation, which is a subspecies of ψ-ontic interpretations in which the quantum state is the *only* ontology.

have access to the same information. Other interpretations, such as Healy's quantum pragmatism and Rovelli's relational quantum mechanics, argue that quantum state assignments are objective.

Note that this is *not* the same as the distinction between objective and relativist interpretations. Relational quantum mechanics is relativist in that the properties (or outcomes) assigned to a system depend on the external system relative to which it is being described. But given a choice of external system, there is an objectively correct quantum state assignment relative to that system.

For present purposes, the question of whether state assignments are objective or subjective is not relevant. Even if there is an objectively correct state, all Copenhagenishts agree that it is not an *intrinsic* property of an *individual* quantum system. To quote Bub, the quantum state is not a "truthmaker of propositions" (Bub and Pitowsky 2010). Although the quantum state encodes our predictions about quantum experiments, it is not part of the essential nature of the individual system involved in a single run of the experiment.

This implies that, unlike the orthodox interpretation, Copenhagenish interpretations cannot and do not endorse the eigenvalue–eigenstate link, since this link implies that the quantum state is an intrinsic property of a system. We discuss this further in section 7.2.2.1.

Indeed, as we will discuss further in the next section, the Copenhagenisht need not assign objective intrinsic properties to *any* of the systems that are being described according to quantum mechanics. Therefore, two different quantum state assignments (which may have resulted from two different descriptions of a measurement process) need not be contradictory. They may simply represent the descriptions of two different observers, and the two descriptions need not be unifiable into a single global picture.

7.2.1.4. *Completeness*

Completeness is the assumption that there is no deeper description of the world to be had than that provided by quantum theory. Taken together with universality and anti-ψ-ontology, this has far-reaching implications. Universality implies that any system can be given a quantum description, anti-ψ-ontology implies that such a description does not assign objective intrinsic properties to the system, and then completeness implies that we cannot add anything further to that description. This may seem to imply that systems can have no objective intrinsic properties at all, but this is not quite the case. It merely implies that, if there are any such properties, then physics cannot provide a description of them when we are describing the system according to quantum mechanics. In other words, any such properties are *ineffable*[12]. Therefore, any such properties do not provide a hidden variable theory or ontological model, since in those frameworks it is assumed that we can supply a physical theory that explains our experiments in terms of the properties ascribed by the hidden variables or ontic states.

When we are describing a system according to quantum mechanics, the three assumptions imply that there is *no* description of the objective intrinsic properties of an individual system to be had. Therefore, Copenhagenish interpretations must be antirealist to a certain extent, that is, when we are describing a system according to quantum mechanics, then we cannot also speak meaningfully about its objective intrinsic physical properties.

However, observers observe implies that some systems do have physical properties that we would presumably like to be able to talk about, namely, the results of observations made on quantum systems. In order to talk about them, those results must be described classically, which entails that we must split the world into a part we are describing according to quantum mechanics, which includes the system that is being observed, and a part we are describing classically, which includes the measurement outcomes. Depending on the Copenhagenish interpretation under consideration, these classical properties may be assigned to the observer alone, the system relative to the observer or the interaction between the system and observer. For simplicity, we will tend to use the shorthand of describing measurement results as properties "of the observer", but this is meant to encompass the other two possibilities as well. What's important is that the measurement results cannot be intrinsic properties of the system alone because the system is being described according to quantum mechanics.

In this sense, Copenhagenish interpretations need not be wholesale antirealist, that is, they need not be antirealist about systems for which classical mechanics provides an adequate account, or about the subject matter of higher level sciences like biology, but they are antirealist about intrinsic properties of systems that are being described quantum mechanically.

Another consequence of completeness is that most Copenhagenishts reject counterfactual reasoning, for example, about what would have happened had I made a different measurement from the one I actually chose, as this would entail the

12 Ineffable is usually defined as "incapable of being expressed in words". Here, we mean that it is literally indescribable in any terms. In a slogan: the moon may be there when you are not looking at it, but it is fundamentally impossible for you to describe its properties in language, pictures, mathematics, computer code or anything else.

existence of properties that go beyond the predictions of quantum mechanics. Also, any concrete claims about events that are in principle unobservable are typically viewed as meaningless on the same grounds.

Note that the term ψ-completeness was defined in the ontological framework to be the conjunction of completeness and ψ-ontology (Harrigan and Spekkens 2010). Now, ψ-completeness is a much stronger assumption that is rejected in Copenhagenish interpretations, since they reject ψ-ontology. However, some authors use the term completeness to mean ψ-completeness, that is, they define completeness to mean that the quantum state specifies all of the physical properties of a system. We do not think that a different meaning of completeness is intended by these authors. Rather it seems that ψ-ontology is taken for granted by them. Two factors that contribute to this are that the orthodox interpretation, which is ψ-complete, is often conflated with the Copenhagen interpretation, and that the best known realist interpretations – Everett/many-worlds, de Broglie–Bohm theory and spontaneous collapse theories – are all ψ-ontic. So it is fairly easy to fall into the trap of believing that, apart from *wholesale* antirealism, ψ-ontic interpretations are the only serious possibility[13]. However, as we explain in section 7.2.2.1, Copenhagenish interpretations are not the orthodox interpretation, and they are not wholesale antirealist, so the assumptions of ψ-ontology and completeness should be separated. This can lead to confusion[14], as some important works use completeness to mean ψ-completeness. For example, Maudlin uses the word "completeness" in this way in his well-known formulation of the measurement problem as a trilemma (Maudlin 1995).

7.2.2. *What Copenhagenish interpretations are not*

Since the Copenhagen interpretation is often thought to be the default interpretation of quantum mechanics, it is often conflated with what is found in quantum mechanics textbooks. Most textbooks on quantum mechanics are not primarily concerned with interpretation, and they usually describe something closer to what we call the *orthodox* interpretation of quantum mechanics, or try to avoid the

13 This impression may be exacerbated by the Pusey–Barrett–Rudolph theorem (Pusey et al. 2012) and other ψ-ontology theorems, but these are all proven in the ontological models framework, which is rejected by Copenhagenishts as well as realist interpretations with exotic ontology, such as Everett/many-worlds, interpretations that involve retrocausality (Wharton and Argaman 2020; Price 2012; Adlam 2022) or interpretations that involve providing nonclassical causal explanations (Schmid et al. 2021; Ormrod and Barrett 2024).

14 For instance, conflating the two led Okon (2022a, 2022b) to argue that the conjunction of completeness, universality and the objective version of observers observe (used in some extended Wigner's friend arguments) are already ruled out by Maudlin's trilemma, but this is not true without an implicit assumption of ψ-ontology.

issue altogether by adopting the "shut up and calculate" attitude. Neither of these views is at all close to the Copenhagen interpretation or its modern descendants. Many issues that are thought to be problems for Copenhagenish interpretations, or for quantum mechanics per se, such as the measurement problem and Wigner's friend, are in fact only problems for the orthodox interpretation. We explore this in sections 7.3 and 7.4.

7.2.2.1. *The orthodox interpretation*

The key postulate of the orthodox interpretation (also known as the *textbook* or *Dirac–von Neumann* interpretation) is the *eigenvalue–eigenstate link*[15]. This rule specifies intrinsic properties for a quantum system, and, as such, it directly contradicts the Copenhagenish principle (alluded to above and discussed further in section 7.2.3) that systems being described quantum mechanically should not be assigned intrinsic properties. It also renders the orthodox interpretation ψ-ontic (in fact ψ-complete), so it directly contradicts the Copenhagenish principle of anti-ψ-ontology.

The eigenvalue–eigenstate link states that an observable, represented by a self-adjoint operator A, corresponds to an intrinsic property of a quantum system if and only if the quantum state $|\psi\rangle$ is an eigenvector of A. Furthermore, the corresponding eigenvalue is the value of the property that obtains. Finally, echoing completeness, these are the *only* intrinsic properties of the system[16]. For example, a particle has an intrinsic property of spin in the z-direction when the quantum state is an eigenstate of the Pauli σ_Z operator, but, in this case, it does not have an intrinsic property corresponding to spin in the x-direction.

In addition, the orthodox interpretation assumes that measurements have a single unique outcome, that is, it is not a many-worlds theory, and it endorses the textbook postulates of quantum mechanics, including the fact that measurements must generally change the quantum state of the system and that they can be implemented in such a way that the projection postulate holds.

The eigenvalue–eigenstate link renders the orthodox interpretation ψ-ontic. To see why, suppose that the system is assigned the quantum state $|\psi\rangle$ and consider the observable represented by the rank-1 projector $|\psi\rangle\langle\psi|$. Now, $|\psi\rangle$ is an eigenvector of $|\psi\rangle\langle\psi|$ with eigenvalue 1, so this observable is an intrinsic property of the system that takes value 1. Furthermore, up to an global phase, $|\psi\rangle$ is the *only* quantum state vector

15 The term was first coined by philosopher Fine (1973). Gilton (2016) presents the case that the principle, although unnamed, is present in early textbook accounts of quantum mechanics.

16 Apart, perhaps, from things that do not vary as we change the quantum state, such as the dimension of the Hilbert space of the system.

that has eigenvalue 1 for this observable. Knowing which rank-1 projector is a property of the system with value 1 suffices to determine the quantum state of the system, but, by definition, a ψ-ontic interpretation is one in which knowing the intrinsic properties of the system determines the quantum state uniquely.

Furthermore, the orthodox interpretation is ψ-complete because once you have specified the quantum state $|\psi\rangle$, all of the other properties of the system are determined by it. For any observable represented by an operator A, you simply have to check whether $|\psi\rangle$ is an eigenvector of A and, if so, what the corresponding eigenvalue is.

The intuition behind the eigenvalue–eigenstate link is that if it is possible to determine the outcome of a measurement with probability one, and moreover possible to do so without changing the quantum state of the system, then that is grounds for asserting that the quantity being measured was already a property of the system beforehand. The lack of change of the quantum state is taken as an indication that the system is not disturbed by the measurement (although this implicitly assumes the ψ-complete view). This intuition is similar to motivation for the Einstein–Podolsky–Rosen (EPR) criterion for reality, except that for EPR, the lack of disturbance of the system is supposed to be enforced by spacelike separation rather than lack of change of the quantum state.

Copenhagenish interpretations reject the eigenvalue–eigenstate link for several reasons. First, this link requires that there be a unique, objectively correct, quantum state to assign to a system. Up to a global phase, each distinct state vector is either an eigenstate of a different set of observables or has a different eigenvalue for at least one observable. Thus, either the definite properties or the values of those properties are necessarily different for different state vectors. But since these are the intrinsic physical properties of the system, they must be uniquely specified, so only one state assignment can be correct, and this violates the anti-ψ-ontology. This means that the measurement problem poses an extreme difficulty for the orthodox interpretation. It can avoid inconsistency only by rejecting universality. We will discuss this more in section 7.3. Second, the orthodox interpretation requires the existence of superluminal influences via the EPR argument, whereas a supposed virtue of Copenhagenish interpretations is that they avoid the need for nonlocality.

Part of the reason that the Copenhagen interpretation became conflated with the orthodox interpretation is that the Copenhagen founders often used ontological and epistemological language interchangeably and inconsistently. The epistemological counterpart of the eigenstate–eigenvalue link is: the outcome of a measurement of an observable can be predicted with certainty if and only if the state vector assigned to the system is an eigenstate of the corresponding self-adjoint operator, so in this sense, these are the only physical quantities that can be "known" with certainty. This much would be accepted as fact by every physicist, up to minor quibbles about terminology (e.g. whether "known" has the implication that the system possessed the physical quantity before the measurement was made, or whether we should rather say that the measurement "created" the outcome). All of this is compatible with the Copenhagenish principles, as well as any other interpretation of quantum mechanics, because it only refers to the operational predictions of the theory, which everybody agrees upon.

Things only become problematic when we replace talk of "predictions" and "knowledge" with talk of "properties" and "definite values". We are then dealing with ontology, which is the realm of the eigenvalue–eigenstate link.

In a historical study, Gilton (2016) traces the increasing significance given to the eigenvalue–eigenstate link in the major quantum mechanics textbooks from Heisenberg to Messiah. Gilton argues that a version of the link is stated in Heisenberg's book, although not in modern terminology. The link is stated in Dirac's textbook, although much more explicitly in the second edition than the first. It is stated in a more mathematical language by von Neumann (whence the name Dirac-von Neumann interpretation). By the time we get to Messiah, the eigenstate–eigenvalue link is clearly highlighted as an important principle of quantum mechanics, as it appears in italics in a separate paragraph. Based on this, Gilton argues that the eigenvalue–eigenstate link is part of "standard" quantum theory. Due to Heisenberg's statement, we may even think that it is part of the Copenhagen interpretation.

The problem with Gilton's analysis is that, with the exception of von Neumann, all of the authors Gilton studies state both something similar to the eigenvalue–eigenstate link *and* something similar to its epistemic counterpart. They do so without noting that there is a distinction, and often in the same section of the textbook. Gilton takes this as evidence that they intended that *both* principles should hold, but you could just as easily argue that it means they had not clearly understood the difference between the ontological and epistemic principles, and so which one they intend to endorse is ambiguous. The exception to this is von Neumann, who only states the eigenvalue–eigenstate link in terms of the presence or absence of properties, which is clearly ontological.

In conclusion, we believe that the orthodox interpretation arose from an ontological reading of a principle that should only have been used as an epistemological principle by the Copenhagen founders. It was conflated with Copenhagenishism only because the two readings of the principle were not clearly distinguished at the time (this is good news for the Copenhagenishts, since the orthodox interpretation is either inconsistent or incomplete, as we will see in section 7.3).

7.2.2.2. *Shut up and calculate*

David Mermin (2004) once stated "If I were forced to sum up in one sentence what the Copenhagen interpretation says to me, it would be 'Shut up and calculate!'". This term (often misattributed to Feynman) is meant to invoke the attitude of many practicing physicists who are satisfied with the predictive power granted by quantum theory and uninterested in the interpretative question of what it teaches us about reality.

It is a mistake to associate this view with Copenhagenish interpretations. Indeed, the "shut up" part is an extremely inaccurate description of the works of founders like Bohr and Heisenberg, who wrote voluminously on the subject of interpretation and talked about it at great length. Bohr famously argued with Schrödinger at such length that Schrödinger became ill and took to his bed, which Bohr failed to take as a signal to stop talking to him about quantum mechanics. Modern Copenhagenishts also fail to "shut up" in their voluminous works[17]. Indeed, Mermin later disowned his "shut up and calculate" statement, calling it a "childishly brusque dismissal of such an exquisitely subtle point of view" (Mermin 2004). He subsequently came up with his own Copenhagenish interpretation (Mermin 1998), before becoming a latter day QBist (Fuchs et al. 2014).

Mermin's conflation of Copenhagenishism with "shut up and calculate" is understandable from a historical point of view. As Kaiser (2012) has argued, prior to the 1940s most quantum mechanics textbooks were published in Europe and contained lengthy discussions of the philosophical implications of the theory, usually along Copenhagen lines (although perhaps with an undue emphasis on the eigenvalue–eigenstate link). During the second world war, physics took a much more pragmatic and practical turn, partly because the center of gravity of physics moved from Europe to the United States, which has always taken a more practical attitude toward science, and partly due to the implications of the Manhattan project, which suddenly made fundamental physics relevant to urgent political, ethical and technological issues.

Regardless of the reason, in the post-war period, textbooks vastly reduced their discussion of the interpretation of quantum mechanics and work on the foundations of quantum mechanics became taboo in physics departments. The standard view became that it is pointless to worry about such issues because Bohr (1963) had sorted everything out back in the 1930s, so it is easy to see how Copenhagen became conflated with "we don't worry about it".

---

17 One of them famously posts their entire email correspondence on the subject on the internet (Fuchs and Mermin 2001; Fuchs et al. 2015).

7.2.2.3. *Operationalism*

The "shut up and calculate" view is a kind of agnosticism about the interpretation of quantum mechanics, underpinned by the claim that differences in interpretation will never make a difference to the predictions of the theory. However, if this point of view is taken to its logical conclusion, then we arrive at strict operationalism or instrumentalism, which claims that it is not the job of physics to give an account of the physical world independent of our experiments. Instead, concepts like "preparation", "measurement" and "transformation" are taken as primitives and the job of physics is simply to predict the probabilities of measurement outcomes in experiments composed of these primitives. Given that "shut up and calculate" gets conflated with Copenhagen, it is understandable that operationalism does as well.

But Copenhagenish interpretations are *not* strict operationalism. They do not forbid you from believing that quantum systems have properties independently of whether they are measured in an experiment. They only assert that you can say very little about those properties when you are describing a system according to quantum mechanics. Copenhagenishts are also usually realist about the macroscopic world of our experience, even though they do forbid us from giving a straightforwardly realist description of a system that is being treated quantum mechanically.

7.2.3. *Shifty splits*

Here, we detail an important consequence of the four postulates of Copenhagenish interpretations.

On the first encounter, the two postulates of special relativity seem to contradict each other. How can the laws of physics be the same in all inertial reference frames if there is also an absolute velocity, which should pick out a preferred frame? The apparent contradiction only holds if we keep all other structures of classical physics held fixed. In particular, revising the Newtonian concepts of space and time allows the postulates to peacefully coexist. Likewise, the defining postulates of Copenhagenish interpretations initially seem to be in conflict. How can quantum theory be complete, universal and have the straightforward existence of measurement outcomes if it denies that the things being described by quantum theory (including observers and measuring devices) can have intrinsic properties? How can state-update rules *and* unitary dynamics both be universally applicable, given that the latter describes deterministic and continuous dynamics, while the former describes indeterministic and discontinuous dynamics? Such apparent conflicts can again be resolved by making a radical change to the structure of physics elsewhere, that is, by rejecting the requirement that the goal of physics is to provide a unique objective account of the properties of the world, independent of observers. Rather, the goal of physics is more focused on epistemology than ontology, i.e. more focused on giving a prescription for reasoning about what observers may observe, rather than on giving a description of what actually happens independently of observation (including on the microscale).

In this section, we discuss how this follows from the interplay between and consequences of the four defining postulates.

On the face of it, you may be surprised that we list universality as a Copenhagenish principle, given that Bohr insisted on the need to describe our experimental arrangements in the language of classical physics, which necessitates a split (usually called the *Heisenberg cut*) between the quantum system we are investigating and the apparatus that we describe classically. However, this split is supposed to be movable, so we can give a quantum description of any system in the universe, so long as it is embedded in a larger system, the rest of which is described classically. The movability of this split has led to it being termed "the shifty split" (Bell 1990), but – to be clear – this shiftiness is deliberate. Normally, if a theory is built on operational principles, we hope to later give it a realist underpinning in terms of a more fundamental physical theory. Such is the relationship between equilibrium thermodynamics and statistical mechanics for example. But the Copenhagen founders thought that quantum theory was a fundamental theory of physics, not an operational expedience to be replaced by something deeper. Since they viewed it as fundamental, they had little choice but to think that it applies to everything in principle. However, due to their empirical leanings, they still needed something outside the quantum mechanical description to which their empirical records could refer. The shifty split is thus an attempt to have their cake and eat it, that is, for the theory to be fundamental, but also partly operational.

The cut cannot be moved arbitrarily, however. When you attempt to describe an experiment, there is a highest level and a lowest level at which you can put the cut. The lowest level arises because if you coherently interfere a degree of freedom which is viewed as being on the classical side of the cut, then you will get the predictions wrong (today, we can use decoherence theory (Zurek 2003) to decide where the lowest level for the cut is). The highest level arises because analyzing an experiment requires us to reason about the immediate sensory experiences of observers. As those experiences are only describable in classical terms, we must place those observers on the classical side of the cut. That is, if I want to account for the particular definite outcomes that someone observed in a given run of an experiment, then that is ascribing a property *to that observer*, and so I cannot include them in my quantum description.

For each different placement of the cut, we get a different description of a given experiment. The fact that different quantum states are assigned at different levels is not a problem, because Copenhagenishts are anti-ψ-ontologists. All that matters is that the different levels agree on the predictions for experiments that are *actually performed*. Whether or not it is the case that these different descriptions are consistent, and whether they can be combined in any nontrivial ways, is not immediately clear, and is the subject of debate in the literature. We will elaborate on these questions in the context of some specific examples from section 7.3 onwards.

It is quite subtle to reconcile the fact that there are different incompatible descriptions of any given physical experiment. Different Copenhagenish interpretations disagree on if or how these descriptions can be combined or seen to be consistent. These problems ultimately arise because there is a highest level for the Heisenberg cut, as this prevents us from simply treating *all* systems quantum mechanically. However, the existence of a highest level is necessary, both as a consequence of the Copenhagenish idea that all personal experience must ultimately be described classically, and also to ensure consistency of such interpretations. As we will see more explicitly when we discuss variations of Wigner's friend thought experiments, modeling an observer unitarily while also reasoning about the outcomes they observed leads to inconsistencies. As such, it is useful to expand on what it means to reason about observations.

There are multiple distinct senses in which you might "reason about the observations of an observer", all of which are subject to some degree of suspicion in Copenhagenish interpretations:

1) You could assert that the observer saw an outcome.

2) You could assert that the relative frequency of these observed outcomes is distributed according to the Born rule.

3) You could condition your reasoning on the observer having seen a particular outcome.

4) You could apply a state-update rule to the state assigned by an agent to assign them a new quantum state based on which outcome they observed.

This is not to say that when we chooses to model an observer as a quantum system, we must believe that the observer will *not* observe an outcome. Indeed, to do so would be to reject observers observe. It is just that in any argument or calculation about what can be expected to occur in some experiment, we cannot combine facts about the observer's observed outcomes together with facts that follow from treating the observer as a quantum system. The two perspectives are incompatible (or complementary, in the sense of Bohr), and so are not to be combined.

This is awkward from the standpoint of a straightforward realist picture of the world. If we believe that there is an absolute fact about what the observer observed (as in objective Copenhagenish interpretations), and about the frequencies with which each observation occurs, then why should we ever be forbidden from using this knowledge in ours reasoning? But it is perhaps sensible from a Copenhagenish perspective wherein quantum theory is more of an epistemological tool than a picture of the world that extends to the microphysical level. And in a realtivist Copenhagenish interpretation, the impossibility of combining these distinct accounts makes great sense, because the "facts" to be combined are not generally both true relative to any single observer.

7.3. The measurement problem

There are several aspects to what is normally called "the measurement problem". The first is that the postulates of quantum mechanics refer to the notion of "measurement", but they give no precise criterion for when a measurement occurs. This is apparently problematic because the theory says that an experiment produces data only when a measurement occurs, and the rules for updating the quantum state over time differ depending on whether or not a measurement has occurred. In the orthodox interpretation, where the quantum state determines which properties exist, differing state assignments imply different properties for the system, so the two rules for updating the state are contradictory. Therefore, the orthodox interpretation can only be made consistent if we provide a precise criterion for when a measurement has occurred. This means that the orthodox interpretation cannot endorse universality, which implies that the interaction between a system and an observer can always be described by a unitary interaction.

In Copenhagenish interpretations, there is no question of dropping universality, as it is one of the postulates of this class of interpretation. Hence, a Copenhagenish interpretation cannot endorse any precise criterion for when a measurement has occurred. The reason that this does not immediately lead to inconsistency with the two different rules for updating a quantum state over time (depending on whether or not a measurement has occurred) is because of anti-ψ-ontology. Having two different state assignments does not necessarily imply contradictory statements about the properties of the system.

Let us consider the quantum description of a measurement in a little more detail. We will follow the presentation in Schmid et al. (2023). Suppose a qubit system, labeled S, has an associated Hilbert space $\mathcal{H}_S$ and is prepared in the state $|+\rangle_S = \frac{1}{\sqrt{2}}(|0\rangle_S + |1\rangle_S)$. The system is measured in the computational basis $\{|0\rangle_S, |1\rangle_S\}$ in such a way that the projection postulate applies. According to the measurement postulates of quantum mechanics, the $|0\rangle_S$ outcome will be obtained with probability $1/2$, and if it is obtained, the state is updated to $|0\rangle_S$. Similarly, the

$|1\rangle_{\mathrm{S}}$ outcome will be obtained with probability $1/2$, in which case the state is updated to $|1\rangle_{\mathrm{S}}$. The net result is that after the measurement, the state will be either $|0\rangle_{\mathrm{S}}$ or $|1\rangle_{\mathrm{S}}$, with probability $1/2$ each.

However, a measurement device is just a physical system, made of atoms like everything else, so according to universality, we should be able to associate it with a Hilbert space $\mathcal{H}_M$, so that the interaction between the qubit and the measuring device can be described by a unitary on $\mathcal{H}_S \otimes \mathcal{H}_M$ according to the Schrödinger equation. Note that $\mathcal{H}_M$ should be taken to be a large enough subsystem of the universe such that the interaction between the system and the measuring device can always be modeled by a unitary. If we think that environmental decoherence is necessary for the device to register a definite measurement outcome, then a significant portion of the environment beyond the measurement device itself should also be included in $\mathcal{H}_M$.

Suppose that the measurement device starts in a ready state $|R\rangle_{\mathrm{M}}$, a state in which the measurement device is ready to start a new measurement, for example, it is calibrated and the pointer needle points to zero. Note that there will typically be many possible $|R\rangle_{\mathrm{M}}$ states, for example, because parts of the apparatus may be in thermal contact with the environment. All that matters is that the measurement device is in one such state, as they will all behave similarly in the measurement interaction.

Assuming that the measurement device works correctly when the system is prepared in either the $|0\rangle_{\mathrm{S}}$ state or the $|1\rangle_{\mathrm{S}}$ state, the unitary evolution U_{SM} describing the measurement must satisfy:

$$\begin{aligned} U_{\mathrm{SM}} |0\rangle_{\mathrm{S}} |R\rangle_{\mathrm{M}} &= |0\rangle_{\mathrm{S}} |0\rangle_{\mathrm{M}} , \\ U_{\mathrm{SM}} |1\rangle_{\mathrm{S}} |R\rangle_{\mathrm{M}} &= |1\rangle_{\mathrm{S}} |1\rangle_{\mathrm{M}} , \end{aligned} \qquad [7.1]$$

where $|0\rangle_{\mathrm{M}}$ and $|1\rangle_{\mathrm{M}}$ are states in which the measurement device is registering the outcome 0 and 1, respectively. Note that, as with $|R\rangle_{\mathrm{M}}$, there will typically be many possible $|0\rangle_{\mathrm{M}}$ and $|1\rangle_{\mathrm{M}}$ states, but all that matters is that the evolution takes $|0\rangle_{\mathrm{S}} |R\rangle_{\mathrm{M}}$ and $|1\rangle_{\mathrm{S}} |R\rangle_{\mathrm{M}}$ to one such pair of states. Typically, we want to have $\langle 0|1\rangle_{\mathrm{M}} \approx 0$, so that the outcome can be determined with good accuracy by measuring the measurement device, but this is actually immaterial for the argument.

For some ways of implementing a measurement, $|R\rangle_{\mathrm{M}} := |0\rangle_{\mathrm{M}}$, in which case the unitary evolution for the measurement satisfying equation [7.1] is the CNOT gate[18]. From now on, we assume this convention, so equation [7.1] becomes the following:

---

18 For example, suppose the system is an ion in an ion trap that is confined to be in its ground state $|0\rangle_{\mathrm{S}}$ or first excited state $|1\rangle_{\mathrm{S}}$. We assume that a selection rule prevents the spontaneous transition from $|1\rangle_{\mathrm{S}}$ to $|0\rangle_{\mathrm{S}}$, so that we can reliably encode a qubit in the $|0\rangle_{\mathrm{S}}, |1\rangle_{\mathrm{S}}$ subspace.

$$U_{\mathrm{SM}} |0\rangle_{\mathrm{S}} |0\rangle_{\mathrm{M}} = |0\rangle_{\mathrm{S}} |0\rangle_{\mathrm{M}} ,$$
$$U_{\mathrm{SM}} |1\rangle_{\mathrm{S}} |0\rangle_{\mathrm{M}} = |1\rangle_{\mathrm{S}} |1\rangle_{\mathrm{M}} . \qquad [7.2]$$

Now consider the case where the qubit is prepared in the $|+\rangle_{\mathrm{S}}$ state, and thus the joint state of the qubit and the measuring device is $|+\rangle_{\mathrm{S}} |0\rangle_{\mathrm{M}}$. We apply U_{SM} to $|+\rangle_{\mathrm{S}} |0\rangle_{\mathrm{M}}$ and obtain, by linearity, the joint state after the measurement:

$$\begin{aligned} U_{\mathrm{SM}} |+\rangle_{\mathrm{S}} |0\rangle_{\mathrm{M}} &= \frac{1}{\sqrt{2}} \left(U_{\mathrm{SM}} |0\rangle_{\mathrm{S}} |0\rangle_{\mathrm{M}} + U_{\mathrm{SM}} |1\rangle_{\mathrm{S}} |0\rangle_{\mathrm{M}} \right) \\ &= \frac{1}{\sqrt{2}} \left(|0\rangle_{\mathrm{S}} |0\rangle_{\mathrm{M}} + |1\rangle_{\mathrm{M}} |1\rangle_{\mathrm{M}} \right) =: |\Phi^+\rangle_{\mathrm{SM}} . \end{aligned} \qquad [7.3]$$

This is clearly not the same quantum state as we obtain in the first description of the measurement process (either $|0\rangle_{\mathrm{S}}$ or $|1\rangle_{\mathrm{S}}$). From the point of view of the orthodox interpretation, these two descriptions are in flat-out contradiction since the orthodox interpretation endorses the eigenvalue–eigenstate link. This implies that, if the state of the system is either $|0\rangle_{\mathrm{S}}$ or $|1\rangle_{\mathrm{S}}$, then the system has a property corresponding to the projector $|0\rangle\langle 0|_{\mathrm{S}}$, which takes either value 1 or 0 depending on whether the state is $|0\rangle_{\mathrm{S}}$ or $|1\rangle_{\mathrm{S}}$. If the state of the system and the measurement device is $|\Phi^+\rangle_{\mathrm{SM}}$, then no such property exists because $|\Phi^+\rangle_{\mathrm{SM}}$ is not an eigenstate of $|0\rangle\langle 0|_{\mathrm{S}} \otimes I_{\mathrm{M}}$. From the orthodox view, one of these descriptions must be right and the other wrong. There must be a fact of the matter about whether a measurement, and hence a state collapse, has occurred. So the orthodox interpretation is inconsistent with universality, and it is incomplete unless it gives a definite prescription for when a measurement has occurred.

However, this reasoning does not apply to Copenhagenish interpretations. The two different descriptions can both be correct, but applicable in different circumstances, depending on where you choose to place the shifty split. Which description you should use depends on whether you choose to treat the measurement device according to quantum mechanics or not. Because of anti-ψ-ontology, and the fact that Copenhagenishts reject the idea that state assignments imply *anything* about

---

One way of measuring the energy of the ion is to excite the ion to a higher energy state $|2\rangle_{\mathrm{S}}$ with a laser pulse, tuned so that the transition $|1\rangle_{\mathrm{S}}$ to $|2\rangle_{\mathrm{S}}$ is on resonance, but the transition $|0\rangle_{\mathrm{S}}$ to $|2\rangle_{\mathrm{S}}$ is blocked by a selection rule. The measurement consists of applying this pulse and then detecting any fluorescence of the ion on a screen. The state of the screen at the beginning of the measurement, and at the end if the outcome is $|0\rangle_{\mathrm{S}}$, will be the same: a state $|0\rangle_{\mathrm{M}}$ showing no fluorescence on the screen. However, when the outcome is $|1\rangle_{\mathrm{S}}$ the screen will transition to a state $|1\rangle_{\mathrm{M}}$ showing fluorescence on the screen. Although this is not the only way of measuring the ion's energy, it is one possible way of doing it, and so any general problems with our understanding of quantum measurement ought to apply to this example in particular.

the intrinsic properties of the system, having two different quantum state assignments in two different descriptions of the experiment is not a contradiction. Given that counterfactual reasoning about alternative possible measurements is precluded in Copenhagenish interpretations, the only thing that matters is whether the two descriptions agree on the predictions for measurements that are *actually performed.*

For example, suppose that we want to know what an observer, who is external to the measuring device, will observe about the outcome of the measurement described above. To reason about the outcomes of this observer, we must place the observer on the classical side of the cut. However, there remains a choice about whether to place the measurement device M on the classical or quantum side of the cut. If the split is placed between the system and the measuring device, the collapse is applied as soon as the measurement device interacts with the system. The observer simply sees the outcome already registered on the measuring device, the probability that they see each outcome is $1/2$, and afterward they will assign the state $|0\rangle_S$ or $|1\rangle_S$ to the system, as they now know which of the states the system collapsed to in its earlier interaction with the measuring device. On the other hand, if we put the split between the measurement device and the observer, then the collapse is only implemented when the observer looks at the measurement device. However, they will still see each outcome with equal probability, since both terms in the entangled state in equation [7.3] have equal amplitude of $\frac{1}{\sqrt{2}}$. Moreover, this will collapse the state of the system and measuring device to $|0\rangle_S|0\rangle_M$ or $|1\rangle_S|1\rangle_M$ afterward, at which point the descriptions for either placement of the cut assign the same state to the system, and so both descriptions would make the same predictions about any later measurements on the system.

From this, we see that the split can be placed at a range of *levels*, where a higher level description places more of the world on the quantum side of the split, without necessarily leading to different predictions for actually performed experiments.

7.4. Wigner's friend

The Wigner's friend thought experiment (Wigner 1995) can be thought of as a variant of the Schrödinger's cat experiment where a human observer (the friend) takes the place of the cat, or as a dramatization of the measurement problem where the friend plays the role of the measurement device. The friend, system and measuring device are sealed in a box, which is an isolated system. We take the "friend" system F to consist of the friend, the measuring device and enough of the environment that the interaction between the system S and the friend F can be modeled by a unitary interaction from Wigner's point of view. The friend performs a computational basis measurement on the system, which is prepared in the initial state $|+\rangle_S = \frac{1}{\sqrt{2}}(|0\rangle_S + |1\rangle_S)$. Just as in the measurement problem, we will model this experiment in two ways: one that places the split between S and F, and another that places the split between SF and Wigner. We will refer to these as "the friend's point of view" and as "Wigner's point of view", respectively, but it should be understood that this is just shorthand (since any agent can reason in the same manner as any other agent). We again follow the presentation in Schmid et al. (2023).

The friend observes an outcome, based on which they update the state using the projection postulate. If they observe the 0 outcome, then they update the state of the system to $|0\rangle_{\mathrm{S}}$. If they observe the 1 outcome, then they update the state of the system to $|1\rangle_{\mathrm{S}}$. Each possibility is equally likely.

From Wigner's perspective, the friend is a quantum system, and the system and friend together form an isolated system that evolves unitarily. Wigner models his friend in the same way as we modeled the measurement device in the previous section. Let $|0\rangle_{\mathrm{F}}$ and $|1\rangle_{\mathrm{F}}$ denote the (coarse-grained) quantum states in which the friend has observed the outcome 0 or 1, respectively. Like before, we let $|0\rangle_F$ also be the ready state of the friend before the measurement. Then, the unitary evolution U_{SF} describing the measurement must satisfy:

$$U_{\mathrm{SF}}\,|0\rangle_{\mathrm{S}}\,|0\rangle_{\mathrm{F}} = |0\rangle_{\mathrm{S}}\,|0\rangle_{\mathrm{F}}\,, \quad U_{\mathrm{SF}}\,|1\rangle_{\mathrm{S}}\,|0\rangle_{\mathrm{F}} = |1\rangle_{\mathrm{S}}\,|1\rangle_{\mathrm{F}}\,.$$

Wigner models the friend's measurement by applying U_{SF} to the initial state $|+\rangle_{\mathrm{S}}\,|0\rangle_{\mathrm{F}}$, obtaining one of the Bell states:

$$\begin{aligned} U_{\mathrm{SF}}\,|+\rangle_{\mathrm{S}}\,|0\rangle_{\mathrm{F}} &= \frac{1}{\sqrt{2}}\left(|0\rangle_{\mathrm{S}}\,|0\rangle_{\mathrm{F}} + |1\rangle_{\mathrm{S}}\,|1\rangle_{\mathrm{F}}\right) \\ &= \left|\Phi^{+}\right\rangle_{\mathrm{SF}}\,. \end{aligned} \qquad [7.4]$$

Since Wigner is describing the state of another observer, he is often referred to as a *superobserver*. This term will be even more appropriate in later sections, where we imagine that superobservers also have the extreme technological capability to apply coherent quantum operations to macroscopic systems including other observers.

From the point of view of the orthodox interpretation, this implies a contradiction in exactly the same way as the standard measurement problem of the previous section. Also as in section 7.3, there is no contradiction within a Copenhagenish interpretation: since Copenhagenishts reject ψ-ontology, the different state assignments for the system made in the two different perspectives do not reflect different states of reality, just different choices of where to place the shifty split.

The two can easily come into agreement, for example, if Wigner subsequently opens the box and asks his friend which outcome they saw. Both Wigner and his friend are modeling Wigner classically, so at this point, Wigner will collapse the state of the friend and the system, and they will agree on the final state of the system.

While the box was closed, Wigner assigned a quantum state to the system and friend, which is something that is impossible to even formulate in the friend's description. If the friend wants to reason about their own observations, they necessarily must place themself on the classical side of the split, so the friend cannot adopt this description. But for the purposes of predicting what the friend will say they saw after Wigner opens the box and asks, Wigner could adopt the friend's description, which indeed leads to the same prediction about what he will see when he opens the box. If he does so, then the two will make the same predictions about subsequent measurements made on the system as well.

In general, an observer, Alice, will have a range of possible levels at which they can place the split without changing the predictions for the actually performed experiment. If they want to reason about some observer's outcomes (be it themself or some other observers, Bob, Charlie, etc.), then they must necessarily put those observers on the classical side of the split. Just in front of these observers is the highest possible level at which Alice can place the split.

On the other hand, if Alice is going to do an experiment that checks for coherent superpositions of some set of systems, then those systems must be placed on the quantum side of the split, otherwise Alice's predictions will be wrong. This is the lowest possible level at which Alice can place the split. Any level between these two will make the same predictions for the actually performed experiments[19].

We say that two observers, Alice and Bob, have *range compatibility* if they can reach *level agreement*. This occurs when Alice's highest possible level is higher than Bob's lowest possible level and Bob's highest possible level is higher than Alice's lowest possible level. In this case, they can agree upon a level that is in the allowed range for both of them. This is the case in the Wigner's friend experiment. Although Wigner could start off with his friend on the quantum side of the split, putting the friend on the classical side instead will not change Wigner's predictions, so Wigner and his friend can mutually decide to adopt this description for the sake of reaching level agreement.

On the other hand, there is *range conflict* if Alice and Bob cannot reach level agreement. This happens if Alice's lowest possible level is higher than Bob's highest possible level or vice versa. If range conflicts never occur, then Alice and Bob will

19 In realistic experiments, there may be small differences between the theoretical predictions made at different levels. Provided these are below the threshold of experimental precision we can include levels that make slightly different predictions within the allowed range. This will lead to some fuzziness about the location of the lowest and highest levels of the range.

never make different predictions[20]. However, the next two experiments show that range conflicts do in fact occur in some experiments.

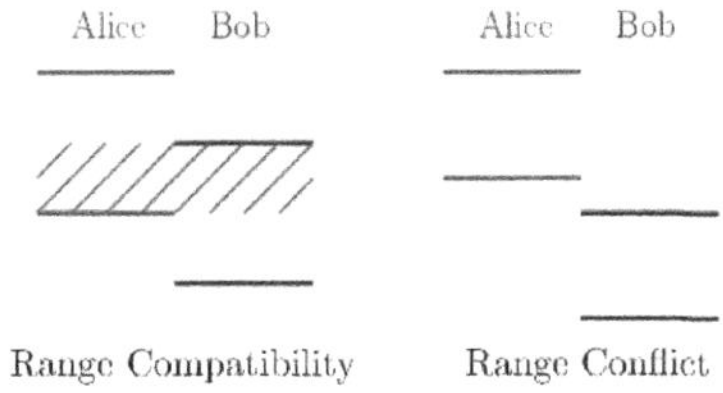

Figure 7.1. *Illustration of range compatibility and range conflict. For a color version of this figure, see: www.iste.co.uk/zwirn/quantum.zip*

7.4.1. *Wigner's enemy*

Thought experiments of the sorts described in this and in the next subsection were first introduced by Deutsch (1985), Albert (1983) and Albert and Putnam (1995).

In the Wigner's friend experiment, Wigner and his friend have range compatibility, so they can always agree on a common location for the shifty split that is compatible with both of their predictions for the actually performed experiment. Things would be simple for the Copenhagenisht if every set of observers always had range compatibility. However, there are situations in which range conflicts necessarily occur.

Suppose Wigner is a superobserver with the (extreme) technological capability to apply unitary operations to the friend and the system while they are inside the box. For example, he could apply $U_{\mathrm{SF}}^{\dagger}$, the inverse of the unitary U_{SF}, on the joint system of the friend and the system S, effectively reversing (or "undoing") the friend's measurement. According to Wigner, undoing the measurement returns the state of the friend and system S to their initial state, as:

$$U_{\mathrm{SF}}^{\dagger} \left|\Phi^{+}\right\rangle_{\mathrm{SF}} = \left|+\right\rangle_{\mathrm{S}} \left|0\right\rangle_{\mathrm{F}} . \qquad [7.5]$$

In this description, neither Wigner nor the friend has any knowledge of what the friend's outcome was after Wigner's undoing operation. If Wigner were to

20 Since the highest and lowest levels for a given observer are slightly fuzzy, it may not be possible to make a sharp distinction between range conflict and compatibility when the lowest level of one agent is close to the highest level of another. But it will be clear that there is compatibility when there is a large amount of overlap and conflict when the highest level of one observer is much lower than the lowest level of another.

subsequently open the box and make a measurement on the system in the $|\pm\rangle_S$ basis, he would obtain the $+$ outcome with certainty. We call this experiment *Wigner's enemy*, since friends do not recohere friends.

Inside the box, before Wigner's application of $U^{\dagger}_{SF}$, the friend is aware of the outcome of their measurement by observers observe. If they want to take account of the outcome they have seen, then they necessarily have to place the split before themselves. They will collapse the quantum state of the system to either $|0\rangle_S$ or $|1\rangle_S$ depending on which outcome they saw in the measurement. We suppose that the friend knows that Wigner is going to implement a procedure on them that Wigner describes as $U^{\dagger}_{SF}$, so how should the friend update the state they assign to the system when this happens? The difficulty is that the unitary $U^{\dagger}_{SF}$ is a quantum interaction between the system and friend, but the friend has no quantum description of themselves. So it appears that they cannot say anything about how the system changes during this interaction. But this is not entirely true.

It is permissible for the friend to neglect their own measurement outcome (i.e. to not reason about it) and to reason about Wigner's undoing exactly as Wigner would, placing the cut above themself and so predicting that Wigner will obtain the $|+\rangle_S$ outcome with certainty. Alternatively, the friend can reason with their own measurement outcome, in which case they must forgo all reasoning about Wigner's undoing, as we just established. However, what the friend *cannot* do is combine their reasoning in the two different descriptions into a single picture. On the objective Copenhagenish point of view, this is because Wigner's outcome is ineffable when the split is placed before the friend, and the friend's outcome is ineffable when the split is placed above the friend. So there is no point of view from which we can combine our reasoning about the two descriptions. This is not to say that Wigner's outcome and the friend's outcome do not both exist. Observers observe states that they do. In an objective Copenhagenish interpretation, both facts are true in an absolute sense, and for all observers, but there is an epistemological preclusion on reasoning about them together[21]. On the relativist view of observers observe, both facts are true in one or

21 If both Wigner's outcome and the friend's outcome exist in an absolute sense, then after many runs of the experiment, there are facts about the joint relative frequencies for both observations. Then why don't they immediately determine the probabilities that the observers ought to assign, which would then determine how they should reason about the observations? This only applies to a strict frequentist account of probability. On any other account, the actual frequencies are related to the probabilities, but do not completely determine them. For example, on a subjective Bayesian account, probabilities represent the subjective degrees of belief of an observer, which are updated by conditioning on the observed data. Provided the prior distribution is generic, and there are a large number of data samples, then the conditioned probabilities will be close to the relative frequencies. However, if the observer necessarily

another perspective, but there is no single perspective from which they are both true, and this relativism is what enforces the epistemological preclusion.

From some perspectives (such as hard-headed realism), these restrictions on reasoning may seem overly convoluted and suspect. What would go wrong if we were to remove the restriction on observers assigning quantum states to themselves while also reasoning about their own measurement outcomes? If this were allowed then, presumably, before Wigner's undoing of the measurement, the friend ought to assign either the state $|0\rangle_{\mathrm{S}} |0\rangle_{\mathrm{F}}$ or the state $|1\rangle_{\mathrm{S}} |1\rangle_{\mathrm{F}}$ to the system and themselves, depending on whether they observed the outcome 0 or 1. Consequently, after the application of the unitary, the friend would assign:

$$U_{\mathrm{SF}}^{\dagger} |0\rangle_{\mathrm{S}} |0\rangle_{\mathrm{F}} = |0\rangle_{\mathrm{S}} |0\rangle_{\mathrm{F}} \quad [7.6]$$

or

$$U_{\mathrm{SF}}^{\dagger} |1\rangle_{\mathrm{S}} |1\rangle_{\mathrm{F}} = |1\rangle_{\mathrm{S}} |0\rangle_{\mathrm{F}} . \quad [7.7]$$

At this point both Wigner and his friend agree that the system is uncorrelated with the friend, but they disagree on the state of the system. If Wigner were to open the box and make a measurement on the system in the $|\pm\rangle_{\mathrm{S}}$ basis, then the two would make different predictions for that measurement. Wigner would predict that the outcome will be $|+\rangle_{\mathrm{S}}$ with certainty and the friend would predict that each outcome will occur with probability $1/2$.

---

cannot become aware of some of the measurement outcomes, then they cannot condition on them, and so the conditioned probabilities will not reflect their relative frequencies. For example, suppose an experiment involves observing two variables, A and B. The objective version of observers–observe requires that definite values of A and B are obtained in each run of the experiment and hence that a relative frequency distribution $P(A, B)$ exists in nature. However, suppose that Alice observes the values of A and Bob observes the values of a B but, by virtue of the way the experiment is conducted, Alice can never become aware of the values of B and Bob can never become aware of the values of A. Then, Alice's marginal probability distribution for A should converge to $P(A)$ over time and Bob's marginal for B should converge to $P(B)$, but nobody's joint probability distribution need converge to $P(A, B)$. Since the correlations in $P(A, B)$ are unobservable, they may (with no empirical consequences) differ from what we would expect them to be from a naive application of some physical theory. From an epistemological point of view the marginal $P(A)$ constrains how we should reason about Alice's observations and $P(B)$ constrains how we should reason about Bob's, but the correlational structure of $P(A, B)$ has no such import. Although this point is easiest to make for a Bayesian approach to probability, it is not limited to this context, since the direct connection between relative frequency and probability is severed in almost all seriously advocated accounts of probability.

Note that, unlike in our treatment of the shifty split in the measurement problem, Wigner and his friend would here make different predictions for an *actually performed* experiment, and they would not have a level disagreement at the time the measurement is performed. They could both put the split between system and friend for the purposes of analyzing Wigner's measurement. Their disagreement on the predictions could also persist for an arbitrarily long time, since Wigner could delay performing his measurement for as long as he wants. So, although the difference of predictions arises because of the range conflict that happened when the friend was inside the box, it would persist even after the conflict was resolved.

How can this be resolved from a Copenhagenish point of view? First, we may observe that a difference of probabilistic predictions for an actually performed experiment still does not constitute a contradiction provided we take a sufficiently liberal view of the nature of probabilities (see footnote 21). For example, on a subjective Bayesian account, there are never any objective facts about what probability must be assigned to an event, so observers may make arbitrarily different probability assignments. However, a better response (as noted earlier) is to point out that this argument is illegal from the Copenhagenish point of view, which does not allow the friend to reason about their own measurement outcome while also describing their perception of it according to quantum mechanics at the same time.

7.4.2. *Wigner's stalkee and penpal*

Next, imagine the following alternative to the Wigner's enemy experiment. Suppose that the friend measures the system in the $|0\rangle_{\mathrm{S}}$, $|1\rangle_{\mathrm{S}}$ basis as before, which, according to Wigner, leaves the joint system comprised of the friend and system in the entangled state $|\Phi^{+}\rangle_{\mathrm{SF}}$. Now imagine that, instead of undoing the measurement and then measuring S in the $\pm$ basis, Wigner performs a measurement on S and F in a basis containing the projector onto the entangled state $|\Phi^{+}\rangle_{\mathrm{SF}}$. As Wigner's state assignment for the system SF is $|\Phi^{+}\rangle_{\mathrm{SF}}$, he predicts that he will obtain this outcome with certainty.

Wigner can implement this measurement in a minimally disturbing manner, so that he applies the projection postulate and so his state assignment $|\Phi^{+}\rangle_{\mathrm{SF}}$ is unchanged by the measurement. It may seem natural to think that if Wigner's state assignment has not changed then he has not changed anything about the SF system. Then, if the friend's measurement outcome existed before Wigner's measurement, it should still exist, unperturbed, afterward. We call this experiment "Wigner's stalkee" because Wigner is attempting to keep track of the state of his friend without doing anything that would alert the friend to his presence, so he is effectively stalking his friend.

The difference between this experiment and Wigner's enemy is that, because the friend's measurement is never undone, both the friend's outcome and Wigner's outcome exist at the same time. You may be willing to accept restrictions on what can be said about outcomes that never coexist (like the two outcomes in the Wigner's enemy scenario), but a bit more uneasy about doing this for things that exist at the same time.

In addition to coexisting, it would seem that we can even arrange the experiment so that a single agent has access to both outcomes at once. We simply have Wigner write the outcome of his measurement on a piece of paper and hand it to the friend through a hole in the box that is small enough and open for a short enough time that no other interaction between the inside and outside of the box can occur[22]. With this modification, it seems that the friend could now become aware of both their own measurement outcome and Wigner's at the same time.

Since the friend is allowed to reason from Wigner's point of view, presumably he agrees with Wigner that the only possible thing that can be written on the paper is, "I got the Φ^+ outcome". The friend is also aware of their own outcome, 0 or 1. We could artificially impose the restriction that the friend is not allowed to reason about these outcomes together, but that is difficult to make sense of when both outcomes are simultaneously available in the friends consciousness awareness[23].

However, this reasoning is a little too quick. To argue that the friend's outcome exists (and is unperturbed by Wigner's measurement), we relied upon the fact that in Wigner's description, Wigner's state assignment was unchanged by Wigner's measurement, which leads to the intuition that *nothing* has changed. But this reasoning is not warranted under anti-ψ-ontology. The state is not an intrinsic property of the system, nor does it determine any instrinsic properties that the system does have. The intuition behind this reasoning really stems from the orthodox interpretation, and is not consistent with Copenhagenishism. Therefore, we cannot say that the friend's outcome is unperturbed by Wigner's measurement. In fact, any reasoning about the friend's measurement outcome requires us to adopt a description where the split is placed before the friend, and in that description there is no joint state of SF that can be said to remain undisturbed.

22 The genre of thought experiment in which Wigner and his friend exchange notes to each other through holes in the box should be called "Wigner's penpal". For those who are fed up with the cutesy names for thought experiments, rest assured that this is the last one. Note that the version of Wigner's penpal described here is different from the one described by Deutsch (1985), where the friend passes a note to Wigner saying "I observed an outcome".

23 OK, we lied: "Wigner's friend's multiple personality disorder" anyone?

Wigner's measurement may or may not have changed the friend's measurement outcome, but the question of which is the case has no answer in the Copenhagenish point of view. It is ineffable, because there is no single description that includes both the friend's outcome and Wigner's measurement.

So, providing one understands the Copenhagenish attitude on the meaning of the quantum state, there is in the end not much difference between the Copenhagenish responses to Wigner's enemy and Wigner's stalkee/penpal.

7.5. Conclusion

In this work, we proposed a principled characterization of Copenhagenish interpretations of quantum theory, identifying four core postulates – observers observe, universality, anti-ψ-ontology and completeness – that capture the commitments common to a broad family of views inspired by the Copenhagen tradition. While proponents of interpretations in this family often emphasize their differences, our aim was to distill the shared assumptions that distinguish Copenhagenish approaches from other competing interpretations.

We then examined several Wigner's friend-type scenarios: Wigner's friend, enemy, stalkee and pen pal. These demonstrate a core feature of Copenhagenish interpretations: that they often provide multiple complementary descriptions of an experiment, where each description is valid, but not able to be combined with the other descriptions into a single description. To remain consistent, a Copenhagenisht must impose epistemological constraints on combining certain types of information. While these constraints can be imposed in both objective and relativist Copenhagenish interpretations, the advantage of relativist approaches is that they ground the epistemological constraints in the metaphysics of the interpretation: the complementary descriptions cannot be combined into one description because they are not *both true* in any one perspective. Of course, this added simplicity comes at a steep metaphysical price, as these interpretations hold that different observers live in distinct, fragmented realities rather than a single, shared, absolute reality.

These analyses illustrate how Wigner's friend-type arguments engage more meaningfully with the commitments of Copenhagenish views than do many standard no-go theorems in quantum foundations. We hope that our analyses in these simple scenarios will serve as a starting point for a careful exploration of the implications of Copenhagenish views in more complex scenarios, such as extended Wigner's friend scenarios.

7.6. Acknowledgments

This work was supported by Perimeter Institute for Theoretical Physics. Research at Perimeter Institute is supported in part by the Government of Canada through the Department of Innovation, Science and Economic Development and by the Province of Ontario through the Ministry of Colleges and Universities. Y.Y. was also supported by the Natural Sciences and Engineering Research Council of Canada (Grant No. RGPIN-2024-04419). M.L. was supported by Grant 63209 from the John Templeton Foundation. The opinions expressed in this publication are those of the authors and do not necessarily reflect the views of the John Templeton Foundation.

7.7. References

Adlam, E. (2022). Two roads to retrocausality. *Synthese*, 200(5), 422.

Albert, D.Z. (1983). On quantum-mechanical automata. *Physics Letters A*, 98(56), 249252.

Albert, D.Z. and Putnam, H. (1995). Further adventures of Wigner's friend. *Topoi*, 14, 17–22. doi: 10.1007/BF00763474.

Baghramian, M. and Carter, J.A. (2025). Relativism. In *The Stanford Encyclopedia of Philosophy (Spring 2025 Edition)*, Zalta, E.N. and Nodelman, U. (eds). Stanford University, Stanford [Online]. Available at: https://plato.stanford.edu/archives/spr2025/entries/relativism/.

Bell, J.S. (1964). On the Einstein Podolsky Rosen paradox. *Physics*, 1, 195–200. doi: 10.1103/PhysicsPhysiqueFizika.1.195.

Bell, J.S. (1990). Against 'measurement'. *Physics World*, 3(8), 33–40. doi: 10.1088/2058-7058/3/8/26.

Bohr, N. (1963). Quantum physics and philosophy – Causality and complementarity. In *Essays 1958–1962 on Atomic Physics and Human Knowledge*. John Wiley & Sons, New York.

Bong, K.-W., Utreras-Alarcón, A., Ghafari, F., Liang, Y.-C., Tischler, N., Cavalcanti, E.G., Pryde, G.J., Wiseman, H.M. (2020). A strong no-go theorem on the Wigner's friend paradox. *Nature Physics*, 16(12), 1199–1205.

Brukner, Č. (2015). On the quantum measurement problem. *arXiv:1507.05255*. doi: 10.48550/arXiv.1507.05255.

Brukner, Č. and Zeilinger, A. (2003). Information and fundamental elements of the structure of quantum theory. In *Time, Quantum and Information*, Castell, L. and Ischebeck, O. (eds). Springer, Berlin, Heidelberg. doi: 10.1007/978-3-662-10557-3_21.

Brukner, Č. and Zeilinger, A. (2005). Quantum physics as a science of information. In *Quo Vadis Quantum Mechanics?*, Elitzur, A.C., Dolev, S., Kolenda, N. (eds). Springer, Berlin, Heidelberg. doi: 10.1007/3-540-26669-0_3.

Bub, J. (2005). Quantum mechanics is about quantum information. *Foundations of Physics*, 35(4), 541–560. doi: 10.1007/s10701-004-2010-x.

Bub, J. (2016). *Bananaworld: Quantum Mechanics for Primates*. Oxford University Press, Oxford. doi: 10.1093/acprof:oso/9780198718536.001.0001.

Bub, J. and Pitowsky, I. (2008). Two dogmas about quantum mechanics. *arXiv:0712.4258*. doi: 10.48550/arXiv.0712.4258.

Bub, J. and Pitowsky, I. (2010). Two dogmas about quantum mechanics. In *Many Worlds? Everett, Quantum Theory, & Reality*, Saunders, S., Barrett, J., Kent, A., Wallace, D. (eds). Oxford University Press, Oxford.

Cavalcanti, E.G. and Wiseman, H.M. (2021). Implications of local friendliness violation for quantum causality. *Entropy*, 23(8), 925. doi: 10.3390/e23080925.

Deutsch, D. (1985). Quantum theory as a universal physical theory. *International Journal of Theoretical Physics*, 24(1), 1–41.

Di Biagio, A. and Rovelli, C. (2021). Stable facts, relative facts. *Foundations of Physics*, 51(1), 1–13.

Fine, A. (1973). Probability and the interpretation of quantum mechanics. *British Journal for the Philosophy of Science*, 24(1), 1–37.

Frauchiger, D. and Renner, R. (2018). Quantum theory cannot consistently describe the use of itself. *Nature Communications*, 9(1), 3711. doi: 10.1038/s41467-018-05739-8.

Fuchs, C.A. (2011). *Coming of Age with Quantum Information*. Cambridge University Press, Cambridge.

Fuchs, C.A. (2017). On participatory realism. In *Information and Interaction*, Durham, I. and Rickles, D. (eds). Springer, Cham. doi: 10.1007/ 978-3-319-43760-6_7.

Fuchs, C.A. and Mermin, N.D. (2001). Notes on a Paulian idea: Foundational, historical, anecdotal and forward-looking thoughts on the quantum. *arXiv:quant-ph/0105039*. doi: 10.48550/arXiv.quant-ph/0105039

Fuchs, C.A. and Schack, R. (2009). Quantum-Bayesian coherence. *arXiv:0906.2187*. doi: 10.48550/arXiv.0906.2187.

Fuchs, C.A. and Schack, R. (2013). Quantum-bayesian coherence. *Reviews of Modern Physics*, 85(4), 1693 [Online]. Available at: https://journals.aps.org/rmp/abstract/10.1103/RevModPhys.85.1693.

Fuchs, C.A. and Stacey, B.C. (2019). Qbism: Quantum theory as a hero's handbook. *arXiv:1612.07308*. doi: 10.48550/arXiv.1612.07308.

Fuchs, C.A., Mermin, N.D., Schack, R. (2014). An introduction to QBism with an application to the locality of quantum mechanics. *American Journal of Physics*, 82(8), 749–754. doi: 10.1119/1.4874855.

Fuchs, C.A., Schlosshauer, M., Stacey, B.C. (2015). My struggles with the block universe. *arXiv:1405.2390*. doi: 10.48550/arXiv.1405.2390.

Gilton, M.J.R. (2016). Whence the eigenstate-eigenvalue link? *Studies in History and Philosophy of Science Part B: Studies in History and Philosophy of Modern Physics*, 55, 92–100.

Harrigan, N. and Spekkens, R.W. (2010). Einstein, incompleteness, and the epistemic view of quantum states. *Foundations of Physics*, 40(2), 125–157. doi: 10.1007/s10701-009-9347-0.

Healey, R. (2012). Quantum theory: A pragmatist approach. *British Journal for the Philosophy of Science*, 63(4), 729–771.

Healey, R. (2017). *The Quantum Revolution in Philosophy*. Oxford University Press, Oxford. doi: 10.1093/oso/9780198714057.001.0001.

Healey, R. (2018). Quantum theory and the limits of objectivity. *Foundations of Physics*, 48(11), 1568–1589. doi: 10.1007%2Fs10701-018-0216-6.

Healey, R. (2020). Pragmatist quantum realism. In *Scientific Realism and the Quantum*, French, S. and Saatsi, J. (eds). Oxford University Press, Oxford. doi: 10.1093/oso/9780198814979.003.0007.

Heisenberg, W. (1925). Über quantentheoretische Umdeutung kinematischer und mechanischer Beziehungen. *Zeitschrift für Physik*, 33(1), 879–893.

Howard, D. (2004). Who invented the Copenhagen interpretation? A study in mythology. *Philosophy of Science*, 71(5), 669–682.

Kaiser. D. (2012). *How the Hippies Saved Physics: Science, Counterculture, and the Quantum Revival*. W. W. Norton & Co., New York.

Kastner, R. (2020). Unitary-only quantum theory cannot consistently describe the use of itself: On the Frauchiger–Renner paradox. *Foundations of Physics*, 50(5), 441–456. doi: 10.1007/s10701-020-00336-6.

Maudlin, T. (1995). Three measurement problems. *Topoi*, 14(1), 7–15. doi: 10.1007/BF00763473.

Mazurek, M.D., Pusey, M.F., Kunjwal, R., Resch, K.J., Spekkens, R.W. (2016). An experimental test of noncontextuality without unphysical idealizations. *Nature Communications*, 7(1), ncomms11780. doi: 10.1038/ncomms11780.

Mermin, N.D. (1998). What is quantum mechanics trying to tell us? *American Journal of Physics*, 66, 753–767. doi: 10.1119/1.18955.

Mermin, N.D. (2004). Could Feynman have said this? *Physics Today*, 57(5), 10–11. doi: 10.1063/1.1768652.

Nielsen, M.A. and Chuang, I.L. (2010). *Quantum Computation and Quantum Information*, Cambridge University Press, Cambridge.

Okon, E. (2022a). On the objectivity of measurement outcomes. *The British Journal for the Philosophy of Science*, 76(2). doi: 10.1086/719046.

Okon, E. (2022b). Reassessing the strength of a class of Wigner's friend no go theorems. *Entropy*, 27(6), 563. doi: 10.3390/e27060563.

Ormrod, N. and Barrett, J. (2024). Quantum influences and event relativity. *arXiv:2401.18005*. doi: 10.48550/arXiv.2401.18005.

Peres, A. (1978). Unperformed experiments have no results. *American Journal of Physics*, 46(7), 745–747. doi: 10.1119/1.11393.

Peres, A. (1995). *Quantum Theory: Concepts and Methods*. Springer, Dordrecht.

Price, H. (2012). Does time-symmetry imply retrocausality? How the quantum world says, maybe? *Studies in History and Philosophy of Science Part B: Studies in History and Philosophy of Modern Physics*, 43(2), 75–83 [Online]. Available at: https://www.sciencedirect.com/science/article/pii/S1355219811000736.

Pusey, M.F., Barrett, J., Rudolph, T. (2012). On the reality of the quantum state. *Nature Physics*, 8(6), 475–478 [Online]. Available at: https://www.nature.com/miscs/nphys2309.

Renes, J.M. (2021). Consistency in the description of quantum measurement: Quantum theory can consistently describe the use of itself. *arXiv:2107.02193*. doi: 10.48550/arXiv.2107.02193.

Rovelli, C. (1996). Relational quantum mechanics. *International Journal of Theoretical Physics*, 35, 1637–1678. doi: 10.1007/BF02302261.

Rovelli, C. (2018). Space is blue and birds fly through it. *Philosophical Transations of the Royal Society A*, 376(2123), 20170312. doi: 10.1098/rsta.2017.0312.

Schmid, D., Selby, J.H., Spekkens, R.W. (2021). Unscrambling the omelette of causation and inference: The framework of causal-inferential theories. *arXiv:2009.03297*. doi: 10.48550/arXiv.2009.03297.

Schmid, D., Yīng, Y., Leifer, M. (2023). A review and analysis of six extended Wigner's friend arguments. *arXiv:2308.16220*. doi: 10.48550/arXiv.2308.16220.

Schrödinger, E. (1935). Die gegenwärtige Situation in der Quantenmechanik. *Naturwissenschaften*, 23(48), 807–812.

Wharton, K.B. and Argaman, N. (2020). Colloquium: Bell's theorem and locally mediated reformulations of quantum mechanics. *Rev. Mod. Phys.*, 92, 021002. doi: 10.1103/RevModPhys.92.021002.

Wigner, E.P. (1995). Remarks on the mind-body question. In *Philosophical Reflections and Syntheses. The Collected Works of Eugene Paul Wigner*, Mehra, J. (ed.). Springer, Berlin, Heidelberg. doi: 10.1007/978-3-642-78374-6_20.

Wood, C.J. and Spekkens, R.W. (2015). The lesson of causal discovery algorithms for quantum correlations: Causal explanations of bell-inequality violations require fine-tuning. *New Journal of Physics*, 17(3), 033002. doi: 10.1088/1367-2630/17/3/033002.

Zurek, W.H. (2003). Decoherence, einselection, and the quantum origins of the classical. *Rev. Mod. Phys.*, 75, 715–775. doi: 10.1103/RevModPhys.75.715.

Zwirn, H. (2016). The measurement problem: Decoherence and convivial solipsism. *Found. Phys.*, 46, 635–667. doi: 10.1007/s10701-016-9999-5.

Zwirn, H. (2024). Convivial solipsism as a maximally perspectival interpretation. *Found. Phys.*, 54(3), 1–16. doi: 10.1007/s10701-024-00773-7.

8

Dark Energy

Nathalie PALANQUE-DELABROUILLE[1,2]
[1]*Lawrence Berkeley National Laboratory, California, USA*
[2]*CEA-Saclay, Paris, France*

8.1. The expanding universe

The early 20th century saw considerable advances in our understanding of the cosmos. In 1908, American astronomer Henrietta Levitt, employed at Harvard Observatory as a "human computer" to analyze the luminosity and distance of stars from photographic plates, published a decisive discovery: a law relating the intrinsic luminosity of cepheids, a category of stars of variable intensity, to their pulsation period. This relationship makes it possible to calculate the intrinsic luminosity of these stars by analyzing the evolution of their intensity over time and thus to accurately assess their distance by comparing this intrinsic luminosity with the observed luminosity. On the basis of this work, American astronomer Edwin Hubble announced the existence of galaxies beyond our own Milky Way Galaxy in 1924. Then, in 1929, he demonstrated a fundamental result: most galaxies are moving away from us at a speed that increases with distance. Meanwhile, on the other side of the Atlantic, the Belgian abbot Georges Lemaître published an almost unnoticed paper in 1927, proposing a solution to Einstein's equations of general gravitation describing an expanding universe. The recession of galaxies observed by Hubble proved to be a direct manifestation of this expansion, constituting one of the fundamental pillars of the Big Bang model. In homage to Hubble, the constant of proportionality between the recession velocity of galaxies and their distance is called the Hubble constant, or H. Its value today is on the order of H_0 = 70 km/s/Mpc, where 1 parsec = 3.26 light years = 3×10^{16} m and 1 mega-parsec is the order of magnitude of the distance between two galaxies. In other words, owing to the

Quantum Physics and Cosmology,
coordinated by Hervé ZWIRN.

expansion of the Universe, the distance between two galaxies is increasing by approximately 7% per billion years.

The idea of an expanding universe was not easy to accept. For millennia, the cosmos was perceived as a vast collection of fixed stars in unchanging space. Albert Einstein himself adhered to this static conception of the Universe. If the laws of general relativity he had established in 1915 allowed the Universe to expand, it would inevitably end up collapsing under its own gravity. To restore the Universe to its static state, Einstein introduced a cosmological constant Λ into his equations, a form of vacuum energy whose repulsive gravitational effect compensates for gravitational attraction. However, he considered this modification artificial and contrary to the mathematical elegance of his theory. As a result, he was quick to abandon this cosmological constant when the expansion of the Universe was demonstrated, going so far as to call it "the biggest blunder of his life".

Once we accept that the Universe is expanding, we have to assume that it had a beginning (the famous "Big Bang") and could come to an end. Its age is directly linked to its expansion rate: the higher the expansion rate is, the younger the Universe. Any variation in the expansion rate over time would therefore influence its age and history (see Figure 8.1).

The relationship between cosmic expansion and the age of the Universe has long been the subject of serious controversy. A team of researchers led by Alan Sandage estimated an H_0 value on the order of 50 km/s/Mpc on the basis of a selection of what they considered to be the best "standard candles". On the other hand, the French astronomer Gérard de Vaucouleurs, who preferred to dilute the risk by using several experimental techniques, obtained a much higher value of approximately 100 km/s/Mpc. The latter presented a problem, since it led to an age of the Universe lower than that of certain stars in the Milky Way. In particular, "globular clusters" – stellar structures so named because of their spherical appearance – were attributed an age of between 11 and 18 billion years, suggesting paradoxically that they were older than the Universe itself! This disagreement between two great figures in astronomy persisted for almost 30 years. It was resolved only through the efforts of the collaboration known as the Hubble Space Telescope Key Project.

The new collaboration identified systematic effects affecting cepheid measurements and applied calibrations to overcome them, resulting in an estimated H_0 value of approximately 72 km/s/Mpc to within 10%, bringing the debate close. The age of the Universe is now estimated at 13.8 billion years, and everything seems to be in order again…or almost.

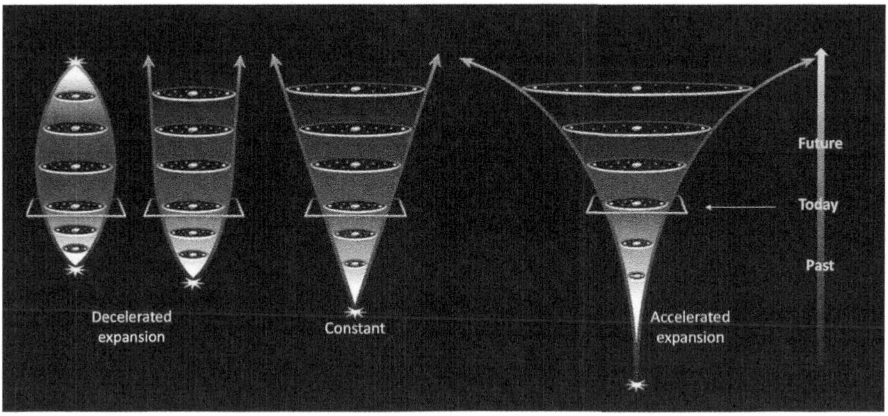

Figure 8.1. *Age and fate of the Universe according to evolution in its expansion rate over time. If expansion is accelerating today, the rate of expansion was lower in the past, implying an origin of time (symbolized by the star at the bottom of each diagram) further back in the past, meaning an older Universe today. The fate of the Universe is also affected. A decelerating expansion could eventually lead to a recontraction of the Universe ("Big Crunch" as in the diagram on the left), whereas an accelerating expansion rate would lead to an ever larger, ever colder Universe ("Big Chill", diagram on the right). Figure adapted from the NASA website (https://starchild.gsfc.nasa.gov/docs/StarChild/questions/question28.html). For a color version of this figure, see www.iste.co.uk/zwirn/quantum.zip*

8.2. First clues to the existence of dark energy

Although we are talking about Hubble's "constant", it is important to understand that it is a priori constant in space, in other words, at any point of observation, but not necessarily in time! We conclude this historical overview by presenting the work of two teams of researchers who, at the end of the 20th century, set out to measure the evolution of the expansion rate of the Universe. Their aim was to detect any slowdown in cosmic expansion caused by the gravitational force. This force is exclusively attractive for all forms of matter and radiation and is supposed to oppose the expansion of space–time.

Hubble's law establishes a relationship between the recession velocity of galaxies (v) and their distance (d): v = H d. The distance is obtained from photometric data by studying objects of known intrinsic luminosity. Their apparent luminosity is then used to determine their distance. Speed, on the other hand, is measured by an effect analogous to the acoustic Doppler effect, but in this case, it is applied to light. In the same way that sound appears more highly pitched when a sound source (such as a car) approaches us and lower pitched when it moves away, the color of a galaxy undergoes a redshift – that is, toward longer wavelengths – as it moves away from us. More precisely, this phenomenon reflects the increase in space–time between the galaxy and the observer. This redshift, known as z, quantifies the increase in the dimensions of the Universe between the emission of a photon by the galaxy and its reception in the telescope by a scaling factor $a = 1/(1 + z)$, with $z = (\lambda_{obs} - \lambda_e)/\lambda_e = (1\text{-}a)/a$, where λ_e and λ_{obs} express, respectively, the wavelengths of the photon at emission and at observation.

Cepheids, although invaluable for measuring cosmic distances, are stars that are too faint to be visible beyond Milky Way's neighboring galaxies. To study Hubble's law at distances of several billion light-years, it is necessary to turn to much brighter objects: supernovae. Supernovae are cataclysmic stellar explosions that occur in binary systems when a compact star progressively accretes matter from its companion. Once the star reaches a certain critical mass, it can no longer counterbalance its own gravitational pressure, triggering a thermonuclear explosion of exceptional intensity. In the days following the explosion, the luminosity peaks before gradually diminishing. As the phenomenon always occurs under similar conditions and at a precise mass reached by the compact star, it is possible to precisely calibrate the intensity of this luminous peak by taking into account the color of the supernova and the duration of the radioactive decay of the light emitted. In this way, the maximum luminosity of these supernovae can also be used as a reference source of known intrinsic intensity. This property makes it possible to calculate the supernova's distance with a high degree of reliability, extending the possible study of cosmic distances.

On the basis of this phenomenon, two teams – one led by Brian Schmidt and Adam Riess and the other by Saul Perlmutter – set out to measure the rate of expansion of the Universe and its evolution over the last few billion years. For each supernova that explodes in a distant galaxy, researchers have estimated its distance by measuring its peak luminosity and determined its redshift by analyzing its spectrum (the decomposition of its light as a function of wavelength). Mirroring Hubble's work, they in turn produced a graph representing these two quantities, which they compared to the relationship expected if the Universe was to expand at a constant rate. A supernova brighter than that predicted by this relationship would be closer, indicating that less time would have elapsed since the emission of its light. This would indicate a slowdown in the expansion, as expected. The question was, by how much? However, to their great surprise, they observed the opposite: supernovae appear systematically fainter, and therefore further away, than anticipated! In 1998, the two teams thus discovered that the expansion of the Universe was not slowing down but rather accelerating! This unexpected discovery, made simultaneously by two independent teams, earned them the Nobel Prize in Physics in 2011. This completely counterintuitive result can be explained by the existence of an unknown form of energy, known as "dark energy", on which gravity would have a repulsive effect. The observed acceleration implies that 70% of the Universe is in this new form of energy. When we also consider that most of the matter itself is not made up of ordinary matter, such as that found in stars or the intergalactic medium, but that 85% of it corresponds to dark matter of as yet unknown nature, this reduces the truly understood fraction of our Universe to a modest 5%!

As Michael Turner put it when he dubbed this new component "dark energy": "it is the most profound mystery in all of science". This was true when it was first discovered in 1998, and it remains true today, almost 30 years later. Dark energy remains one of the greatest enigmas of contemporary physics.

8.3. What could the nature of dark energy be?

The equations of general relativity establish the relationship between the energy content of the Universe and its geometry. On the one hand, they describe the density of matter, radiation, relativistic particles and other forms of energy via the stress–energy tensor $T_{\mu\nu}$. On the other hand, they determine the space–time curvature via tensor $G_{\mu\nu}$. This fundamental relationship is expressed by the equation $G_{\mu\nu} = (8\pi G/c^4)\ T_{\mu\nu}$. As summarized by physicist John Archibald Wheeler, "Spacetime grips mass, telling it how to move… Mass grips spacetime, telling it how to curve". How can dark energy be taken into account in this representation?

The simplest mathematical approach is to introduce a cosmological constant Λ, as Einstein did to model a static universe, before eliminating this term as inappropriate when the discovery of the Universe's expansion rendered it useless. The acceleration of cosmic expansion is now putting Λ back on the agenda. The equation linking the energy content of the Universe to its geometry is then modified:

$$G_{\mu\nu} = (8\pi G/c^4)\ T_{\mu\nu} - \Lambda g_{\mu\nu}$$

where $g_{\mu\nu}$ represents the "metric", or in other words, the coefficients encrypting the calculation of distances in four-dimensional space–time.

This approach has been shown to describe the vast majority of observations remarkably well and is the framework most commonly used to explain the acceleration of expansion. It has thus led to what is known as the standard model of cosmology, which we will discuss below.

On a theoretical level, the cosmological constant can be interpreted as the energy density of the vacuum, which is a simple and appealing physical explanation. In fact, the vacuum is not so "empty" as it is teeming with energy fluctuations, expressed in the form of pairs of particles and antiparticles that are constantly appearing and disappearing. The larger the mass of these virtual particles is, the shorter their lifespan, in accordance with the laws of quantum mechanics and Heisenberg's uncertainty principle. By taking into account all of the particles likely to contribute to this phenomenon, we can, in principle, evaluate the energy of the vacuum. This would be an elegant explanation for dark energy...if it were not for the fact that the energy density of the vacuum is 120 orders of magnitude greater than that measured experimentally! Certain theories of particle physics, such as supersymmetry, may be able to mitigate this disagreement. According to this model, each known particle is associated with a new particle, its "supersymmetric" partner. The properties of these partners would allow a large proportion of their contribution to vacuum energy to be simply cancelled out by each other. However, even with this adjustment, the gap between the theoretical prediction and experimental values remains enormous, on the order of 60 orders of magnitude. Supersymmetry, which has not been demonstrated experimentally, is therefore not a sufficient explanation. What remains to be discovered is a new theory capable of accounting for the astonishingly low value of the vacuum energy density, if indeed this is the origin of the cosmological constant.

An alternative approach is to consider dynamic dark energy with properties that could evolve over time. Like any component of the Universe, this fluid can be characterized by its equation of state, which relates its pressure P to its density ρ: $w = P/\rho$. Matter, whether ordinary or dark, is described by $w = 0$, due to the absence of pressure. A cosmological constant, as discussed above, could also be interpreted, in this framework, as a fluid whose pressure would be exactly opposite to its energy density. Note that this means a component with negative pressure! What does this mean? If we place a membrane in front of an ordinary gas with positive pressure, the latter applies a force to the membrane, pushing it away. In the case of dark energy with negative pressure, the membrane would, on the contrary, be "sucked in". A cosmological constant would thus be equivalent to a fluid characterized by the equation of state $w = -1$. However, in the more general case, it is conceivable that w is not constant and varies over time, paving the way for more complex dark energy models.

Finally, a third approach is to focus not on the energy content of Einstein's equations (the term $T_{\mu\nu}$) but on their geometric aspect ($G_{\mu\nu}$). While general relativity is now largely validated on the scales of the solar system and in the gravity domains typical of planetary systems, what about cosmological scales? The cosmos is extremely tenuous. The Universe is largely made up of vast expanses of vacuum, where the mass is extremely diluted. Galaxies are extreme concentrations of mass relative to the average density of the Universe. In the absence of gravitating matter, testing the laws of gravity is difficult. This raises a fundamental question: can we conceive of modifications to the laws of gravitation which, while similar to the laws of general relativity in the regimes already tested, would offer an explanation for the accelerating expansion of the Universe? Evidence of modifications to the laws of gravity can be found in the history of the galaxy and cluster formation and, more generally, in all of the large gravitational structures of the Universe. Interestingly, modified gravity models have different impacts on trajectories, depending on whether the object is a massive or a massless particle such as a photon. As such, complementary conclusions can be drawn from studies of the velocities of galaxies (massive sources) and of the impact of gravity on the trajectory of light, as in the case of gravitational lenses. We will return to this aspect in section 8.5, which covers observational approaches.

8.4. The standard model of cosmology

In the 20 years since the discovery of the accelerating expansion of the Universe in 1998, up to the start of the major so-called fourth-generation surveys in the 2020s (see section 8.7 on DESI), a consensus has gradually emerged within the scientific community to describe our Universe by a standard model of cosmology, the lambda cold dark matter model (ΛCDM). In this framework, the Universe is made up of approximately 5% ordinary matter (atoms, the vast majority of which are hydrogen, whether in the intergalactic space or trapped in stars), nearly 25% cold dark matter (only sensitive to gravitation and necessary to explain the formation of galaxies or the rotation curves of spiral galaxies, for example, see Chapter 9), and finally 70% dark energy, described by the cosmological constant Λ.

Importantly, this is the composition of the Universe "today". In the course of cosmic expansion, the various components of the Universe have evolved in distinct ways. The matter, whether ordinary or dark, is diluted as distances increase, so its density decreases inversely with increasing volume. The energy density in the form of radiation decreases more rapidly because of the loss of photon energy through its redshift in addition to the dilution effect. On the other hand, if dark energy is expressed as a cosmological constant, its density remains constant despite the expansion of the Universe. Therefore, while the early Universe was a bath of radiation, after just 50,000 years, it became composed predominantly of matter and remained so for over 10 billion years, or approximately 70% of its history. Only in the last 3.5 billion years has dark energy taken over. The “dark energy” domination is therefore quite recent!

Although the ΛCDM model describes a universe whose main components are as yet unknown in nature, it allows the vast majority of observations to be interpreted with remarkable accuracy. One of the best examples is the cosmic microwave background, measured by the Planck satellite. This is the oldest observable light in the Universe, emitted when the latter was 380,000 years old. At that time, the Universe was still remarkably homogeneous, marked only by minute temperature variations of approximately 0.001% relative to the average. The distribution of temperature fluctuations over the entire sky as a function of the size considered, as illustrated in Figure 8.2, is in good agreement with the ΛCDM model. James Peeble noted when he was awarded the Nobel Prize in 2019 for his work on modern cosmology: “I remain surprised and impressed by the ability of the ΛCDM model to pass ever more demanding tests. However, I continue to hope that we will find problems that will guide us toward an even more complete theory”. The standard model is spectacularly robust ...but full of unknowns!

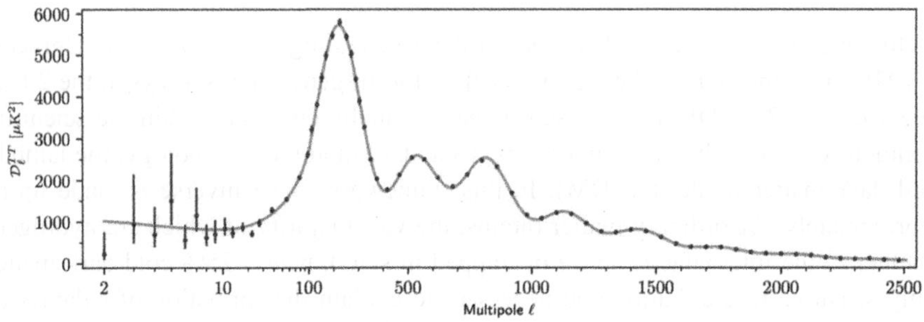

Figure 8.2. *Temperature fluctuations of the cosmic microwave background on the sky as a function of angular scale expressed in multipole (a type of Fourier transform on the sphere). Large scales are on the left (l = 2 corresponds to scales of approximately 90°), small scales are on the right, and the main peak at approximately l = 200 corresponds to scales on the order of 1°. Figure taken from Planck Collaboration (2018). For a color version of this figure, see www.iste.co.uk/zwirn/quantum.zip*

A few tensions occasionally cloud the situation but without seriously shaking the edifice. The main divergence concerns the determination of the current value H_0 of the Hubble constant. There are two independent approaches to accessing this information, which, to date, lead to results that are in clear disagreement.

The first method is to measure H_0 directly, as Edwin Hubble did a century earlier. However, this approach encounters practical difficulties: the galaxies themselves move in the expanding universe, as they are drawn toward each other, gradually forming the cosmic web (see Figure 8.3). Galaxies thus have their own velocity, linked to the local gravitational field, which is added to the redshift due to cosmic expansion. The total redshift measured in the galaxy spectrum is the combined result of these two effects. For cosmic expansion to be the dominant effect and for peculiar velocities to contribute only to negligible disturbance, it is necessary to observe galaxies that are sufficiently distant since the velocity due to expansion increases with distance. The determination of H_0 is then based on a succession of distance indicators.

The distance to the stars in our Galaxy is determined by the so-called parallax effect. This is based on the angle at which the star appears to move relative to the "fixed" stars – those that are sufficiently far away – between two distinct observation positions. This is the principle of stereoscopic vision, which provides a perception of depth. Owing to the Gaia satellite in orbit since 2014, this technique enables precise measurements to be made up to approximately 30,000 light-years from Earth. Beyond this distance, researchers have used cepheids and a law that relates their pulsation period to their distance. This provides relative values, which are then determined absolutely by anchoring this distance scale using cepheids in our own Galaxy, for which parallax measurements are also possible. Finally, for distant galaxies, beyond some ten million light-years, distance measurements are obtained via supernovae, which in turn are calibrated relative to cepheids on the basis of galaxies containing both types of objects. The value of H_0 thus determined is 73.0 ± 1.0 km/s/Mpc according to the most recent data (Riess et al. 2022).

The second technique involves studying the Universe as a whole. This can be done, for example, by analyzing the properties of the cosmic microwave background and determining the parameters of the ΛCDM model that best reproduce the observations. It is then possible to deduce the value of the Hubble constant from these parameters. This approach leads to a value of 67.4 ± 0.5 km/s/Mpc from Planck satellite data.

The situation is far from the conflict of the last century, when H_0 estimates varied by a factor of two depending on the technique used. The current situation, however, is not much better. Considering their low uncertainties, the two measurements are incompatible. Scientists are looking for a bias in the interpretation of measurements from either approach, which would allow the two results to be reconciled but thus far without success.

8.5. Observational approaches to deciphering dark energy

Only new experimental data will enable us to decide between the various possible natures of dark energy, as previously described. In 2006, a consortium of researchers – the Dark Energy Task Force – took up this issue. The report they produced proclaimed the urgent need for a vigorous experimental program to explore this dark energy, which defies our understanding of the fundamental laws of physics. It also identifies the most promising observational means of enabling the scientific community to make progress on this subject (Albrecht et al. 2016).

The report classifies cosmological probes suitable for tackling the question of dark energy into four broad categories: thermonuclear supernovae (the class of stellar explosions mentioned above), large-scale surveys of galaxies for the measurement of baryonic acoustic oscillations (or BAOs – to which we will return at length later in this chapter), gravitational lenses and the properties of galaxy clusters. The latter approach remains less developed than the other three approaches and has not yet reached a comparable level of robustness. We therefore do not discuss this further. Gravitational wave sources from the merger of two neutron stars, which are not mentioned in the report but are nonetheless promising, could also be used as standard sirens. The first case was observed in 2017 and led to an estimate of H_0 with about 10% precision. However, eight years later, this is still the only such event to have been detected. Many more will be needed to extract precise cosmological constraints.

Large-scale gravitational lens studies are used to analyze the distortion of images of cosmic sources in the background caused by the presence of galaxies or clusters of galaxies in the line of sight. These massive structures generate a gravitational field that bends the trajectory of light rays passing nearby, creating characteristic distortions in the images observed. These distortions result from the action of gravity on light and allow us to study this facet of possible modifications to the laws of gravity.

The large-scale study of gravitational lenses is the main objective of the European Euclid survey. The satellite, launched in July 2023, began observations in February 2024 and will run for six years. It maps the shapes of more than 100 million galaxies over approximately 35% of the sky to study the effects of the action of gravitational fields on the propagation of light. Gravitational lens surveys have already been carried out by experiments based on observatories located in different parts of the globe: in Chile and in Hawaii. However, fluctuations in the Earth's atmosphere make such shape measurements difficult. A space-based observatory, by contrast, offers not only significantly better sharpness (image resolution is essentially limited by the diffraction of the telescope, rather than by atmospheric disturbances) but also greater optical stability, since it is unaffected by atmospheric variations. Until the launch of the Euclid satellite, gravitational lens surveys carried out from the ground obtained results that, on the whole, supported the ΛCDM model.

The final approach proposed in the Dark Energy Task Force report is based on the study of large galaxy structures, as illustrated in Figure 8.3. The distribution of matter in the Universe is not uniform. Galaxies are grouped along planes, filaments and nodes and separated from each other by immense cosmic voids. These cosmic structures are caused by the gravitational collapse of matter in the densest regions. Even more interesting for the study of dark energy than the formation of this cosmic web is the fact that these cosmic structures bear the imprint of a characteristic distance that can be used as a standard ruler. This ruler is set in the fabric of space–time and evolves only as the Universe expands, in proportion to the scale factor (a) mentioned at the beginning of this chapter. When measured at different epochs in the history of the Universe, this reference ruler, and more precisely, the evolution of its length over time, is an ideal tracer of the Universe's expansion history.

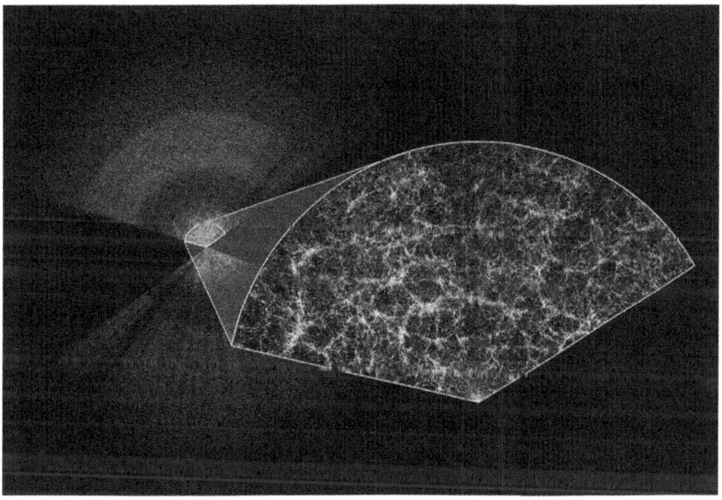

Figure 8.3. *Map showing the distribution of galaxies in a thin slice of the Universe. Each point represents a galaxy. The observer is located at the tip of the cone. The variations in brightness in the background map are the mark of the observational strategy, according to which the density of objects probed depends on the distance studied. The foreground image is a zoom on the region closest to Earth (covering a distance of approximately two billion light-years), in which the structures of the cosmic web are clearly visible (credit: Claire Lamman/DESI collaboration). For a color version of this figure, see www.iste.co.uk/zwirn/quantum.zip*

8.6. The origin of a standard ruler for measuring cosmic expansion

As with reference sources such as cepheids or supernovae, researchers now have a standard ruler! This is of immense interest to cosmology. On the one hand, doubt is inherent to the scientific process: a result, especially a spectacular one, only becomes truly credible if it is confirmed by multiple experiments, and all the more so as these experiments are based on independent approaches. Dark energy clearly falls into the category of first-rate mysteries, and the need to understand its nature amply justifies the use of several complementary techniques. On the other hand, a reference ruler has a considerable advantage over a reference light source such as a supernova. Indeed, while the brightness of a cosmic object can be altered by many factors, whether instrumental (such as an uncontrolled loss of efficiency of an optical lens) or astronomical (such as light absorption by cosmic dust), the determination of a distance via the measurement of angles and redshifts is far more robust and less subject to such perturbations.

The origin of this standard ruler returns to the primordial Universe. A few 10^{-32} s after the Big Bang, the Universe emerged from a hypothetical inflationary phase. It was perfectly homogeneous, with the exception of a few tiny lumps of quantum fluctuations, which formed seeds of over density around which matter could gradually aggregate. This gave rise, billions of years later, to the great structures we observe today – galaxies, clusters of galaxies – all aggregated within the cosmic web. Three minutes after the Big Bang, the Universe, which was still extremely dense and hot, was a plasma of photons, protons, electrons and a few light nuclei. In this environment, there is competition between gravitational attraction, which tends to pull matter toward the initial small overdensities, and the radiation pressure of photons, which opposes it. As in a laboratory plasma, this interplay of forces generates pressure waves, known as acoustic waves, which persist and propagate beyond their point of origin for as long as the plasma remains. These waves are called "baryon acoustic oscillations" (BAOs) – baryons are particles of ordinary matter, such as protons, neutrons and atomic nuclei.

Approximately 380,000 years after the Big Bang, the expanding Universe cooled sufficiently to allow electrons and atomic nuclei to combine into neutral atoms. The Universe became transparent to photons, which then almost completely ceased to interact with matter. The speed at which pressure waves spread suddenly decreases. This event played a fundamental role in cosmic history. On the one hand, the photons, now free, can be detected and offer us a snapshot of the Universe as it was at that time; these are the images of the cosmic microwave background, an invaluable source of information about our primordial Universe. On the other hand, matter, now decoupled from photons, can begin to accumulate in the densest zones, amplifying the initial fluctuations and giving rise to the first stars and galaxies.

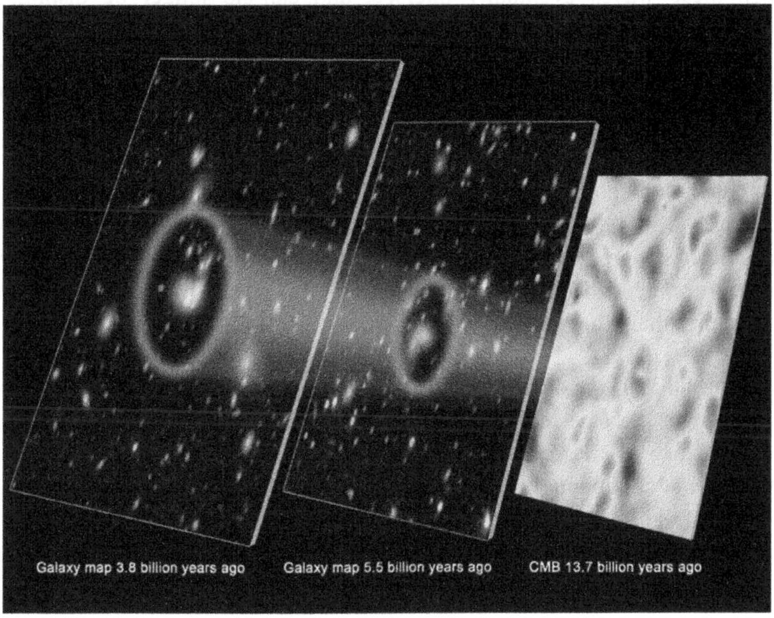

Figure 8.4. *Large-scale galaxy surveys measure the 3D distribution of galaxies at different epochs, revealing their precise position, the age of the Universe at that moment and thus the history of the expansion of the Universe. The study of dark energy relies on the analysis, in this 3D map, of a reference ruler that can be linked to the density fluctuations in the Cosmic Microwave Background (CMB, shown on the right) that later gave rise to the galaxies and large structures of the Universe. (Credit: Eric Huff, the SDSS-III team, and the South Pole Telescope team. Diagram by Zosia Rostomian). For a color version of this figure, see www.iste.co.uk/zwirn/quantum.zip*

The situation for our acoustic waves is comparable to that of a lake in which many small gravels fall simultaneously, each generating ripples that propagate from the point of impact. The surface of the water would be covered with waves, such as those in the primordial plasma. Now, imagine that a sudden drop in temperature instantly freezes the lake. The ripples stall at the maximum distance they can reach. In the same way, primordial acoustic waves are frozen at that instant. At the center of these waves is dark matter, which, since it does not interact with photons, has not been dragged along by them. Around the location of these initial fluctuations, a spherical shell formed. Its radius can easily be determined from the propagation speed of the acoustic waves (approximately 60% of the speed of light) and their propagation time (380,000 years). A more precise calculation reveals that the radius of these waves is then 150 kiloparsecs (kpc), or approximately 450,000 light-years. Although these waves eventually intersect and can no longer be distinguished individually – imagine the surface of the lake under a deluge of gravel! – the distance of 150 kpc leaves a characteristic imprint on the distribution of galaxies that will later form in these regions of over density, as illustrated schematically in Figure 8.4.

From the moment these waves froze, the Universe expanded by a factor of approximately 1,000. As a result, the imprint of acoustic waves in today's Universe has a characteristic size 1,000 times larger, or approximately 450 million light-years. We therefore expect to observe a slight excess of galaxy pairs separated by this distance, rather than, for example, 400 or 500 million light-years.

Measuring the physical length of this BAO feature at different times is one of the major challenges of large-scale three-dimensional galaxy surveys. This enables us to study the history of the expansion of the Universe and the role played by dark energy. The first convincing measurements were obtained in 2005 from observations of the Sloan Digital Sky Survey (SDSS) collaboration. A succession of technical developments has gradually led to a new generation of projects, including the emblematic DESI (dark energy spectroscopic instrument) experiment.

8.7. DESI and large-scale galaxy surveys

DESI is the major galaxy survey of the 2020s, following the SDSS project mentioned above. As a fourth-generation experiment (according to the nomenclature of the Dark Energy Task Force report) to address the question of dark energy, DESI is a much larger survey with much more powerful instrumentation.

From a scientific perspective, logic is simple. To improve the statistical accuracy of cosmic expansion measurements, there are two possible approaches. The first is to increase the number of sources whose redshift can be determined. The second is to increase the volume of the observed Universe, which means both wider coverage of the extragalactic sky and deeper redshift exploration to observe even the most distant sources. The DESI program is thus designed to observe over 40 million extragalactic sources over approximately 30% of the sky, avoiding regions contaminated by emission from stars and dust in our own galaxy. 3D mapping is achieved in two stages. First, a photometric survey provides a 2D image of the Universe, from which researchers select specific sources to be supplied to the DESI program. The third dimension of these objects is then obtained by their redshift, which is determined via spectroscopy. The DESI focuses on four major classes of galaxies to cover the history of the Universe homogeneously over the last 12 billion years or so. In the near Universe, these include old galaxies, which are sufficiently luminous to be observed even in the presence of the moon; red galaxies, which are also very bright and visible at distances of approximately eight billion light-years; and young galaxies, which are still in the star-forming phase, located at distances of approximately six to nine billion light-years. These latter galaxies, although fainter, exhibit spectra with intense emission lines, evidence of ongoing star formation, enabling them to be identified and their redshift measured. Finally, to explore the most distant universe, DESI targets quasars, which are galaxies housing a supermassive black hole a few million to a billion times the mass of the sun. When the surrounding matter is heated before being engulfed, these black holes emit intense radiation, making them visible to the farthest reaches of the cosmos. Like galaxies, quasars are used as direct tracers of matter up to a distance of approximately ten billion light-years.

Beyond this distance, the density of quasars in the Universe becomes too low to allow precise mapping of matter. Researchers are therefore using a new approach in which a few rare quasars act as cosmic beacons that illuminate the intergalactic matter in their line of sight. This matter, essentially neutral hydrogen, absorbs light from the distant quasar at the wavelengths of transitions between energy levels of the hydrogen atom. The most likely transition, called Lyman-α, takes place at a wavelength of 121.6 nm. When originating from distant hydrogen clouds, this absorption of light is observed at a longer wavelength because of cosmic expansion. The intergalactic medium is responsible for such absorptions all along the line of sight of background quasars. It thus generates a "Lyman-α forest", consisting of a series of absorptions whose intensity at each wavelength depends on the gas density at any point along the line of sight. Figure 8.5 shows a quasar spectrum, in other words, the decomposition of its light as a function of wavelength, and its Lyman-α forest. Like galaxies for closer distances, the Lyman-α forest is used in cosmology as a probe of the Universe's matter for the farthest distances, from approximately 10–12 billion light-years.

From an instrumental perspective, DESI represents a major advance over previous experiments. Its 4-m-diameter primary mirror offers a collecting surface that is six times larger than that of the telescope used by the SDSS. In addition, DESI's focal plane is equipped with 5,000 optical fibers mounted on robotized actuators, a considerable improvement over the previous system, which was based on aluminum plates unique to each field studied. Prior to any observation, each plate had to be perforated with 1,000 holes at the precise positions of the galaxies in the focal plane, and then an optical fiber had to be manually inserted into each prepared location. Owing to its robotized technology, which enables all of the fibers to be repositioned on a new set of targets in less than two minutes, DESI is able to measure the spectrum of over 150,000 sources in a single night of observation. This marked a considerable leap forward compared with the SDSS generation, which could at best measure 9,000 sources per night.

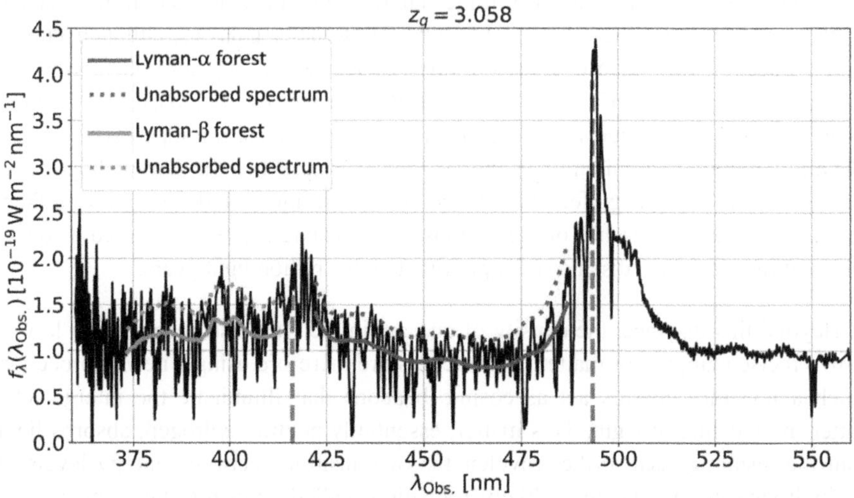

Figure 8.5. *Light intensity versus wavelength distribution of a quasar located at a redshift of 3.058, or a distance of approximately 11.4 billion light-years from Earth. The region highlighted in blue is marked by the succession of absorptions caused by the Lyman-α transition of the hydrogen in the intergalactic medium on the quasar's line of sight. The dotted line indicates the level this spectrum should have in the absence of absorption. In orange, the same phenomenon occurs not only for the Lyman-α transition but also for the Lyman-β transition (figure adapted from eBOSS Collaboration (2021)). For a color version of this figure, see www.iste.co.uk/zwirn/quantum.zip*

8.8. A new turning point for dark energy?

The DESI experiment began mapping the Universe in May 2021. Among the data from the first year of observations, the collaboration analyzed nearly six million galaxies – around three times more than all previous spectroscopic surveys in 20 years – and 420,000 Lyman-α forests, twice as many as SDSS.

The DESI collaboration published its first results in April 2024. Despite the cautious terms of the publication, they went viral in the scientific community, and the echoes from the international press were staggering. What was so revolutionary about DESI?

From the discovery of the accelerating expansion of the Universe in 1998 to the publication of DESI's first results, the majority of experiments have, over the years, confirmed and refined the standard model of cosmology and determined its parameters ever more precisely. The DESI is the first generation IV experiment (according to the Dark Energy Task Force report) to obtain new results on dark energy.

The DESI experiment studies the expansion of the Universe by measuring the characteristic size of the BAO footprint at different epochs. In particular, with its 3D mapping, DESI measures the evolution of cosmic expansion over seven distinct epochs covering a broad period of the Universe's history, from the present to approximately ten billion years in the past. This study highlights the role played by dark energy in cosmic history. Although the DESI results allow interpretation in line with the ΛCDM model, according to which dark energy is described by a cosmological constant Λ, it nevertheless favors an alternative in which dark energy would be dynamic and would have varied over time (DESI Collaboration 2024). If this result is confirmed by future data, it will revolutionize the scientific community, and the standard model of cosmology will have to be revised.

Although the effect is minimal, DESI's distance measurements are accurate to within fractions of a percent, whether taken from Lyman-α forest data for more distant regions or from galaxy mapping for closer regions. In addition, the collaboration worked with exemplary rigor: the entire analysis was carried out initially on data that had been deliberately "blurred" to avoid any influence, be it conscious or otherwise, linked to the conformity of the results with a particular model. Only after fixing all options and parameters was the analysis applied to the real 3D map. This "blind" analysis approach lends exceptional robustness to the conclusions obtained and adds to their credibility.

Figure 8.6 illustrates the DESI measurements, highlighting the imprint of primordial acoustic oscillations in the distribution of matter and the preference for a dynamical dark energy model (dashed curve) over the standard ΛCDM model (horizontal straight line at value 1).

By 2025, barely a year after the first results were published, the DESI experiment had already analyzed more than twice as many data. With a mapping of the Universe extending to almost 15 million galaxies and quasars and over 800,000 Lyman-α forests, DESI confirms, with even greater precision, its initial conclusions. The collaboration has carried out a whole range of tests to verify the stability of the results and has explored numerous avenues in search of a possible, more natural explanation, without identifying the slightest reason to suspect a problem in the data or their interpretation. Yes, dark energy does appear to vary over cosmic time, weakening in the last billion years!

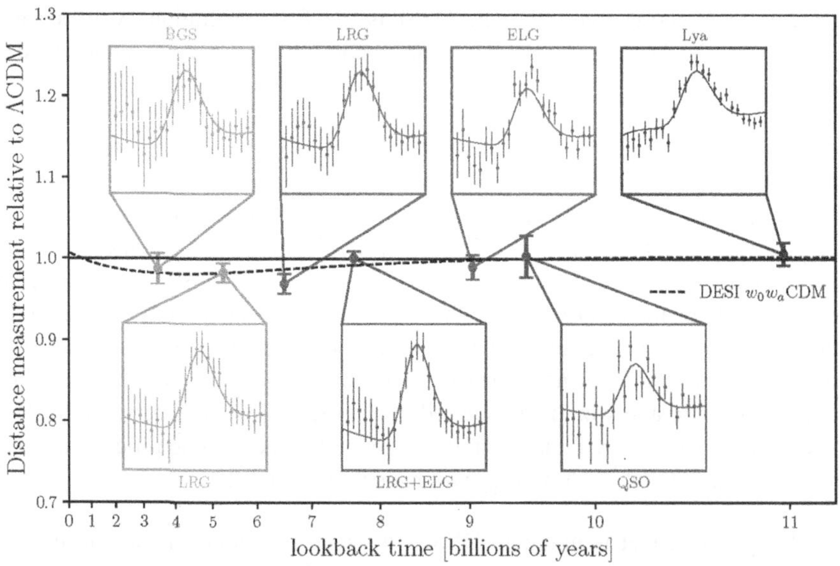

Figure 8.6. *Measurement of the size of the cosmic standard ruler by the DESI experiment from the first year of data. Acoustic oscillations (or BAOs) leave a clearly visible imprint on the mapping of matter, translated here by a peak in the distribution of the probability (vertical axis in each of the insets) of having two tracers (galaxies or Lyman-α absorptions) separated by a given distance (x-axis in each of the insets). Each thumbnail represents a different type of galaxy or a measurement in the Lyman-α forest, probing the history of the expansion of the Universe at different times. The horizontal black line at value 1 corresponds to the size of the BAO footprint in the context of a cosmological constant Λ. The dashed curve shows the prediction for a dynamical dark energy model, marked w0waCDM (Credit: DESI press release, https://newscenter.lbl.gov/2024/04/04/desi-first-results-make-most-precise-measurement-of-expanding-universe/). For a color version of this figure, see www.iste.co.uk/zwirn/quantum.zip*

What do these results tell us about the history of the expansion of the Universe? Initially, the expansion followed the classic scenario. For approximately seven billion years after the Big Bang, when the Universe was made up mainly of dark matter, its expansion gradually slowed. This behavior is predicted both by the ΛCDM and by the dynamic dark energy models. Over time, the dark energy density became predominant, reversing the trend and causing the expansion to accelerate. If dark energy was translated into a cosmological constant, this acceleration would continue indefinitely, becoming ever faster, as illustrated in the rightmost diagram of Figure 8.1. However, DESI measurements indicate that dark energy seems to obey an equation of state whose parameter w has evolved: from a value $w = -1$ for a cosmological constant, or even more negative, it now tends toward a value closer to zero. This evolution indicates a decrease in the repulsive influence of dark energy over time. Over the past billion years or so, this weakening of dark energy has made the current expansion slower than that predicted by a model based on a cosmological constant.

Do these results have any impact on the fate of the Universe? It is premature to draw conclusions on this question without greater knowledge of the nature of dark energy. The future evolution of the Universe will depend on the relative dynamics of dark energy and matter. As long as the parameter w of dark energy remains sufficiently close to -1, the expansion will continue to be governed by the repulsive effect of dark energy, and the rate of expansion will continue to accelerate. On the other hand, if w maintains its evolution toward less negative values, dark energy could have an effect increasingly similar to that of matter. The fate of the Universe is then uncertain and will depend on the origin of such an evolution of the dark energy equation of state. Cosmic expansion could gradually slow down under the combined action of the gravitational attraction of matter and dark energy without ever canceling out. This scenario of eternal but ever-slowing expansion is illustrated in the second image from the left in Figure 8.1. In contrast, some models, according to which dark energy would decay into dark matter, could lead to a new epoch of matter domination, potentially reversing the current expansion and resulting in a recontraction of the Universe, as illustrated in the leftmost diagram.

While DESI is the first experiment to announce a preference for time-varying dark energy rather than the expansion described by the ΛCDM model, other projects using supernovae or gravitational lenses have also revealed a trend in this direction, albeit less marked. The most precise results come from the combination of these different experimental approaches. By analyzing the DESI data in conjunction with the cosmic microwave background data, a model based on dynamical energy seems to be preferable to a cosmological constant by approximately 3σ, resulting in an error probability of approximately 0.3%. With the addition of supernova measurements, the probability of error considerably decreases. The tradition in particle physics is to announce a discovery only after an effect at 5σ, when the probability that the result is due to statistical fluctuation is less than one chance in a million. Additional data are needed to cross this threshold and reach a definitive conclusion.

Can the observed deviation from the expected impact of a cosmological constant be explained by a theory of gravitation other than general relativity? The evolution of the size of the acoustic oscillation footprint is no longer sufficient to answer this question. We need to look at all of the information coming from the formation of the Universe's large structures on all scales. The DESI collaboration is also carrying out analyses along these lines and announced its first results at the end of 2024. The analysis of such data is complex and delicate. This requires an understanding of the many possible sources of bias that can impact the measurement, whether of theoretical, instrumental or astrophysical origin, and not just on a single characteristic scale, but also on scales covering approximately 40–400 mega-parsec (i.e. approximately 120–1,200 million light-years). DESI's results from the first year of observation show no evidence to challenge the principles of general relativity. This study would benefit greatly from additional data to refine the conclusions. With a larger number of galaxies, analysis of the formation of large-scale structures on any scale will provide complementary information to the BAO, enabling us to improve the precision of the equation of state for dark energy and to test the hypothesis of gravitational laws beyond general relativity in greater detail. It should be remembered, however, that modifications to the laws of gravitation are often boiled down to two parameters: one captures the changes to which light is sensitive, and the other is devoted to those affecting the trajectories of massive objects.

The DESI project can only constrain the latter, since it measures the effects of gravitation on massive objects. Gravitational lensing analyses therefore provide complementary information that is fundamental to tackling the question of the laws of gravitation from all angles.

8.9. Conclusion

Dark energy appeared at the very end of the 20th century, following the unexpected discovery that the expansion of the Universe was accelerating. Since then, the scientific community has gradually developed a standard model of cosmology, which describes the vast majority of observations with remarkable accuracy, albeit on the basis of the astonishing notion that dark energy of an unknown nature accounts for 70% of all matter and energy in our Universe. While an interpretation of this dark energy in the form of a cosmological constant has long been favored, the results announced in 2024 by the DESI collaboration call this proposition into question, suggesting, on the contrary, that dark energy may have varied over time!

The question of dark energy is often referred to as "the greatest mystery of contemporary physics". Numerous experiments are already underway or planned for the near future to shed new light on this enigma. To this end, several experimental approaches are being considered to examine different facets of dark energy. Some methods are based on 3D mapping of matter in the Universe and use tracers such as galaxies, quasars or even hydrogen gas in the intergalactic medium. The aim is to measure the evolution of a predetermined characteristic distance under the effect of cosmic expansion. Other approaches exploit thermonuclear supernovae and their known intrinsic luminosity to track the evolution of expansion over time. Other studies have examined the effects of gravitation on light from distant galaxies by analyzing their apparent shapes, thus testing the laws of gravity on cosmological scales. The complementary nature of these various techniques is essential for understanding the full range of properties of dark energy and determining whether it is a cosmological constant, a new form of energy whose nature needs to be determined, or whether the observed effects stem from an incomplete model of the laws of gravitation. The coming years will be rich in new experimental data based on one or more of these techniques. The mapping of the Universe by the DESI experiment will continue until at least 2028 and possibly longer. A DESI-II project is planned to take over from DESI and extend it to denser coverage to improve the accuracy of the maps. At the same time, it will explore new tracers of matter in the distant Universe, which will find their way into the next generation of projects. The Euclid satellite begins measuring the deformations caused by gravitational lensing on background galaxies. Other observatories, both on Earth and in satellites, will soon follow, such as the Nancy Roman satellite, the Vera Rubin Observatory in Chile, or the Prime Focus Spectrograph on the Subaru telescope in Hawaii. Studies on the design of the next generation of instruments are already underway and will, in turn, contribute to our knowledge of the cosmos, its composition and evolution.

8.10. References

Albrecht, A., Bernstein, G., Cahn, R., Freedman, W.L., Hewitt, J., Hu, W., Huth, J., Kamionkowski, M., Kolb, E.W., Knox, L. et al. (2016) Report of the dark energy task force. *arXiv*. doi: 10.48550/arXiv.astro-ph/0609591.

DESI Collaboration (2024). DESI 2024 VI: Cosmological constraints from the measurements of Baryon acoustic oscillations. *Journal of Cosmology and Astroparticle Physics*. doi: 10.1088/1475-7516/2025/02/021.

eBOSS Collaboration (2021). The completed SDSS-IV extended Baryon oscillation spectroscopic survey: Baryon acoustic oscillations with Lyman-α forests. *Physics Review D*, 103, 083533.

Planck Collaboration (2018). Planck 2018 results. I. Overview and the cosmological legacy of Planck. *arXiv*. doi: 10.1051/0004-6361/201833880.

Riess, A.G., Yuan, W., Macri, L.M., Scolnic, D., Brout, D., Casertano, S., Jones, D.O., Murakami, Y., Breuval, L., Brink, T.G. et al. (2022). A comprehensive measurement of the local value of the Hubble constant with 1 km/s/Mpc uncertainty from the Hubble space telescope and the SH0ES Team. *arXiv*. doi: 10.48550/arXiv.2112.04510.

9

Dark Matter

Françoise COMBES[1,2]
[1]*LUX – Observatoire de Paris – PSL, France*
[2]*Collège de France, CNRS, Sorbonne University, Paris, France*

The existence of invisible matter in the Universe remains one of the greatest mysteries of modern cosmology. It has defied our standard cosmological model for decades. It has been named "dark matter" because it neither produces nor absorbs electromagnetic radiation and is, in fact, more transparent than dark matter. Despite all of the direct and indirect detection experiments, the progressive exclusion of several candidates and the exploration of multiple hypotheses, its identity remains elusive. While research in recent years has provided a better understanding of what dark matter is not, the nature of this invisible component is still unknown.

9.1. A brief historical overview

The first person to suggest the presence of "dark" matter in the Milky Way was Henri Poincaré in 1906, when he investigated the equilibrium of the galactic system. He noted that if a large number of bright stars were visible, there must also be an unknown number of dark stars circulating in space. At the time, the existence of galaxies had not yet been established. This would only be after the Great Debate of 1920 between the astronomer Heber Curtis, who defended the hypothesis of a universe made up of a large number of galaxies such as our own, and the astronomer Harlow Shapley, who, on the contrary, did not place the Sun at the center of the Milky Way and considered the spiral nebulae he observed to be mere clouds of gas from our galaxy. However, this had not yet become the dark matter of the modern cosmological model.

Quantum Physics and Cosmology,
coordinated by Hervé ZWIRN. © ISTE Ltd 2026.

One of the true pioneers of dark matter was the astronomer Fritz Zwicky, who was born in Switzerland but spent most of his career in California. He was a visionary thinker with a brilliant and inventive mind, although often considered eccentric. He is best known for proposing that stars at the end of their lives explode into supernovae, giving rise to neutron stars. He also discovered over a hundred supernovae. In 1934, while studying the velocities of the galaxies making up the Coma cluster, Zwicky reported that these galaxies were moving at speeds of approximately 1,000 km/s, far too fast to be held back by the gravity of the "visible" mass alone. According to his calculations, the mass required to explain this cohesion had to be approximately 100 times greater than the cumulative mass of the galaxies making up the cluster. Today, this high factor can be explained by the existence of an immense reservoir of hot gas (at a temperature of approximately ten million degrees), which was still unknown at the time, and whose presence would only be confirmed 30 years later by X-ray imaging. The total mass of this gas exceeds that of the matter that forms the galaxies themselves by a factor of 10. Owing to these observations, Zwicky was the first to suspect the existence of an invisible mass influencing the galaxy dynamic. However, his dark matter hypothesis aroused little interest at the time, all the more so as he defended other controversial ideas, such as that of tired light as an alternative to the expansion of the Universe.

At the scale of individual galaxies, the imbalance between visible and dynamic masses is much smaller and remains shrouded in quantification uncertainties. This was notably the case with Horace Babcock's study of the speed of stars in the Andromeda galaxy in 1938 (Babcock 1939). The approach, however, is simple and elegant. Let us, for example, take our solar system. According to the laws of Newtonian gravity, the orbital velocity of planets depends on the mass of the Sun and decreases with distance from it. This same principle should make it possible to determine the mass of galaxies by measuring the velocity at which stars revolve around the galactic center as a function of their distance. Zwicky, on the other hand, did not believe in this approach, believing that the viscosity of the interstellar medium caused stars to rotate globally, similar to a solid body.

It was not until more than 30 years later that the question of dark matter resurfaced. In the 1970s, Vera Rubin and her colleagues published a series of rotation curves for spiral galaxies, classified according to their morphological type. Contrary to expectations, these curves showed no significant decrease in velocity far from the galactic center, suggesting the presence of an invisible mass extending beyond the optical radius. Shortly afterward, similar measurements, notably by Morton Roberts and Albert Bosma on atomic gas using hydrogen hyperfine line emission at a wavelength of 21 cm, confirmed these flat rotation curves beyond the visible component of galaxies.

Gradually, astronomers became convinced of the existence of this unknown matter, now known as dark matter. Its proportion relative to visible matter seems to increase with scale: almost non-existent at the center of spiral galaxies, it becomes predominant at the periphery and makes an even greater contribution at the scale of galaxy clusters. All that remained was used to measure the proportion of this dark matter on very large scales, namely, to determine its universal value.

In the second half of the 20th century, the scientific community was also fascinated by another subject. Following the revolutionary discovery by Penzias and Wilson (1965) of a persistent and ubiquitous cosmological signal (known as the cosmic microwave background (CMB)), numerous experiments have attempted to map this radiation to detect faint fluctuations. These were considered to be the seeds of the galaxies, galaxy clusters and cosmic voids (globally referred to as "large structures") that we observe in the Universe today. In the 1980s, however, the anisotropies of this diffuse background remained elusive at relative amplitude levels on the order of 10^{-3} to 10^{-4}, posing a serious problem for cosmological models.

In an expanding universe, the formation of galactic structures under the effect of gravity is a particularly slow process. In contrast to a non-expanding environment, such as the interior of a galaxy, where density fluctuations become unstable and grow exponentially to give rise to stars, cosmic expansion is opposed to gravitational collapse. In this context, fluctuations grow only linearly: their relative density $\delta\rho/\rho$ evolves in proportion to the characteristic size of the Universe, a quantity that can be measured. Indeed, cosmic expansion stretches the wavelengths of light emitted by galaxies, causing a redshift, denoted z. The relationship between the observed wavelength λ_{obs} and the emitted wavelength λ_e is given by $\lambda_{obs} = \lambda_e(1 + z)$, reflecting the increase in the size of the Universe since the emission of radiation. Overdensities thus evolve according to the law $\delta\rho/\rho \propto 1/(1 + z)$. We can then determine the epoch, or redshift, at which time galaxy formation begins.

As the Universe continues to expand, the energy of photons in the CMB decreases. This energy is expressed by a temperature T, which evolves with a redshift as $T = T_0(1 + z)$, where T_0 represents the current temperature, measured at 2.7 K. Just after the Big Bang, the Universe was composed of a hot plasma of ionized matter and photons. When the temperature decreased to 3,000 K, some 380,000 years later, the first atoms were formed, releasing photons and allowing matter to begin to coalesce under the influence of gravity. This moment, which marks the start of structure formation, corresponds to a redshift of z = 1,000. Since then, density fluctuations have only been able to be amplified by this same factor of 1,000. If their initial amplitude is less than 10^{-3}, they will not have had time to reach a nonlinear regime (corresponding to $\delta\rho/\rho > 1$). In other words, without sufficient instability from the outset, galaxy formation seems impossible.

Faced with this dead end, astronomers have considered the existence of exotic, invisible dark matter, which does not interact with photons and is therefore not trapped in primordial plasma. This dark matter offered a solution to the problem of galaxy formation, as it could collapse long before the baryonic matter (i.e. ordinary matter made up of protons and neutrons) that makes up stars and galaxies. As soon as the Universe ceases to be dominated by radiation (around the redshift $z = 4{,}000$), dark matter can begin to condense and form gravitational potential wells. As the primordial plasma disappears, ordinary matter accumulates around these preexisting structures, facilitating the formation of galaxies. This advance in dark matter has enabled the emergence of the Universe's large-scale structures.

During 1984–1985, cosmological models experienced a major boom. An important advance was the recognition of dark matter as being made up of exotic particles, namely, different from ordinary matter. These particles must be electrically neutral since they do not interact with light, that is, with the electromagnetic force. In many models, these particles are their own antiparticles. They can therefore annihilate each other when the Universe is still dense. With cosmic expansion, the average density gradually decreased, and there was a time when collisions become so rare that these particles stopped annihilating each other. From then on, only a fraction of them remained. Current measurements indicate that this relic density of dark matter is approximately five times greater than the density of ordinary matter.

Three main families of dark matter are considered, depending on the velocity dispersion of dark matter particles at the moment when they stop interacting with other particles and can begin to form structures under the effect of gravity. If they are light particles, such as neutrinos, they are moving at relativistic speeds. This is known as hot dark matter (HDM). Conversely, if their mass is much greater than that of a proton, they have a low velocity and are referred to as cold dark matter (CDM). The difference between these two types of dark matter has a major impact on structure formation in the Universe.

In the HDM scenario, the pressure exerted by these relativistic particles prevents the growth of small galactic structures, slowing the formation of the first galaxies. In the CDM case, on the other hand, the slower particles allow the formation of many small clusters, such as dwarf galaxies, in the vicinity of large galaxies. However, these structures are not observed in as many numbers as expected, which leads us to consider a third category: warm dark matter (WDM).

The dark matter model that emerged as the most robust model was proposed by Blumenthal et al. (1984) and is based on CDM. It quickly became the standard model of cosmology. This was also the period when several experiments were launched to directly detect these dark matter particles.

Let us take a little tour of the Universe in search of this invisible matter. We start with galaxies and move onto larger scales, such as galaxy groups or clusters. Finally, we conclude with a study of dark matter on the scale of the entire Universe.

9.2. On a galaxy scale: rotation curves of spiral galaxies

The dynamics of galaxies can be studied using several tracers: stars, neutral gas and ionized gas. The most accurate rotation curves are obtained from neutral gas, whether atomic or molecular. Unlike stars, gas is a dissipative medium that is cooled by radiation. Its velocity dispersion is relatively small, on the order of 10 km/s, which limits its impact on determining the overall rotational velocity of spiral galaxies, typically on the order of 200 km/s, as in the case of the Milky Way. As a result, the rotational velocity measured for neutral gas is close to the circular velocity, offering direct access to the gravitational potential of the galaxy. Atomic gas is the probe of choice, as it extends radially some four times further than stars and molecular gas.

Stars, on the other hand, follow a different dynamic. As collisionless media, their kinematics are very difficult to slow down. The oldest stars in the galaxy plane have a high velocity dispersion, reaching 70–100 km/s, which reduces their interest in determining the rotational velocities of the galaxy. Moreover, this velocity dispersion is not necessarily isotropic, making modeling even more complex. In the case of galactic bulges and elliptical galaxies, whose rotation is weak, it is even more difficult to estimate the gravitational potential from their kinematics. Not only is velocity anisotropy marked, but the three-dimensional structure of these objects can also be inaccurately determined. This flattening is not due primarily to rotation but rather to velocity anisotropy inherited from earlier galaxy mergers. Furthermore, their shape may be triaxial and vary with radius, adding further complexity to their modeling.

Finally, the rotation curves obtained from the study of ionized gas, particularly the Hα hydrogen line (electronic transition of the Balmer series), reflect the dynamics of young stars and are closer to the actual rotation speed of galaxies. Although regions composed of ionized gas have an intrinsically high velocity dispersion on the order of 30 km/s, they are invaluable for studying the center of galaxies, where atomic gas is often scarce and molecular gas is not always sufficient.

To date, over a thousand precise rotation curves of nearby galaxies have been established. These data have led to the identification of universal rotation profiles by normalizing radii with respect to a characteristic scale and velocities with respect to the maximum velocity. Figure 9.1 illustrates this series of universal curves, classified according to the luminosity or velocity of the galaxy. The evolution of these curves is striking in dwarf galaxies (at the bottom of the figure), and the rotational velocity continues to increase right up to the limit of observations; whereas in massive galaxies (at the top), the curve eventually decreases. Moreover, when we model these curves with visible and dark matter components, we observe that the fraction of dark matter is much greater in dwarf galaxies. The more massive a galaxy is, the less dark matter it requires, proportionally speaking. Thus, massive galaxies are essentially made up of ordinary matter, whereas dwarf galaxies are dominated by dark matter right from their core.

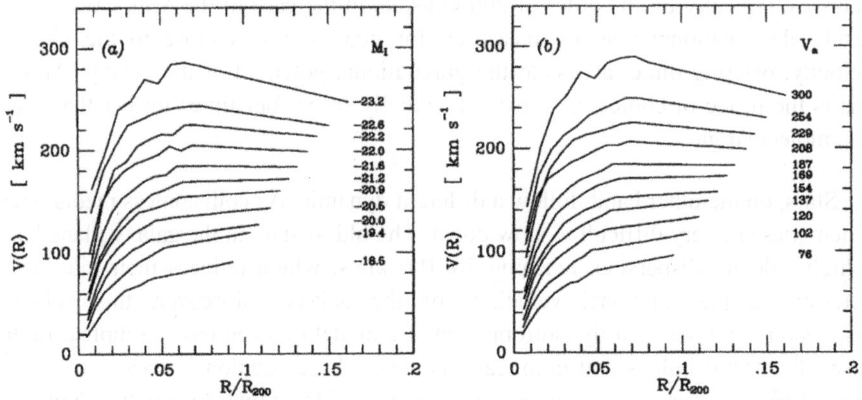

Figure 9.1. *Universal rotation curve of spiral galaxies for different luminosities (a) or velocities (b). Radii are in units of R_{200}, the radius encompassing an average overdensity δρ/ρ of 200 and representing the characteristic scale of dark matter distribution (according to Persic et al. 1996).*

9.3. Scaling relationships for rotation curves

Scaling relationships link various physical properties of galaxies, such as their rotational velocity, mass and size. These relationships are crucial to understanding galaxy dynamics and the nature of dark matter.

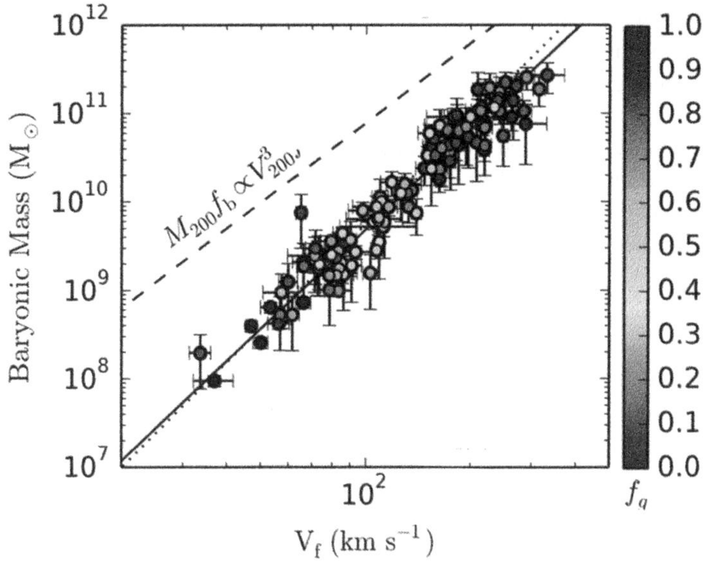

Figure 9.2. *Baryonic Tully–Fisher relationship between the velocity of the rotation curve plateau and the baryonic mass (stars + gas) of spiral galaxies. The color indicates the gas fraction (fg, ratio of gas mass to baryon mass). The data line up on a straight line with a slope of 4. The dashed line illustrates the predictions of the standard CDM model, with the universal baryon fraction f_b = 0.17 and a slope of 3 (after Lelli et al. 2016). For a color version of this figure, see www.iste.co.uk/zwirn/quantum.zip*

One of the most famous relationships is the Tully–Fisher relation, established in 1978, which links the maximum rotational velocity and luminosity of spiral galaxies. Initially, this relationship was used primarily to estimate galactic distances. In 2000, McGaugh realized that including gas-rich dwarf galaxies – in which the mass of gas is often greater than the stellar mass – in this framework was not enough to consider the luminosity of stars. The total baryonic mass (Mb) was considered. This extended version, known as the baryonic Tully–Fisher relation, shows that Mb increases proportionally to the fourth power of the maximum rotational velocity, as illustrated in Figure 9.2. Furthermore, according to the standard CDM model, the initial baryonic mass of a galaxy can be estimated as a fraction f_b (approximately 17%) of its total mass M_{200} (mass within the radius R_{200} defined in Figure 9.1). This result, illustrated by the dashed curve in Figure 9.2, strongly disagrees with the experimental Mb data and reveals that a large proportion of baryons must have been expelled from galaxies during their evolution. With respect to the size of the Milky Way, approximately 80% of baryons must have been expelled, and this fraction is even greater for lower-mass galaxies, such as dwarf galaxies. To reproduce this evolution, numerical simulations must incorporate a powerful feedback mechanism: in dwarf galaxies, it is mainly supernova explosions that expel baryons, whereas in more massive galaxies, this role is played by the activity of supermassive central black holes.

Since they are dominated by dark matter, dwarf galaxies are an excellent testbed for studying dark matter. Uncertainties in the stellar mass-to-light ratio play a negligible role. The rotation curves of these galaxies show that the distribution of dark matter at their centers is not as concentrated as predicted by the standard model simulations and, in particular, does not show a density peak. Navarro et al. (1997) showed that, according to these simulations, the radial profile of dark matter in galactic halos followed a universal power law in 1/r. However, observations reveal a density plateau at the center of galaxies, reaching a surface density of approximately 150 solar masses per square parsec, regardless of the luminosity of the galaxy. This result, highlighted by Donato et al. (2009), corresponds to a gravitational acceleration on the order of 2×10^{-11} m/s^2.

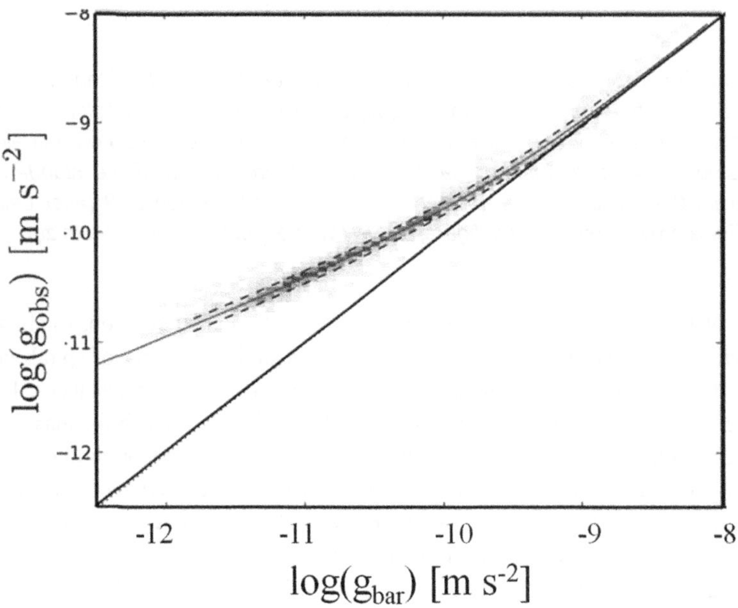

Figure 9.3. *Relationship between baryon gravity acceleration g_{bar} and observed radial acceleration g_{obs}. The solid red line represents the radial acceleration relation (RAR) $g_{obs} = g_{bar}/(1 - \exp - (g_{bar}/a_0)^{1/2})$, where $a_0 = 1.2 \times 10^{-10}$m/s^2. The blue color scale represents 2,694 data points, and the blue dotted lines represent their dispersion (68% of the distribution falls between the two levels) (according to Li et al. 2018). For a color version of this figure, see www.iste.co.uk/zwirn/quantum.zip*

Another fundamental relationship links the radial acceleration measured in spiral galaxies to that predicted by taking baryons alone into account. This relationship, illustrated in Figure 9.3, clearly shows the increasing importance of dark matter in low-mass galaxies. The red line, which represents the observational data, illustrates two asymptotic behaviors: in the high-acceleration range, the observed acceleration (g_{obs}) corresponds to that predicted by baryonic matter (g_{bar}) alone; whereas at low acceleration, the measured acceleration corresponds to $g_{obs} \approx (g_{bar} \times a_0)^{1/2}$, where $a_0 \approx 1.2 \times 10^{-10}$ m/s^2. This relationship, known as the radial acceleration relation (RAR), is not limited to spiral galaxies. It is also valid for irregular, lenticular and elliptical galaxies. This relationship can be used to test various dark matter models.

9.4. Galaxies without dark matter: intriguing exceptions

This well-established relationship becomes more complex when we consider galaxies that evolve within galaxy groups or clusters. In these dense environments, gravitational interactions between galaxies are frequent and generate complex phenomena that make it difficult to study dark matter. Moreover, galaxy clusters are dominated by elliptical galaxies, whose mass distributions and dynamics are difficult to model. Spiral galaxies have generally only recently joined the cluster and are subject to various perturbations linked to their dense environment. These include the stripping of their gas due to the ram pressure of the cluster's hot gas and tidal effects caused by neighboring galaxies or even by the gravitational potential of the cluster itself.

Surprisingly, some galaxies appear to be almost devoid of dark matter, defying conventional models. These exceptions have been discovered owing to modern telescopes, which are able to detect large numbers of ultradiffuse galaxies (UDGs). Some of these galaxies, although low in mass, possess a large number of globular clusters, making it possible to study their dynamics. A prominent example is the galaxy NGC 1052-DF2, of mass 2×10^8 solar masses only, which is a satellite of the giant galaxy NGC 1052. In 2018, a team led by Pieter van Dokkum (van Dokkum et al. 2018) analyzed the motion of its ten globular clusters and concluded that this galaxy contains very little, if any, dark matter. Curiously, other satellite galaxies of NGC 1052 seem to exhibit the same peculiarity, which runs counter to the model according to which dwarf galaxies are normally dominated by dark matter. van Dokkum et al. (2022) proposed an explanation: these dark matter-free galaxies could be the result of a high-speed collision between an outer galaxy and one of the NGC 1052 satellites. The violence of the impact could tear the dark matter away from the rest of the galaxy, leaving behind fragments rich in stars and gas. These fragments would then form an elongated structure, extending over several million light-years. This hypothesis remains to be confirmed with other cases.

Another, more frequent and less extreme scenario could also explain these dark matter-free galaxies: interactions between spiral galaxies. When two large spiral galaxies of comparable mass merge, they form long filaments of gas called tidal tails. This gas can collapse under its own gravity, giving rise to new stars. These gas condensations are particularly important in the outer parts of the filaments, where small galaxies, called tidal dwarfs, are created that contain no dark matter. Unlike baryons, dark matter does not radiate and therefore cannot condense to form dark matter halos in this way.

A similar process can occur when a spiral galaxy crosses a galaxy cluster. Its gas is swept away by the pressure of the surrounding medium, leading to the formation of tidal dwarfs. This process is rarer, however, because of the hostile environment of the cluster's hot gas. These dark matter-free galaxies are characterized by a greater proportion of heavy elements in their stars than would be expected for their mass, testifying to their particular origin. However, their proximity to their parent galaxy makes it difficult to analyze their kinematics, which are still strongly altered by tidal effects, and it has been difficult to demonstrate that they are indeed not surrounded by a dark matter halo.

9.5. Dark matter on the scale of galaxy clusters

There are several methods for determining the amount of matter in a galaxy cluster.

The first method, proposed by Fritz Zwicky, involves measuring the velocity of each galaxy in the cluster. Using the Virial theorem, which relates the kinetic energy of a galaxy to its gravitational potential in the galaxy cluster, the total mass of the cluster can be deduced. If a sufficient number of velocities are known, establishing the radial distribution of its mass becomes possible.

Another approach is to study the hot gas that fills the cluster. This gas, detectable in X-rays, is maintained in equilibrium by the interplay of forces between the thermal pressure and gravitational attraction. By analyzing its density and temperature at different distances from the cluster center and using hydrodynamic equilibrium models, we can reduce the distribution of total mass. This method is particularly useful, as hot gas accounts for a significant fraction of the baryonic mass of clusters. This method generally provides results compatible with the previous method, subject to the limits of the uncertainties, except when the cluster presents a complex structure with several subcomponents.

Finally, a third technique is based on the gravitational lensing effect, which makes it possible to analyze the deflection of light rays under the effect of the cluster's mass. The first gravitational arcs were discovered by Geneviève Soucail and her colleagues in Toulouse (Soucail et al. 1987) in the Abell 370 cluster, a cluster similar to the one shown in Figure 9.4. Two images of the same quasar, a phenomenon known as gravitational mirage, had already been detected by British astronomer Dennis Walsh (Walsh et al. 1979), but little attention had been given to the small structures visible on the photographic plates of galaxy clusters. These structures, which resulted from the same gravitational mirage, had been regarded as mere flaws (like so many others) in the plates. It was not until Lynds and Petrosian (1989) published an observation similar to that of the Toulouse team, albeit achieved earlier but still unexplained.

These gravitational arcs, namely, fine curved structures that reveal distorted images of background galaxies, result from strong amplification of their light. Just as an optical lens focuses light, the mass of the cluster deflects and amplifies light rays. Depending on the relative alignment between the background galaxy, the cluster lens and the observer, amplification can reach a factor of 100. As background galaxies have a different redshift (caused by the expansion of the Universe) to the foreground cluster, they are measured on a different part of their spectrum and therefore have a different color from the cluster galaxies. While the latter are generally more yellow elliptical galaxies, containing old stellar populations, the gravitational arcs appear blue or red depending on their distance and star-forming activity. In Figure 9.4, the presence of a large number of gravitational arcs makes it easy to imagine that virtually all of the small points of light, especially those different in color from the galaxies in the cluster, turn out to be background galaxies that have undergone gravitational deformation.

It may seem difficult to distinguish these deformations from the intrinsic shapes of galaxies, such as spiral galaxies seen edge-on. However, the gravitational field imparts spatial coherence to deformation, and when thousands of galaxies are analyzed in a single region, statistics can be used to reconstruct the gravitational field and thus map dark matter. The Hubble Space Telescope (HST), which provides the spatial resolution needed to detect slight distortions of thousands of background galaxies, has been a powerful tool since the 1990s. Although not without its limitations, not least because it determines the mass contained in a cylinder aligned with the line of sight, this tool is an essential complement to other mass estimation methods. The combination of the galactic velocity measurements and X-ray analyses described above provides a detailed view of the distribution of dark matter in galaxy clusters.

Figure 9.4. *(a) The Abell 2218 galaxy cluster, as seen by the Hubble Space Telescope. Intertwined with the cluster's yellow elliptical galaxies are numerous gravitational arcs, in different colors (red and blue), corresponding to background galaxies whose light rays are deflected by the cluster's mass. Credit HST (NASA-ESA). For a color version of this figure, see www.iste.co.uk/zwirn/quantum.zip*

9.6. Cosmic shear and cluster collisions: revealing dark matter

Weak gravitational lensing, also known as cosmic shear or weak lensing, has been used to map the distribution of total matter in galaxy clusters. By comparing the distribution of dark matter with that of ordinary visible matter, which is mainly contained in the multimillion-degree hot gas that fills the clusters and emits X-rays, it appears that dark matter is strongly correlated with visible matter on a large scale. In the cosmic web, empty zones in galaxies also correspond to dark matter-free regions.

In cluster collisions, however, dark matter can decouple from ordinary matter. The first observation of this phenomenon was made by Clowe et al. (2004): a small cluster, launched at an impressive speed of 4,700 km/s (Mach 3), passed like a cannonball through a more massive cluster, called the bullet cluster (see Figure 9.4(b)). In this event, the very hot gas of the small cluster virtually stopped at the center of the collision, while its much more dispersed galaxies continued their trajectory, unaffected by the impact. Analysis of the cosmic shear has shown that the total mass is associated with the galaxies and not with the gas, whose mass can be more than ten times the visible mass of the galaxies. This proves that an invisible component, which is even more massive than the gas, has passed through the collision zone: dark matter particles, which interact neither with light nor with each other, have followed the galaxies.

This spectacular event is all the more interesting in that the collision occurred almost in the plane of the sky and at an unexpected speed, exceeding the predictions of the standard model. Since then, several other cluster collisions have revealed similar, albeit less spectacular, phenomena.

These interactions have enabled us to refine our understanding of dark matter and conclude that it behaves like a collisionless medium: in particular, it has been possible to establish an upper limit on the effective cross-section of collisions between dark matter particles. Unlike ordinary matter, which interacts with light and with itself, dark matter seems to respond only to gravity.

One of the first conclusions reached by Clowe et al. (2004) on the bulletl cluster was to eliminate the hypothesis of modified gravity. Indeed, if gravity were to be modified to explain the effects attributed to dark matter, there should be no decoupling between total mass and the most massive visible component (gas). However, this argument alone is not enough. In spiral galaxies such as ours, we already observe a separation between visible and dark matter: the center is dominated by baryonic matter, whereas dark matter is confined to a halo surrounding the galaxy and dominating the outer regions.

In giant galaxies, it is even possible to explain the observed dynamics using visible matter alone, without recourse to dark matter or a change in gravity. However, galaxy clusters pose a fundamental problem for alternative theories to dark matter, which are unable to fully compensate for the missing mass, except by adding an additional non-luminous component, which amounts, in a way, to reintroducing a form of dark matter.

Thus, observations of cosmic shear and cluster collisions, such as that of the bullet cluster, provide strong evidence in favor of the existence of dark matter and its non-collisional behavior.

Figure 9.4. *(b) Composite image of the bullet cluster. The gas (pink), measured in X-rays by Chandra, is distinct from the area of mass (blue), determined by the gravitational lensing method. The gas in the small cluster shows a spiky shock wave, yielding the speed of the collision. Credit: X-ray: NASA/CXC/CfA/M. Markevitch et al. (2002); Optical: NASA/STScl; Magellan/U. Arizona/D. Clowe et al. 2004; Lensing Map: NASA/STScl; ESO WFI; Magellan/U. Arizona/D. Clowe et al. 2004. For a color version of this figure, see www.iste.co.uk/zwirn/quantum.zip*

9.7. The cosmic microwave background: a measurement of unparalleled precision

The fraction of dark matter in the Universe is determined with remarkable precision because of the mapping of the CMB. Several satellites have contributed to this mapping.

The first, COBE, which was launched by NASA in 1989 and was operational until 1993, demonstrated that the spectrum of the CMB corresponds to that of a perfect black body. This confirmed that the Universe has passed through a dense, hot phase, in line with the Big Bang model. COBE also revealed for the first time the existence of temperature anisotropies of 10–100 μK, corresponding to fluctuations of approximately 10^{-5} relative to the uniform 2.7K background. This is the hallmark of the primordial fluctuations we mentioned earlier, which serve as the seeds of the formation of the galaxy. With a spatial resolution of 0.7°, COBE was unable to map these anisotropies with great precision but was nevertheless able to distinguish them from signal contamination via the Milky Way because of their different dependences on the observing frequency band.

The WMAP satellite, launched in 2001 and positioned at the Lagrange point L2, improved this resolution to 0.2°, thus enabling the detection of the first Doppler peaks, which are traces of the acoustic waves that traveled through the primordial plasma.

Finally, the Planck satellite, launched in 2009 by the European Space Agency, further refined these observations, with a resolution of up to 10 arcminutes and enhanced sensitivity owing to improved instrument cooling. All Doppler peaks can be measured with such precision that the error bars in the data are barely perceptible, obscured by the line thickness of the theoretical model (see Figure 9.5).

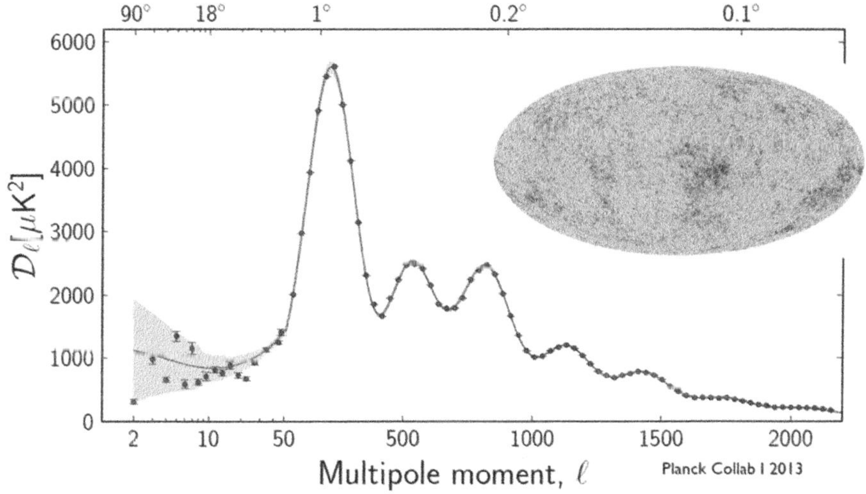

Figure 9.5. *Amplitude of cosmic microwave background anisotropies as a function of angle (top scale) or multipole moment (bottom scale). These amplitudes correspond to the temperature fluctuations in the map shown at the top right, according to Planck Collaboration (2013). For a color version of this figure, see www.iste.co.uk/zwirn/quantum.zip*

Analysis of the CMB provides essential information on the structure and composition of the Universe. The first Doppler peak, located around an angle of 1°, corresponds to the sound horizon, namely, the distance covered by acoustic waves in primordial plasma before the Universe became made up of neutral atoms, approximately 380,000 years after the Big Bang. This allows us to measure the curvature of space, which turns out to be zero with great precision, confirming that the Universe is globally flat. The third Doppler peak is particularly sensitive to the fraction of baryons in the Universe. Owing to Planck's observations, the determination of cosmological parameters has reached unprecedented accuracy, improved by several orders of magnitude compared with earlier estimates. We now know that the Universe is 13.8 billion years old and consists of approximately 5% baryons, 25% dark matter and 70% dark energy.

We have now completed our overview. On the scale of the Universe, we now know that the fraction of matter in baryonic form is 17%, and that of dark matter is thus 83%.

9.8. Dark matter solutions

The existence of a missing mass at all scales and the impossibility of explaining it with ordinary matter have led to a search for exotic particles beyond the standard model for almost 40 years. Among the many hypotheses put forward, that of WIMPs (weakly interacting massive particles) has dominated the scene for more than 30 years. The standard model of particle physics is successful because of its number of symmetries, which make it possible to predict the existence of missing particles. It is therefore natural to turn to supersymmetry theory, developed in the 1970s, to find these exotic particles. Supersymmetry, or SUSY, is associated with a boson superpartner with each fermion, and vice versa. Among all of these supersymmetric particles, the neutralino particles appear to be the most stable, and the one into which the other particles decay. This hypothetical particle, a combination of photino, zino and Higgsino, is an excellent candidate for dark matter, as it would be electrically neutral, massive and stable over billions of years.

It remains to be seen whether the current abundance of the neutralino is indeed equal to 25% of the Universe's content made up of dark matter. In the SUSY model, the neutralino is produced in thermal equilibrium in the primordial Universe. As this particle is neutral and has its own antiparty, it progressively annihilates itself until the Universe has sufficiently cooled, owing to its expansion, for this particle to no longer be created anew. When the density of neutralinos falls to the point where the characteristic interaction time becomes greater than the age of the Universe, the neutralinos no longer annihilate, leaving a residual density that will no longer evolve in relation to the density of matter (both components will, however, be diluted in the same way owing to cosmic expansion). The calculation of this relic abundance depends on the interaction cross section of the WIMPs. Assuming a value typical of weak nuclear interactions, we find a residual abundance on the order of 25%, corresponding to that attributed to dark matter, for a special value of the neutralino mass. This near-miraculous coincidence has long reinforced the idea that WIMPs, with masses on the order of a hundred times that of the proton, offered the most natural solution to the dark matter problem. However, since the 2010s, several experiments have gradually cast doubt on this hypothesis.

On the one hand, the LHC particle accelerator at CERN has reached sufficient energy to produce these particles and has detected nothing. The discovery of the Higgs boson in 2012, with a mass 125 times greater than that of the proton, has also constrained supersymmetric models and reduced the possible range of existence of WIMPs.

On the other hand, direct detection experiments carried out in underground laboratories, such as Gran Sasso in Italy and others in Europe and North America, have considerably lowered the upper limits on the effective cross-sections of WIMPs without obtaining any conclusive signals. Finally, indirect detection by observing gamma rays from the annihilation of dark matter in dense regions, such as the center of the Milky Way or dwarf galaxies, has also been unsuccessful.

Faced with these repeated setbacks, other hypotheses have been explored. Among them are the possibility that dark matter is made up of primordial black holes. This hypothesis has been advanced by the detection of gravitational waves. Alternatives are sterile neutrinos that could form the "warm" dark matter mentioned above, wimpzillas, string-theory branons with a large number of dimensions etc.

Within this great diversity, perhaps one of the most promising avenues is the hypothesis of ultralight boson-like particles, similar to axions. This latest candidate could offer a viable alternative to the standard CDM model, within the framework of models known as fuzzy dark matter (FDM), as simulated in Figure 9.6.

FDM models are named because the mass of these hypothetical particles is extremely low, on the order of 10^{-21} eV, giving them a macroscopic quantum wavelength. Therefore, their wave characteristics outweigh their particle characteristics. They are similar to axions (axion-like particles (ALPs)), although not the same as those introduced by Peccei and Quinn (1977) to solve a CP symmetry violation problem in quantum chromodynamics. True axions should have a mass between 10^{-6} and 10^{-3} eV, whereas the ALPs put forth for dark matter are much lighter particles.

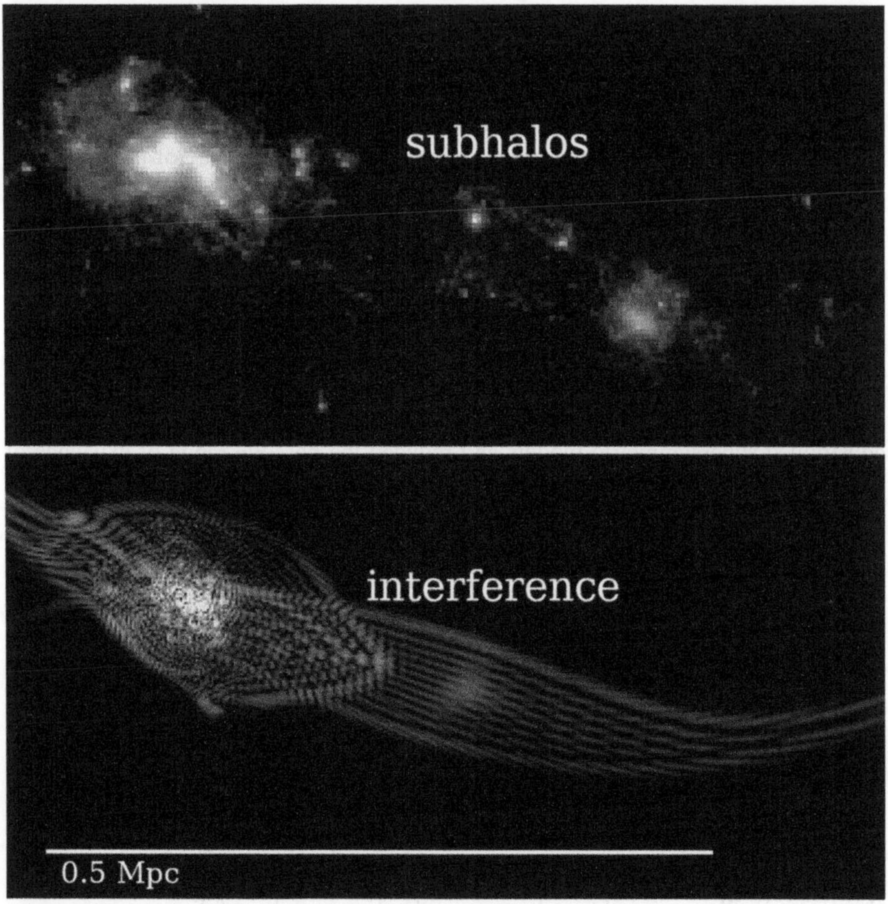

Figure 9.6. *Cosmological simulation of a cosmic filament, within which galaxies form: above the standard CDM model and below the fuzzy dark matter (FDM) model. Quantum interference patterns are visible in the FDM model in both the galaxy and the filament. A core/soliton of size ~1 kpc forms in the galaxy core (according to Mocz et al. 2019). For a color version of this figure, see www.iste.co.uk/zwirn/quantum.zip*

The first ideas linking these particles to dark matter date back to Baldeschi et al.'s (1983) work. These authors suggested that these ultralight bosons could condense at the center of galaxies in a form known as a Bose–Einstein condensate. Unlike the neutralinos of the supersymmetric model, which are fermions with a half-integer spin, these particles are described by a scalar field, which means that they have no spin. Their mass m_a is adjusted to reproduce the distribution of matter observed in galaxy cores. In particular, it avoids the peak of high central density, which is a difficult prediction to reconcile with simulations based on CDM. One of the most striking features of this model is the formation of a soliton at the center of galaxies, namely, a stable structure with a size comparable to the de Broglie length, inversely proportional to the particle's mass. If m_a is between 10^{-22} and 10^{-21} eV, the soliton would have a size of approximately 1 kps, corresponding to observations of dark matter cores in dwarf galaxies. However, this hypothesis is based on an extrapolation to astronomical scales of quantum phenomena, which, to date, are all confined to sub-Angstrom scales. The quantum fluctuations inherent in this model are thus found over several kiloparsecs. These fluctuations cannot be damped since they are quantum in origin and constantly renewed. They create gravitational potential oscillations that could have a measurable effect on ordinary matter, notably by heating and thus thickening the disks of spiral galaxies, which imposes constraints on the mass m_a.

Despite the lack of direct experimental evidence, this model has several theoretical advantages. It reproduces the anisotropies of the CMB, as well as the mass distribution of the Universe's large-scale structures. Moreover, the quantum pressure associated with these particles prevents the formation of very low-mass structures (below 3×10^{-8} solar masses), which also solves the problem of missing satellites encountered in the standard CDM model. Ever since the 2000s, cosmological simulations based on this model have been developed. They are more complex and time-consuming than those of the CDM model, as they require the simultaneous solution of the gravity equation and the Schrödinger equation for the quantum part.

The wave-like nature of these particles gives rise to special phenomena. These include the formation of vortices, which are annular or linear structures of zero density around which the particles circulate. Hui (2021) reported that the central soliton moves randomly around the center of gravity of the dark matter halo, potentially interacting with the supermassive black hole at the galactic core. More recently, Amruth (2023) suggested that certain local anomalies observed in strong gravitational lenses could be better explained by the FDM model than by the standard CDM model.

Although the fuzzy dark matter model still presents experimental and theoretical challenges, it is an interesting alternative that could help resolve certain inconsistencies in the CDM paradigm.

9.9. Experiments for the direct detection of dark matter

Numerous experiments have been conducted to detect WIMPs directly. Detection relies on the elastic scattering of dark matter particles on atomic nuclei.

If WIMPs interact sufficiently with baryons, this process can produce a characteristic signature in dedicated detectors. To minimize the background noise caused by the interaction of cosmic rays in these same detectors, the experiments are arranged in protected environments, such as old underground mines or tunnels dug beneath mountains. Other efforts are aimed at identifying features specific to WIMPs, such as the energy distribution of nuclear recoils, due to possible directional detection. In addition, several signals arising from the interaction of a WIMP in the detector can be detected and distinguished from those arising from background noise: the recoil of the nucleus itself, its deexcitation via a scintillation phenomenon and the production of phonons in a crystal.

The most successful experiments to date are XENONnT in the Gran Sasso underground laboratory (Italy) and Lux-Zeplin in a mine in South Dakota (USA), both of which use several tons of liquid xenon. Enormous progress has been made in the 40 years since the first experiments. Detector sensitivity has been improved by almost ten orders of magnitude, enabling us to exclude certain potential candidates, such as sneutrinos, which are the supersymmetric partners of neutrinos. We are now just an order of magnitude away from reaching the threshold where the signal will be dominated by interactions caused by solar neutrinos, making the detection of possible WIMP interactions more delicate.

9.10. Indirect detection of dark matter

The dark matter particles stop annihilating each other when their density decreases so that the mean time between two collisions exceeds the age of the Universe. However, these interactions can still occur in regions where dark matter is sufficiently concentrated, such as at the center of galaxies. The annihilation of a pair of WIMPs leads to the production of two gamma-ray photons of the same energy, corresponding to the mass of the particles. Annihilation may also give rise to intermediate particles before gamma photons are produced, although this route is less likely and leads to a more diluted energy spectrum.

The preferred sites for this indirect detection are the nearest gravitational potential wells, such as the Sun and the center of our Galaxy. Although the Sun is close by, it is unlikely to contain much dark matter, as it was formed from the gas of the galactic disk. The center of the Milky Way, on the other hand, although further away, is likely to have the highest density of dark matter in the Galaxy. Unfortunately, these regions are both dominated by baryons, which inherently emit significant gamma-ray radiation, thus complicating the identification of a possible dark matter signal. Moreover, the high concentration of pulsars at the galactic center also produces a gamma-ray background that has already led to several false indicators.

Research is now focusing on the dwarf satellite galaxies of the Milky Way. As explained above, these small galaxies are mostly composed of dark matter and contain very few stars; therefore, few baryons are likely to emit stray gamma-ray radiation. Therefore, the detection of a gamma-ray signal would have a high probability of coming from the annihilation of dark matter. Even if the expected signal is weaker, observing a large number of dwarf galaxies increases the sensitivity and improves the chances of detection.

Gamma rays are actively sought by several instruments, notably the Fermi-LAT space telescope, which covers an energy range from 20 MeV to 300 GeV. On the ground, observatories such as HESS and, in the future, the CTA (Cherenkov Telescope Array) take over for higher energies, from 300 GeV to 10 TeV. These telescopes detect the Cherenkov light produced by the secondary particle showers generated when gamma rays from the cosmos interact with the Earth's atmosphere. The results obtained with Fermi methods have already ruled out the existence of thermal WIMPs up to a mass of 300 GeV, and future CTA observations will explore even higher energy ranges.

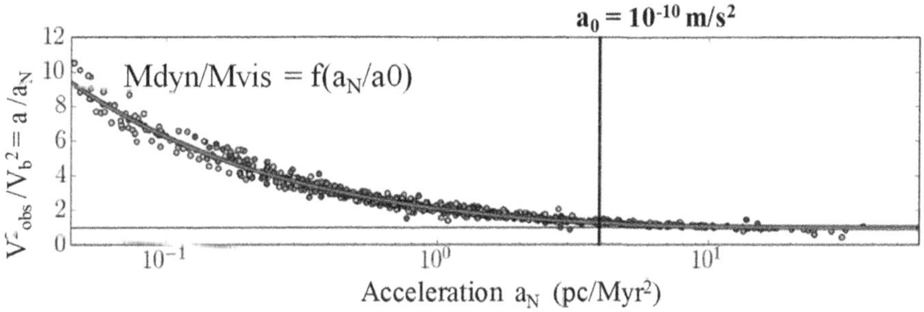

Figure 9.7. *Ratio of the dynamic mass M_{dyn} to the visible mass M_{vis} as a function of Newtonian acceleration a_N. As this ratio increases above 1, which occurs below the limiting acceleration $a_0 = 10^{-10}$ m/s^2, the amount of dark matter increases (adapted from McGaugh 2004). For a color version of this figure, see www.iste.co.uk/zwirn/quantum.zip*

9.11. Modified gravity

An alternative to the standard CDM model is based on the hypothesis of modified gravity. Since hypothetical dark matter particles interact with their environment only via gravity, it is conceivable that they do not exist and that the missing mass results instead of an alteration in large-scale gravitational laws. Observations suggest that this modification of gravity applies only in the weak-field regime, with no direct dependence on spatial or energy scale (i.e. velocities). The MOND (MOdified Newtonian Dynamics) model, proposed by Moti Milgrom in 1983, is based on this idea. Figure 9.7 shows that the dynamic mass only exceeds the visible mass below a critical acceleration of $a_0 = 10^{-10}$ m/s^2. This model postulates that Newtonian gravity a_N, which is valid in strong fields, is gradually replaced by the expression $(a_N a_0)^{1/2}$ as a_N falls below a_0. In this context, the gravitational field no longer decreases at $1/r^2$ but decreases at $1/r$. This mechanism provides a natural explanation for the empirical Tully–Fisher law (Figure 9.2), which relates the mass of galaxies to their rotation speed.

The MOND model successfully reproduces the galaxy rotation curves. In dwarf galaxies, modified gravity dominates from the center, and the rotation curve continues to grow until the last observed radius. Giant galaxies, on the other hand, remain in the Newtonian regime up to the optical radius and only require a modification of acceleration in their halo; their curve rises in the bulge, falls back slightly and then reaches a plateau. Notably, such a simple model, with such a small number of parameters, can account for these observations. Similarly, the radial acceleration relationship (RAR) in Figure 9.3 is remarkably well reproduced. This model also explains why the missing mass appears as soon as the surface density falls below a threshold of approximately 150 solar masses per square parsec. However, MOND encounters several difficulties, not least in explaining the dynamics of galaxy clusters and certain ultradiffuse galaxies. The latter do not seem to be affected by the external gravitational field, although this is supposed to be dominant in their environment. A study carried out by Freundlich et al. (2022) showed that these galaxies do respect MOND dynamics, but as if they were isolated. One possible explanation is that they retain their equilibrium velocity as they enter the cluster, taking several billion years to adapt to their new environment and reduce their internal velocities.

Several formulations of MOND have been proposed. An important advance was made by Bekenstein (2004) with the TeVeS (Tensor–Vector–Scalar gravity) theory, which introduces a relativistic formulation, explaining gravitational lenses, and comprises a metric with a tensor, a scalar field and a vector field. This approach represents an ambitious generalization of Einstein's equations for weak fields. However, it was not without its problems. One of these was that the speed of gravitational waves was predicted to be different from that of light. This hypothesis was invalidated by the 2017 observation of gravitational waves from the merger of two neutron stars by LIGO-Virgo, an event accompanied by a gamma-ray burst detected simultaneously. A more recent version, the AeST (Aether Scalar Tensor) theory, proposed by Skordis and Zlosnik (2021), adjusts the parameters of TeVeS while retaining a time-like vector field, nicknamed "Aether". This modification brought the theory into line with the speed of gravitational waves, the structure of the CMB and the distribution of large-scale structures in the Universe.

However, challenges remain, especially in explaining the dynamics of galaxy clusters. One possible solution lies in taking into account unidentified baryons in the Universe. To date, approximately 60% of the baryons predicted by primordial nucleosynthesis have still not been located. These baryons could be present as hot or cold gas in the cosmic web.

In fact, as Figure 9.8 illustrates, more than 80% of baryons are thought to have been ejected from galaxies and should be found in the intergalactic medium. Even galaxy clusters do not yet contain the expected universal fraction of baryons, suggesting that their distribution could help solve the problem of missing mass in MOND.

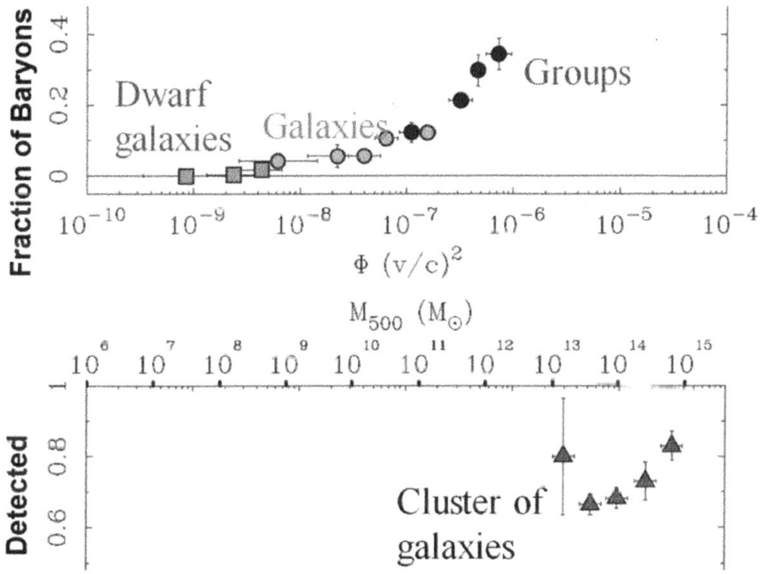

Figure 9.8. *Fractions of baryons detected M_b in various structures, normalized to the universal fraction, that is, $M_b/(0.17\ M_{500})$, with M_{500} being the dynamic mass in a radius where the density is 500 times the average cosmic density. The x-axis gives the kinetic energy of the structures, which increases with their mass (according to McGaugh et al. 2010). For a color version of this figure, see www.iste.co.uk/zwirn/quantum.zip*

9.12. Conclusion

After decades of research, the mystery of dark matter remains unresolved. To make matters worse, it has been accompanied by another fundamental problem: that of dark energy, which is responsible for the accelerating expansion of the Universe. Today, 95% of the Universe's content lies in this still totally unknown dark sector. Nevertheless, cosmology is now a precision science, and many parameters have been discovered and determined with unprecedented rigor.

Hypotheses on the nature of dark matter are narrowing down. Numerous prospects have been ruled out, even though the mass of conceivable particles still covers 34 orders of magnitude. The WIMP (weakly interacting massive particle), which is considered a thermal residue of the young Universe, remains a dominant hypothesis. However, it is based on supersymmetry, a theory that has yet to be confirmed. On the other hand, experimental constraints push the mass of WIMPs to ever higher values, which is difficult to achieve in particle accelerators such as the LHC. Sterile neutrinos have also failed to live up to their promise, concurrently facing severe limitations as candidates for WDM. On the other hand, ultralight bosons, similar to axions, are proposed with characteristics that could solve some of the problems associated with CDM, particularly on galaxy scales. Even so, these particles are subject to quantum oscillations that place further constraints on their mass. In parallel, modified gravity models continue to be developed as alternatives to dark matter, although they still face significant difficulties, both theoretically and observationally.

The future of research now lies with a new generation of telescopes that are capable of testing these different hypotheses with unrivalled precision. Space missions such as Euclid and Nancy Grace Roman, as well as ground-based instruments such as ESO's SKA (square kilometer array) and ELT (extremely large telescope), promise to generate an avalanche of new data. Advances in AI will make it possible to exploit this information and move forward in this field. This door is open to today's young researchers for potentially revolutionary discoveries.

9.13. References

Amruth, A., Broadhurst, T., Lim, J., Oguri, M., Smoot, G.F., Diego, J.M., Leung, E., Emami, R., Li, J., Chiueh, T. et al. (2023). Einstein rings modulated by wavelike dark matter from anomalies in gravitationally lensed images. *Nature Astronomy*, 7, 736–747.

Babcock, H.W. (1939). *The Rotation of the Andromeda Nebula Lick Observatory Bulletin No. 498*. University of California Press, Berkeley.

Baldeschi, M.R., Gelmini, G.B., Ruffini, R. (1983). On massive fermions and bosons in galactic halos. *Physics Letters B*, 122(3–4), 221–224.

Bekenstein, J. (2004). Relativistic gravitation theory for the modified Newtonian dynamics paradigm. *Physical Review Part D*, 70(8), 083509.

Blumenthal, G.R., Faber, S.M., Primack, J.R., Rees, M.J. (1984). Formation of galaxies and large-scale structure with cold dark matter. *Nature*, 311, 517–525.

Bosma, A. (1981). 21-cm line studies of spiral galaxies. II. The distribution and kinematics of neutral hydrogen in spiral galaxies of various morphological types. *Astronomical Journal*, 86, 1825–1846.

Clowe, D., Gonzalez, A., Markevitch, M. (2004). Weak-lensing mass reconstruction of the interacting cluster 1E 0657–558: Direct evidence for the existence of dark matter. *The Astrophysical Journal*, 604, 596–603.

van Dokkum, P., Danieli, S., Cohen, Y., Merritt, A., Romanowsky, A.J., Abraham, R., Brodie, J., Conroy, C., Lokhorst, D., Mowla, L. et al. (2018). A galaxy lacking dark matter. *Nature*, 555, 629.

van Dokkum, P., Shen, Z., Keim, M.A., Trujillo-Gomez, S., Danieli, S., Chowdhury, D.D., Abraham, R., Conroy, C., Diederik Kruijssen, J.M., Nagai, D. et al. (2022). A trail of dark-matter-free galaxies from. a bullet-dwarf collision. *Nature*, 605, 435–439.

Donato, F., Gentile, G., Salucci, P., Frigerio Martins, C., Wilkinson, M.I., Gilmore, G., Grebel, E.K., Koch, A., Wyse, R. (2009). A constant dark matter halo surface density in galaxies. *MNRAS*, 397, 1169.

Freundlich, J., Famaey, B., Oria, P.-A., Bílek, M., Müller, O., Ibata, R. (2022). Probing the radial acceleration relation and the strong equivalence principle with the Coma cluster ultra-diffuse galaxies. *Astronomy & Astrophysics*, 658, A26.

Hui, L. (2021). Wave dark matter. *Annual Review of Astronomy and Astrophysics*, 59, 247–289.

Lelli, F., McGaugh, S., Schombert, J.M. (2016). The small scatter of the Baryonic Tully-Fisher relation. *The Astrophysical Journal Letters*, 816, L14.

Li, P., Lelli, F., McGaugh, S., Schombert, J.M. (2018). Fitting the radial acceleration relation to individual SPARC galaxies. *Astronomy & Astrophysics*, 615, A3.

Lynds, R. and Petrosian, V. (1989). Luminous arcs in clusters of galaxies. *Astrophysical Journal*, 336, 1.

Markevitch, M., Gonzalez, A.H., David, L., Vikhlinin, A., Murray, S., Forman, W., Jones, C. Tucker, W. (2002). A textbook example of a bow shock in the merging galaxy cluster 1E 06557-56. *The Astrophysical Journal*, 567(1), L27–L31.

McGaugh, S. (2004). The mass discrepancy-acceleration relation: Disk mass and the dark matter distribution. *The Astrophysical Journal Letters*, 609, 652.

McGaugh, S., Schombert, J.M., de Blok, W.J.G., Zagursky, M.J. (2010). The Baryon content of cosmic structures. *The Astrophysical Journal Letters*, 708, L14.

Mocz, P., Fialkov, A., Vogelsberger, M., Becerra, F., Amin, M.A., Bose, S., Boylan-Kolchin, M., Chavanis, P.H., Hernquist, L., Lancaster, L. et al. (2019). First star-forming structures in fuzzy cosmic filaments. *Physical Review Letters*, 123(14), 14130118.

Navarro, J., Frenk, C., White, S.D.M. (1997). A universal density profile from hierarchical clustering. *The Astrophysical Journal*, 490, 493.

Peccei, R.D. and Quinn, H.R. (1977). CP conservation in the presence of pseudoparticles. *American Physical Society*, 38, 1440.

Penzias, A.A. and Wilson, R.W. (1965). A measurement of excess antenna temperature at 4080 Mc/s. *The Astrophysical Journal*, 142, 419–421.

Persic, M., Salucci, P., Stel, F. (1996). The universal rotation curve of spiral galaxies – I. The dark matter connection. *Monthly Notices of the Royal Astronomical Society*, 281, 27–47.

Planck Collaboration, Ade, P.A.R., Aghanim, N., Armitage-Caplan, C., Arnaud, M., Ashdown, M., Atrio-Barandela, F., Aumont, J., Baccigalupi, C., Banday, A.J. et al (2013). Planck 2013 results. XVI. Cosmological parameters. *Astronomy and Astrophysics*, 571, A16.

Poincare, H. (1906) The milky way and the theory of gases. *Popular Astronomy*, 14, 475–488.

Roberts, M.S. and Rots, A.H. (1973). Comparison of rotation curves of different galaxy types. *Astronomy and Astrophysics*, 26, 483–485.

Rubin, V.C., Ford, W.K., Thonnard, N. (1980). Rotational properties of 21 SC galaxies with a large range of luminosities and radii, from NGC 4605 (R=4kpc) to UGC 2885 (R=122kpc). *The Astronomical Journal*, 238, 471–487.

Skordis, C. and Zlosnik, T. (2021). New relativistic theory for modified Newtonian dynamics. *Phys. Rev. Lett.*, 127(16), 161302.

Soucail, G., Fort, B., Mellier, Y., Picat, J.P. (1987). A blue ring-like structure in the center of the A 370 cluster of galaxies. *Astronomy and Astrophysics*, 172, L14–L16.

Walsh, D., Carswell, R.F., Weymann, R.J. (1979). 0957+561 A, B: Twin quasistellar objects or gravitational lens?. *Nature*, 279, 381.

Zwicky, F. (1937). On the masses of nebulae and of clusters of nebulae. *The Astrophysical Journal*, 86, 217.

10

Gravitational Waves

Matteo BARSUGLIA[1] and Simone MASTROGIOVANNI[2]

[1]*CNRS, Astroparticule et Cosmologie, Paris, France*

[2]*INFN Sezione di Roma, Rome, Italy*

10.1. Introduction

Gravitational waves are oscillations in the geometry of space-time, propagating as waves. Predicted by Einstein (1916) as a solution to the field equations of general relativity (GR), they were first directly observed in 2015 by the U.S. LIGO instruments as part of the international LIGO–Virgo collaboration (Abbott et al. 2016). Virgo joined LIGO in 2017 and was crucial to precisely localize the neutron star merger of August 17, 2017, which was observed with gravitational waves and electromagnetic waves (GW170817) (Abbott et al. 2017a). Since then, observations of gravitational waves by the network composed of Virgo, LIGO and the Japanese project KAGRA continue, with ~300 candidate sources identified by April 2025, all of which are mergers of compact objects[1].

Gravitational waves represent a new astrophysical and fundamental physics tool to perform new tests of gravity and new tests of cosmological measurements (e.g. the value of the Hubble constant) and explore the extreme states of matter in neutron stars. They also allow us to study compact objects in a completely different and complementary way, compared to studying electromagnetic waves. Unlike electromagnetic waves, which are emitted by a large number of particles, atoms and molecules, gravitational waves are emitted by a coherent displacement of masses. They are not absorbed by matter but travel undisturbed through very long distances. A network of gravitational

1 See: https://gracedb.ligo.org/.

Quantum Physics and Cosmology,
coordinated by Hervé ZWIRN.

wave detectors can observe all of the sky continuously and send alerts to electromagnetic observatories, allowing them to point telescopes and observe potential electromagnetic counterparts (as for GW170817) (Abbott et al. 2017b).

In this chapter, we aim to provide an introduction and overview of gravitational-wave science, which is far from exhaustive and is centered around observations and future prospects with ground-based interferometric gravitational wave detectors. After a description of gravitational wave theory in the framework of GR, we describe ground-based interferometric gravitational wave detectors, their scientific results, the next generation of detectors and scientific future prospects.

10.2. Gravitational waves in the framework of the general relativity theory

After the special relativity theory, published in 1905, Einstein developed the theory of GR, published in its final form in 1915. In GR, gravity is a manifestation of the space–time curvature. Objects move in the simplest trajectory (geodesic) in curved space–time. For example, the Earth orbits around the Sun because the latter curves space–time around the Earth. The theory of relativity is formalized by Einstein's field equations, a system of ten nonlinear equations in which the curvature of space–time, contained in the Riemann tensor, depends on the matter-energy content, contained by the stress-energy tensor. The theory has several experimental confirmations, the first of which is the explanation of Mercury's perihelion precession, an unsolved mystery since the 19th century that was resolved in Einstein's 1915 article. Other confirmations include light deflection, demonstrated by Eddington in 1919, and the gravitational redshift. Other more recent experimental evidence, such as gravitational slowing of time, is obtained through atomic clocks and the global positioning system (GPS). Astronomical tests include experiments performed within the solar system and those with binary pulsars, such as the famous PSR1916+13 discovered by Hulse and Taylor, which was the first demonstration of the existence of gravitational waves.

Einstein's equations admit a wave like solution, the gravitational waves. The waves propagate at the speed of light, have two polarizations and are transverse, similar to electromagnetic waves. In contrast to the latter, which represent vibrations of electric and magnetic fields propagating in space–time, gravitational waves are vibrations of space–time itself, propagating at the speed of light. Gravitational waves are produced by the acceleration of nonsymmetrical mass distributions (technically speaking, having a mass quadrupole moment or a higher mass quadrupole moment). A sphere spinning around itself does not emit gravitational waves, but a system of two bodies orbiting each other emits gravitational waves since it has a mass quadrupole moment. It is clear from Einstein's equations that space–time is extremely rigid, and detectable gravitational waves cannot be generated by humans, but they can be produced only by extremely compact astrophysical bodies, such as binary systems of black holes, neutron stars and white dwarfs.

The effect of gravitational waves on a system of free masses is to change their relative distance (see Figure 10.1).

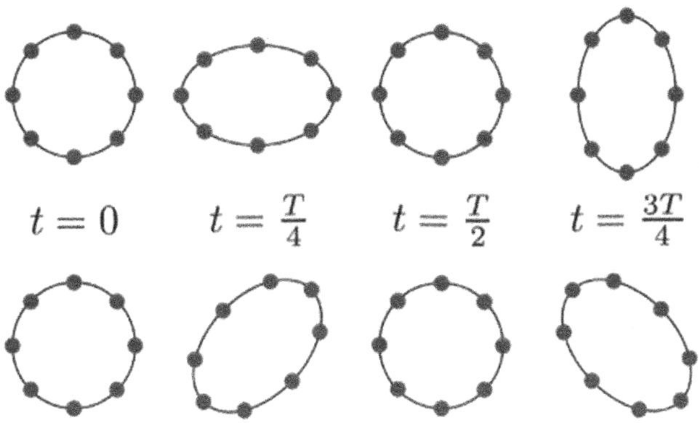

Figure 10.1. *A system of freely falling masses when a gravitational wave passes through the plane of the page. For a color version of this figure, see www.iste.co.uk/zwirn/quantum.zip*

10.3. Gravitational wave sources

As already mentioned, any accelerated distribution of an asymmetric mass (technically speaking, possessing at least one quadrupole moment) produces gravitational waves. However, from Einstein's equations, it emerges that extremely dense and fast-moving mass distributions can emit measurbable gravitational waves. Gravitational wave sources can be continuous or transient, and their waveforms can be modeled through GR or be unmodeled (see Figure 10.2). The sources measured thus far by LIGO–Virgo KAGRA are mergers of compact objects, black hole binary systems of stellar or intermediate mass, neutron stars or mixed systems, consisting of a black hole and a neutron star. Those sources are well modeled since the signal can be superposed with templates obtained by solving GR equations. Current gravitational wave detectors measure the last moments (fractions of a second to several tens of seconds) before the merger, when the frequency of gravitational waves enters the detector's band. For LIGO–Virgo–KAGRA, the band is within 10 Hz–10 kHz, and for future detectors, such as space detectors, it will be extended down to a few Hz (Einstein Telescope and Cosmic Explorer) and to 0.1 mHz–0.1 Hz (the space detector LISA).

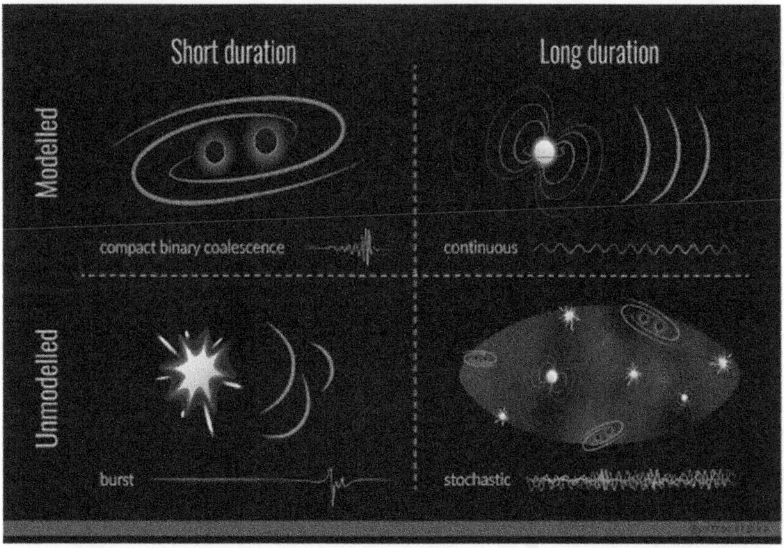

Figure 10.2. *Sources of gravitational waves (credit: Shanika Galaudage)*[2]*. For a color version of this figure, see www.iste.co.uk/zwirn/quantum.zip*

Other examples of gravitational waves that have not been detected thus far, without attempting to be exhaustive, include: (1) gravitational waves produced by rotating neutron stars that possess a small asymmetry in mass distribution. These continuous sources, which also manifest as electromagnetic astrophysics sources in the form of pulsars, produce continuous gravitational waves at a frequency twice the rotational frequency. (2) Gravitational waves produced by the gravitational collapse of massive stars. According to the simulations, gravitational collapses are extremely symmetric and emit a very small fraction (on the order of 10^{-8}) of their mass in gravitational waves, which will be observable only with future detectors and for distances on the order of the size of the galaxy or smaller. These sources are transient and unmodeled. (3) Gravitational waves produced by a cosmological gravitational-wave background produced during cosmic inflation or from phase transitions in the early instants of the Universe.

10.4. Gravitational wave detectors

After an initial attempt made in the 1960s through metal cylinders, which could have been set into vibration by the passage of the gravitational wave (called resonant

2 See: https://shanikagalaudage.github.io/.

bars, or *Weber bars*, named after Joe Weber, the inventor of these devices), the effort on gravitational wave detection quickly turned to Michelson interferometers, instruments that use light to measure the deformation of space–time induced by gravitational waves. In a Michelson interferometer, a light beam is split into two parts by a semireflective mirror, called a *beam splitter*, and sent in two perpendicular directions. The two beams thus obtained are reflected by two mirrors, sent back toward the beam splitter and recombined, producing an interference. The amount of light obtained as a result of interference depends on the optical path difference between the two directions (see Figure 10.3).

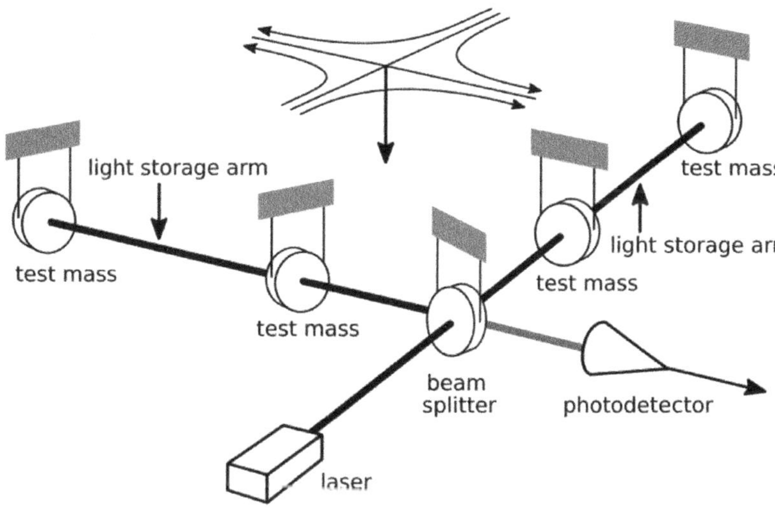

Figure 10.3. *Basic scheme of a Michelson interferometer[3] (credit: LIGO). For a color version of this figure, see www.iste.co.uk/zwirn/quantum.zip*

The passage of a gravitational wave results in an asymmetric deformation of the length of the interferometer arms.

The difficulty of this measurement is related to the extremely small relative length variations produced by gravitational waves, which can be of the order of 10^{-22}. For a ground-based interferometer, such as Virgo or LIGO, this means having to measure length variations on the order of 10^{-18} meters over a few kilometers. The

3 See: https://en.wikipedia.org/wiki/Gravitational-wave_observatory#/media/File:LIGO_schematic_(multilang).svg.

effect of the gravitational wave being to "stretch" spacetime, to increase the signal to a detectable level, kilometric arms are needed: Virgo and LIGO have arms of 3 and 4 km, respectively (Aasi et al. 2015; Acernese et al. 2015). The detector must therefore be perfectly shielded from environmental noises. The mirrors are then suspended from chains of pendula in a cascade, which act as seismic attenuators for frequencies above the resonant frequency of the pendulums themselves. For ground-based interferometers, these frequencies are typically in the hertz range. To detect gravitational waves at lower frequencies, it will be necessary to go to space where there is no seismic noise. This is the goal of the LISA project, which will be launched toward the end of the 2030s.

In addition to seismic noise, the fundamental limitations of an interferometric detector are related to thermodynamics and quantum mechanics. Thermodynamics defines the vibration spectrum of the atoms in the detector mirrors and thus the vibration noise seen by the laser beam. This noise, called *thermal noise*, limits the sensitivity of current detectors to the central part of their frequency range. It can be reduced by decreasing the temperature of the mirrors, enlarging the laser beams, or reducing the mechanical losses of the materials that compose the mirrors. Reducing mechanical losses and dissipations is equivalent to concentrating the thermal energy of vibration at the resonant frequencies of the mirrors, reducing their out-of-resonance component. The ultimate limit of an interferometric detector, once the thermal noise is reduced, is due to the quantum nature of light and is related to the vacuum fluctuations entering the detector.

One manifestation of quantum noise, called *shot noise*, or photon counting noise, decreases when the laser power is increased, but beyond a certain limit, the detector becomes sensitive to another manifestation of quantum noise, called, *radiation pressure noise*, a stochastic force that moves the mirrors. To reduce both noises, a *squeezed vacuum* can be injected into the interferometer (Acernese et al. 2019), which, by creating quantum correlations between the virtual photons in the vacuum, reduces photon noise or radiation pressure noise; these techniques are already being used in Virgo and LIGO. Moreover, particular squeezing techniques, called "frequency-dependent squeezing" make it possible to reduce both manifestations of quantum noise (Ganapathy et al. 2023).

Other noise sources of an interferometric detector are those related to the laser sources (frequency and amplitude noise), the noises related to the electronics, the acoustic and magnetic sources and the scattered light. Every noise should be reduced either by reducing the source itself or by reducing its coupling with the interferometer.

10.5. The first gravitational wave detection, GW150914

The first gravitational wave detection took place on September 14, 2015. The signal was detected by two LIGO detectors in the United States (separated by 3,000 km) and was analyzed within the framework of the LIGO–Virgo scientific collaboration, through a protocol of joint analysis and publication of results in place since 2007.

After five months of checks, the announcement of the first detection was made on February 12, 2016, and the corresponding publication appeared in the journal *Physical Review Letters*, signed by about 1,000 persons from the LIGO and Virgo collaborations (Abbott et al. 2016).

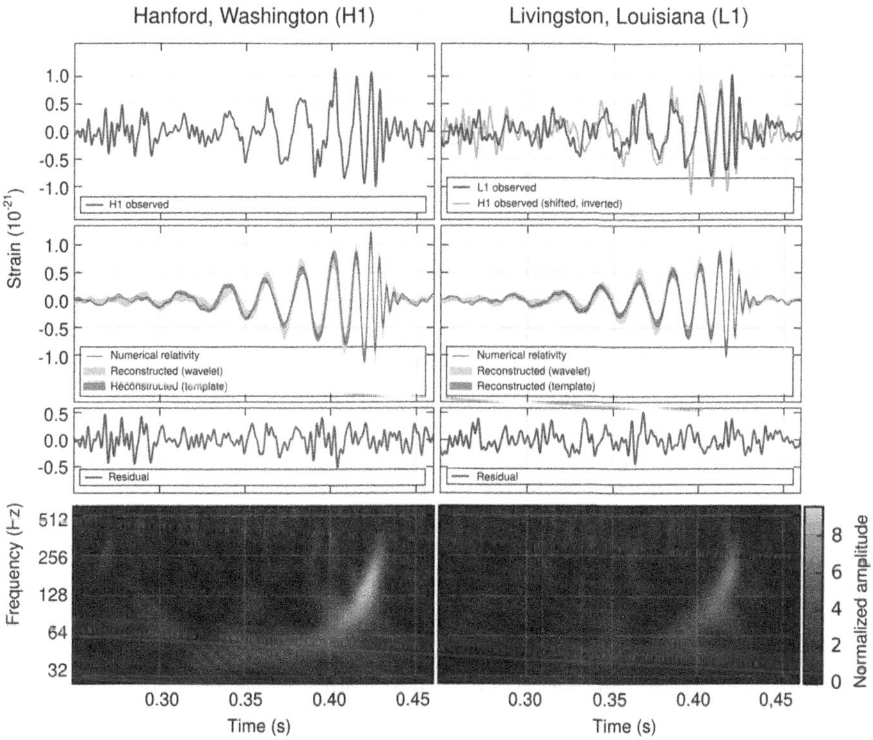

Figure 10.4. *First gravitational-wave detection (figure taken from Abbott et al. (2016)). For a color version of this figure, see www.iste.co.uk/zwirn/quantum.zip*

Following this first detection and the others that followed, the 2017 Nobel Prize in physics was awarded to Rainer Weiss, Kip Thorne and Barry Barish for their theoretical and experimental contributions to the detection of gravitational waves (see The Nobel Prize 2025). The September 14 signal, lasting 0.3 s, was detected by the two detectors LIGO Hanford (near the west coast of the United States) and LIGO Livingston (near the East Coast), with the time separation of the signal consistent with light-speed propagation between the two sites. The signal has the characteristic of a "chirp", a signal that increases in frequency and amplitude, characteristic of the final cycles before the merger of two compact stars. From an analysis of the signal characteristics, applying the general relativity, LIGO and Virgo physicists concluded that the source was a black hole binary of masses of 36 and 29 solar masses and that the event took place at approximately 450 Mpc, corresponding to approximately 1.5 billion light-years. Nevertheless, from the wave analysis, it can be deduced that the two stars orbited at a speed corresponding to 0.3–0.5 times the speed of light (see Figure 10.4), and during the merger, approximately three solar masses were converted into energy, corresponding to a luminosity of approximately 3×10^{36} J/s, which is higher than the total luminosity of the universe in electromagnetic waves. No electromagnetic counterpart was observed.

GW150914, in addition to representing the first direct detection of gravitational waves and thus a test of Einstein's theory in the strong-field regime, also represents the first observation of a binary black hole and the demonstration that these systems can merge in a time comparable to the age of the universe. In addition, the observed source has masses greater than all black holes observed up to that time in the form of X-ray binaries.

10.6. GW170817 multimessenger observation

On August 17, 2017, the LIGO and Virgo detectors observed a merger of two neutron stars, named *GW170817*. At the same time as the gravitational wave signal, the US Fermi and European INTEGRAL satellites reveal a gamma-ray burst (GRB) from the same direction in the sky with a time lag of 1.7 s (Abbott et al. 2017a). Owing to these signals, an alert is sent to approximately 70 telescopes around the world, which observe the region of the sky, which is about 30 square degrees large in the southern hemisphere (Hydra constellation) and at a distance of approximately 40 Mpc, as indicated by Virgo and LIGO.

The electromagnetic telescopes discover an optical transient, a kilonova, in the galaxy NGC 4993, initially in the blue and after a few days in the red. Observations are subsequently performed at all wavelengths, from radio waves to X-rays, through infrared, optical and ultraviolet light. This is the largest multimessenger campaign ever carried out, and the corresponding publications are signed by approximately 3,600 authors (Abbott et al. 2017b). This observation also showed that neutron star systems can merge at a time comparable to the age of the Universe.

Through the joint detection of gravitational waves and GRBs, it is shown that at least some of the short GRBs (those lasting less than 2 s) can be produced by neutron star mergers. By measuring the difference in the arrival time of only 1.7 s between the gravitational waves and the GRB, it is demonstrated that gravitational waves travel at the speed of light with an accuracy of 10^{-15}. Through photometry and spectroscopy of the light emitted from the kilonova, the production of lanthanides and other chemical elements, such as gold and platinum, is observed. This is thus the demonstration that at least some of the heavy nuclei are produced in the fusions of neutron stars. Finally, by measuring the redshift of the NGC 4993 galaxy, a new measure was made of the Hubble constant (or H_0), the expansion rate of the universe today.

10.7. The LIGO–Virgo–KAGRA observations

Figure 10.5. *The LIGO–Virgo–KAGRA network: aerial views of LIGO Livingston (top left), LIGO Hanford (top right), Virgo (bottom left) and an artist impression of the underground detector KAGRA (bottom right). Credit: LIGO, Virgo and KAGRA collaborations. For a color version of this figure, see www.iste.co.uk/zwirn/quantum.zip*

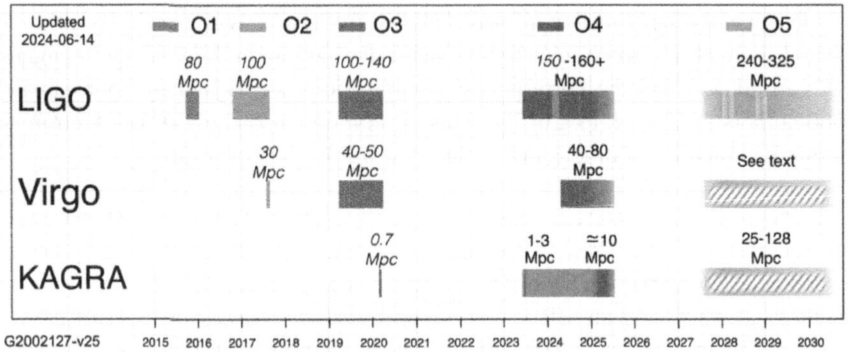

Figure 10.6. *Timeline of observing periods of LIGO, Virgo, KAGRA and their inspiral ranges (average distance at which we can observe a binary-neutron star merger with a signal-to-noise ratio of 8). Credit: LIGO, Virgo and KAGRA Collaborations (see: https://www.ligo.caltech.edu/page/observing-plans). For a color version of this figure, see www.iste.co.uk/zwirn/quantum.zip*

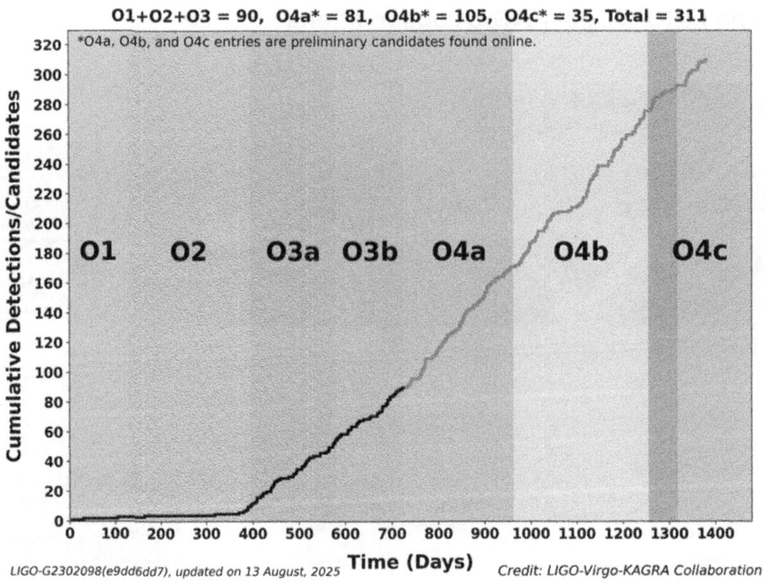

Figure 10.7. *The cumulative number of gravitational wave detections (vertical axis) as a function of observation time (horizontal axis). Among the runs (O1,..., O4C), there were several periods spent to upgrade the detectors that are not reported in this plot. Credit: LIGO-Virgo-KAGRA collaboration (see: https://dcc.ligo.org/public/0190/G2302098/040/cumulative_events%2818%29.png). For a color version of this figure, see www.iste.co.uk/zwirn/quantum.zip*

Since the first gravitational wave detection on September 14, 2015, the LIGO, Virgo and Japanese KAGRA detectors (LVK) have continued observing the sky (see Figure 10.5). The LVK network alternates observational periods, called observing runs (e.g. O1 indicates the first observing run), with periods devoted to the upgrade of the detectors to improve their sensitivity.

The cumulative number of gravitational wave detections from compact binary coalescences (CBCs) is reported in Figure 10.7. At the time of writing of this chapter, LVK reported approximately 300 gravitational wave source candidates from CBC, among which there have been very interesting specific sources. As an example, during the 4th observing run (O4), the LIGO-Virgo-KAGRA collaboration detected the source called GW250114 (Abac 2025a) a loud binary black hole merger with signal-to-noise ratio 80 for which it was possible to validate Hawking's Area theorem, which states that the black holes' surface area can not decrease with time, an important test of the general relativity and of the black-hole nature of the objects observed. As always during O4, we observed the source called GW231123, the most massive binary black hole with a total mass between 190-265 solar masses (Abac et al. 2025b). Most of the detections started with the third observing run (O3), as the sensitivity of the detectors was greatly increased. Gravitational wave detectors are rapidly unveiling a population of merging compact objects that were not previously observable with electromagnetic observations. The study of this population of merging compact objects resulted in several important findings that we briefly summarize below (Abac et al. 2025c)[4].

10.8. The astrophysical properties of compact binary coalescences

Most of the detected signals are consistent with the merging of two black holes, as the masses of the objects span 5 to ~200 solar masses, having a distribution for which the higher black hole masses are less numerous in the overall population. This was an expected result as black holes are likely formed from stars and massive stars are rarer in the Universe with respect to low-mass stars. Furthermore, we discovered that binary black holes were more common when the Universe was younger and that there is probably no mass gap between neutron stars and black holes. Only a few signals are consistent with the masses expected for binary neutron star and neutron star-black hole mergers. Their low number (compared with binary black hole mergers) is because binary black holes produce loud signals that can be detected further away and more easily.

From the detected population of binary black holes, we estimated that in one year of observation, approximately twenty binary black holes merged in a volume of

4 See: https://ligo.org/science-summaries/.

1 Gpc^3. In such a volume, it is possible to find approximately 10 millions of galaxies, meaning that on average, a binary black hole merges in each galaxy every approximately 100,000 years. However, given the age of the universe (~ 14 billion years) and its dimensions, binary black hole mergers are common for LIGO-Virgo-KAGRA detectors.

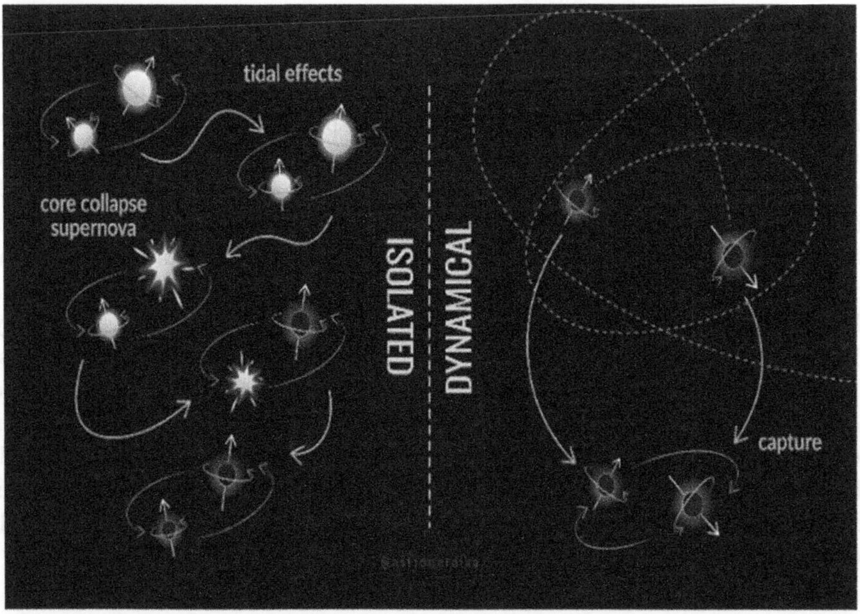

Figure 10.8. *Simplified depiction of two astrophysical formation channels for binary black holes (credit: Shanika Galaudage). On the left: the "isolated" formation channel; on the right: the "dynamical" formation channel. For a color version of this figure, see www.iste.co.uk/zwirn/quantum.zip*

By inspecting the distributions of the masses, spins and distances (redshifts) of binary black holes, we can better understand their astrophysical formation properties. Figure 10.8 depicts two of the most accredited formation channels for binary black holes, the *isolated* and the *dynamical* formation channels. In the isolated formation channel, binary black holes are formed as the natural outcome of an isolated stellar binary system. We expect black holes formed in this channel to have masses less than 40 solar masses (due to a process called the pair instability supernova process) and with spins aligned, as their stellar progenitors exchanged material during their evolution. Instead, in the dynamical formation channel, two independent black holes form a binary when they are close to each other. Typically, we require black holes to be located in very dense stellar environments, such as globular or nuclear clusters, for this formation channel to be effective. Binary black holes formed in this channel are expected to display high masses and high spins (as they can come from previous mergers) that can also be misaligned. However, currently, it is not statistically possible to draw a net line between the presence of the two formation channels in the current data.

10.9. Cosmology with populations of gravitational waves

Gravitational waves from compact binary coalescences can be used to measure cosmic expansion. To do so, we need to observe sources for which we are able to measure the distance and the redshift (recessional velocity with respect to the observer). Gravitational wave sources are the only known astrophysical sources for which it is possible to directly measure the distance from the data. However, they do not provide direct access to the redshift of the source unless an electromagnetic counterpart, as in the case of GW170817, is observed.

Since an electromagnetic counterpart of gravitational wave detection is not common (at the time of writing, only GW170817 is the only such case), the current constraints on the cosmic expansion from gravitational wave detections involve the use of methodologies that are able to *statistically* infer the redshift. These techniques allow us to exploit the hundreds of binary black holes that are observed at cosmological scales. The first technique, proposed by Bernard Schutz (1986), consists of obtaining a statistical redshift estimation from all the galaxies observed by telescopes in the localization volume of the gravitational wave event. With this redshift and the luminosity distance estimated by the gravitational waves, we can obtain a measure of the cosmic expansion. A second technique, proposed by Chernoff and Finn (1993), exploits the fact that when we measure a gravitational wave signal, we do not actually measure the masses of the two objects. Instead, we measure the *redshifted masses*. Therefore, by making astrophysical assumptions on the possible values of the masses of the sources, it is possible to obtain a statistical redshift estimation and measure the cosmic expansion.

The aforementioned techniques, called *dark standard sirens,* have been applied to the population of binary black holes detected in the first three observing runs of the LVK detectors. Interestingly, there is evidence from data that the heaviest object in a binary often has a source mass in the range of 30–40 solar masses. This *mass scale* is used to measure the Hubble constant but, unfortunately, does not provide a value as precise as that of GW170817, as only a few tens of binary black holes were used for the measurement. Even the inclusion of galaxy catalogs to improve the statistical redshift estimation of the sources is not enough to improve the precision of the Hubble constant. This is a consequence of the fact that current galaxy catalogs are highly incomplete at the distances at which we typically observe gravitational wave events. With the term incompleteness, we indicate that a catalog does not report all the galaxies present in the Universe, as many are too dim to be detected. As more GW events are detected, we expect the constraints on the Hubble constant to improve and teach us more about the expansion of the Universe.

10.10 Tests of general relativity with a population of gravitational waves

Gravitational waves are also an unprecedented opportunity to test Einstein's theory of GR. Broadly speaking, the use of gravitational waves for tests of GR can be summarized into two questions:

1) *Are gravitational waves generated from sources as expected from GR? If not, are we sure that what we are observing is a neutron star or a black hole?*

2) *Do gravitational waves propagate as expected in GR?*

Consistency checks on the gravitational waveform attempt to answer the first question, whereas *propagation tests* address the second one. There are a plethora of consistency checks for the gravitational waveform. However, almost all of these methods involve studying the waveform in all its regimes (inspiral, merger and ringdown) and determining whether the waveforms are consistent with each other. As an example, one of the most important tests is the *inspiral-merger-ringdown* (IMR) test, where the mass and spin of the merger remnant are estimated independently from the inspiral and later from the merger and ringdown. If the values of the spins and masses are consistent, then the GR is validated. According to the current data, there is no evidence of GR violations during inspirals, mergers or ringdown. Other consistency checks do not take this agnostic approach and try to parametrize possible deviations from GR either in the inspiral of the two objects or in the relaxation of the final remnant. However, all of these analyses currently agree that GR is still valid, and there is no need for an alternative theory of gravitation to describe the binary black holes that we are observing.

In addition to consistency checks, we can peform propagation tests of general relativity. According to Einstein's GR, gravitational waves propagate at the speed of light in two polarizations, and their amplitude decreases inversely with the luminosity distance to the source. However, this is not the case for many theories of gravity that are devised to explain open questions in cosmology, such as dark energy and dark matter. Here, we do not delve into the plethora of propagation tests of gravitational waves, but we focus on an interesting case closely related to particle physics: massive gravity. If gravity is associated with a massive particle (*the graviton*), we expect gravitational waves to propagate slightly slower than the speed of light, and we expect to observe a different phase evolution of the signal. Current constraints from binary black holes indicate that the mass of the graviton cannot be greater than 10^{-22} ev/c^2, a limit that is ~10^{27} smaller than the mass of the electron (Abbott et al. 2021).

10.11. Future detectors: Einstein Telescope, Cosmic Explorer and LISA

The LIGO, Virgo and KAGRA scientific collaboration plans an additional observing run with increased sensitivity, called O5, which is scheduled to occur between ~2028 and ~2031. For this observing run, LIGO India, another GW detector planned to be built in India near Aundha Nagnath, Hingoli District, Maharashtra, should join the two LIGO detectors, Virgo and KAGRA. Moreover, plans to continue the observations in the years 2030 are under study, with upgraded versions of current detectors (these projects are called A# for an upgraded version of LIGO and Virgo_nEXT for an upgraded version of Virgo).

In parallel, researchers are already working on next-generation successors of current ground-based GW detectors: Einstein Telescope (ET) (ET n.d.), Cosmic Explorer (CE) (https://cosmicexplorer.org/) and the Large Interferometer Space Antenna (LISA) (The European Space Agency n.d.). The concepts of these detectors are shown in Figures 10.9 and 10.10.

LISA is a triangular-shaped GW detector in space. It is a project led by the European Space Agency with NASA as a partner, and it is scheduled to fly in the mid-2030s. The LISA consists of three spacecraft that orbit behind the Earth as we revolve around our Sun. Each spacecraft will contain two lasers that form a Michelson-like interferometer (although we remark that the working concept of LISA is slightly different than that of LIGO, Virgo and KAGRA). Gravitational wave detectors in space have two advantages over ground-based detectors First, they are not limited at lower GW frequencies by seismic noise, and second, they can have an arm length that is orders of magnitude greater than that of ground-based detectors. LISA arms will be approximately 2.5 million km. As such, LISA will be sensitive to all the GW signals from 0.1 mHz to 0.1 Hz. ET and CE are instead the direct successors of the current ground-based detectors. ET is a European project that is planned underground. The original concept of the ET was that of a triangular-shaped gravitational wave detector with 10 km long arms. However, recently, researchers have considered the possibility of building the ET as a network of two L-shaped detectors with 15 km long arms. Building the ET underground will allow us to improve the sensitivity to gravitational wave signals in the frequency band between ~2 and 10 Hz. Cosmic Explorer is an American project of an L-shaped gravitational wave detector with arms expected to be 20–40 km long.

Figure 10.9. *Artistic concept of the space-based gravitational wave detector LISA (credit: NASA)*[5]. *For a color version of this figure, see www.iste.co.uk/zwirn/quantum.zip*

Figure 10.10. *Artistic concept of the ground-based detector Einstein Telescopes (credit: Marco Kraan – Nikhef). For a color version of this figure, see www.iste.co.uk/zwirn/quantum.zip*

5 See: https://commons.wikimedia.org/wiki/File:LISA-waves.jpg.

ET and CE detect all binaries with masses between a few and 1,000 solar masses up to a redshift of two when the universe is approximately three billion years old (Maggiore et al. 2020; Evans et al. 2021). At this epoch, we know that the Universe was very young and active in forming its stars. Detected compact binaries up to that epoch will allow us to study how compact objects form in relation to the very first stars in our universe. ET and CE will also be able to detect binaries with a total mass of ~100 solar masses up to redshift ~20. At this distance, even the first stars were not yet born, and any black hole detected would certainly be of primordial origin.

Instead, LISA will open a new window on a population of black holes that have not been detected before. In fact, the sensitivity of the LISA in the 0.1 mHz–0.1 Hz region allows us to observe binary black holes between 1,000 solar masses and 100 billion solar masses. With LISA, we will be able to observe the merger of supermassive black holes that are in the center of galaxies up to redshifts 6–8. We will also be able to detect binary black holes with a total mass between 1,000 solar masses and 10,000 solar masses, which we have never previously reported. This particular population of black holes is particularly important for astrophysicists, as we expect them to be the seeds that lead to the formation of supermassive black holes that we observe at the center of galaxies.

10.12. References

Aasi, J., Abbott, B.P., Abbott, R., Abbott, T., Abernathy, M.R., Ackley, K., Adams, C., Adams, T., Addesso, P., Adhikari, R.X. et al. (2015). Advanced LIGO. *Classical and Quantum Gravity*, 32, 074001.

Abac, A.G., Abouelfettouh, I., Acernese, F., Ackley, K., Adamcewicz, C., Adhicary, S., Adhikari, D., Adhikari, N., Adhikari, R.X., Adkins, V.K. et al. (2025a). GW250114: Testing Hawking's area law and the Kerr nature of black holes. *Physical Review Letters*, 135, 111403. doi: 10.1103/kw5g-d732.

Abac, A.G., Abouelfettouh, I., Acernese, F., Ackley, K., Adamcewicz, C., Adhicary, S., Adhikari, D., Adhikari, N., Adhikari, R.X., Adkins, V.K. et al. (2025b). GW231123: A binary black hole merger with total mass 190–265. *arXiv:2507.08219*. doi: 10.48550/arXiv.2507.08219.

Abac, A.G., Abouelfettouh, I., Acernese, F., Ackley, K., Adamcewicz, C., Adhicary, S., Adhikari, D., Adhikari, N., Adhikari, R.X., Adkins, V.K. et al. (2025c). GWTC-4.0: An introduction to version 4.0 of the gravitational-wave transient catalog. *arXiv:2508.18080*. doi: 10.48550/arXiv.2508.18080.

Abbott, B.P., Abbott, R., Abbott, T.D., Abernathy, M. R., Acernese, F., Ackley, K., Adams, C., Adams, T., Addesso, P. (2016). Observation of gravitational waves from a binary black hole merger. *Physical Review Letters*, 116, 061102.

Abbott, B.P., Abbott, R., Abbott, T.D., Acernese, F., Ackley, K., Adams, C., Adams, T., Addesso, P., Adhikari, R.X., Adya, V.B. et al. (2017a). GW170817: Observation of gravitational waves from a binary neutron star inspiral. *Physical Review Letters*, 119, 161101.

Abbott, B.P., Abbott, R., Abbott, T.D., Acernese, F., Ackley, K., Adams, C., Adams, T., Addesso, P., Adhikari, R.X., Adya, V.B. et al. (2017b). Multi-messenger observations of a binary neutron star merger. *Astrophysical Journal Letters*, 848, L12.

Abbott, B.P., Abe, H., Acernese, F., Ackley, K., Adhikari, N., Adhikari, R.X., Adkins, V.K., Adya, V.B., Affeldt, C., Agarwal, D. et al. (2021). Tests of general relativity with GWTC-3. *arXiv:2112.06861*. doi: 10.48550/arXiv.2112.06861.

Abbott, R., Abbott, T., Acernese, F., Ackley, K., Adams, C., Adhikari, N., Adhikari, R., Adya, V., Affeldt, C., Agarwal, D. et al. (2023). GWTC-3: Compact binary coalescences observed by LIGO and Virgo during second part of the third observing run. *Physical Review X*, 13, 041039.

Acernese, F., Agathos, M., Agatsuma, K., Aisa, D., Allemandou, N., Allocca, A., Amarni, J., Astone, P., Balestri, G., Ballardin, G. et al. (2015). Advanced Virgo: A 2nd generation interferometric gravitational wave detector. *Classical and Quantum Gravity*, 32(2), 024001.

Acernese, F., Agathos, M., Aiello, L., Allocca, A., Amato, A., Ansoldi, S., Antier, S., Arène, M., Arnaud, N., Ascenzi, S. et al. (2019). Increasing the astrophysical reach of the advanced Virgo detector via the application of squeezed vacuum states of light. *Physical Review Letters*, 123, 231108.

Chernoff, D.F. and Finn, L.S. (1993). Gravitational radiation, inspiraling binaries, and cosmology. *Astrophysical Journal Letters*, 411, L5. doi: 10.1086/186898.

Einstein, A. (1916). Näherungsweise Integration der Feldgleichungen der Gravitation. *Sitzungsber. K. Preuss. Akad. Wiss.*, 1, 688.

ET (n.d.). The Einstein Telescope project [Online]. Available at https://www.einstein-telescope.it/en/home-en/.

Evans, M., Adhikari, R.X., Afle, C., Ballmer, S.W., Biscoveanu, S., Borhanian, S., Brown, D.A., Chen, Y., Eisenstein, R., Gruson, A. et al. (2021). A horizon study for cosmic explorer: Science, observatories, and community. *arXiv:2109.09882*. doi: 10.48550/arXiv.2109.09882.

Ganapathy, D., Jia, W., Nakano, M., Xu, V., Aritomi, N., Cullen, T., Kijbunchoo, N., Dwyer, S., Mullavey, A., McCuller, L. et al. (2023). Broadband quantum enhancement of the LIGO detectors with frequency-dependent squeezing. *Physical Review X*, 13, 041021.

LIGO (2025). Observing plans [Online]. Available at: https://www.ligo.caltech.edu/page/observing-plans.

Maggiore, M., Broeck, C.V.D., Bartolo, N., Belgacem, E., Bertacca, D., Bizouard, M.A., Branchesi, M., Clesse, S., Foffa, S., García-Bellido, J. et al. (2020). Science case for the Einstein telescope. *Journal of Cosmology and Astroparticle Physics*, 2020, 50.

Schutz, B.F. (1986). Determining the Hubble constant from gravitational wave observations. *Nature*, 323(6086), 310–311.

The European Space Agency (n.d.). LISA factsheet [Online]. Available at: https://www.esa.int/Science_Exploration/Space_Science/LISA/LISA_factsheet.

The Nobel Prize (2025). The Nobel Prize in physics 2017. *The Nobel Prize* [Online]. Available at https://www.nobelprize.org/prizes/physics/2017/summary/.

11

What is the Energy of the Vacuum?

Bruno MANSOULIÉ
IRFU, CEA, Université Paris-Saclay, Gif-sur-Yvette, France

11.1. Introduction

In the eyes of an audience far removed from the discipline, the very notion of vacuum energy seems absurd or irrelevant: how could an empty volume have energy? A person who is slightly more familiar with the definition and application of the concept of energy would note that it is difficult, and pointless, to define an absolute scale of energy: in practice, only energy transfers manifest themselves. A possible constant vacuum of energy at any point in space and time would therefore have no practical manifestation and would not require a measurement or even a precise definition. However, the two great scientific revolutions of the early 20th century – general relativity and quantum theory – challenged this simple conception, each in their own way. For apparently very different reasons, vacuum energy cannot be ignored, and it also seems to play an essential role.

In general relativity, it can be linked directly to the "cosmological constant", which is a quantity that determines the shape of our Universe and its evolution. In quantum theory, defining the energy of the vacuum, by definition an unconstrained state with no limits of time or space, becomes a difficult and uncertain exercise. Ultimately, the confrontation of the two theoretical visions has resulted in a shocking quantitative difference, so it has become a witness to the limits of our understanding of the fundamental laws (Martin 2012).

Quantum Physics and Cosmology,
coordinated by Hervé ZWIRN. © ISTE Ltd 2026.

11.2. General relativity

11.2.1. *The foundations of GR and its successes*

For the so-called "classical" physics of Galileo, Newton and Laplace, the world can be understood as a fixed space, where material points can be located on a regular three-dimensional grid. Within this space, uniform time also runs regularly, with all the clocks in the Universe synchronized. Relativity, enunciated by Einstein in 1905 (already present incognito in Maxwell's equations and sensed by Poincaré and Lorentz, but that is another story), changed this vision: distances and times seen by an observer depend on their motion. The event points are thus located on a four-dimensional grid, although a rather unusual one because movement in one dimension, time, is limited (to the future). The consequences of this new description were immense. To mention just two striking aspects: the speed of light is independent of the reference frame in which it is observed (and unsurpassed by any form of matter or energy), and matter and energy become two aspects of the same quantity, as stated by the famous equation $E = mc^2$. Nevertheless, the space–time grid is present, regardless of the presence of matter–energy. We note that Newton's law of gravity, which postulates instantaneous action at a distance, is incompatible with the relativity of 1905, which we have therefore named special relativity. It was largely to include gravity that Einstein further pursued his approach, successfully proposing general relativity (GR) in 1915.

In GR, space–time can be curved, and its shape depends on the presence of matter–energy. Without going into mathematical detail, this relationship between the geometry of space–time and the density and flow of all forms of matter–energy can be read in Einstein's equation below:

$$R_{\mu\nu} - \frac{1}{2} R g_{\mu\nu} + \Lambda g_{\mu\nu} = \kappa T_{\mu\nu}$$

The left-hand term "$R_{\mu\nu} - \frac{1}{2} R g_{\mu\nu}$" represents the parameters of the curvature of space–time, which are purely geometric quantities. The right-hand term "$T_{\mu\nu}$" groups together the matter–energy parameters, and the constant κ simply describes the intensity of the coupling between these two terms.

The term $\Lambda g_{\mu\nu}$ is a simple parameter Λ multiplied by $g_{\mu\nu}$, which is the local neutral element for a relativistic description. We will return to this shortly.

It is easy to see the limit of Einstein's equation for speeds smaller than the speed of light: we find Newton's law of gravitation. Newton's gravitational intensity, the gravitational constant G, is related to the coupling intensity κ via the relation $\kappa = 8\pi G/c^4$. As G is measured in the laboratory, the value of κ can be deduced, so everything in Einstein's equation is known, with the exception of the constant Λ.

At the time of the invention of general relativity, the measurable consequences, such as deviations from Newtonian gravity, were difficult to observe. The motion of planets, for example, involves relative velocities that are small compared with the speed of light, and GR makes only a very small modification. For light itself, of course, it is "relativistic", but its deflection by matter is far too weak to be measured in the laboratory and still very small for deflection by an object as massive as the Sun. However, two famous measurements were made for each of these cases, quickly confirming GR's predictions: Mercury's perihelion precession and the measurement of the deflection of a star's light by the Sun. Since then, theoretical developments and observations have multiplied, offering a multitude of important, very well-measured effects, all of which are in agreement with GR. Among the most striking examples are the existence of black holes, gravitational mirages, GPS corrections on Earth, the existence of gravitational waves, etc.

The most spectacular change in our vision of the world is certainly the notion of the Big Bang, derived directly from GR. Before this theory, the Universe had no beginning or end in time. The Universe could have existed "forever" and should exist "forever". Transformations could take place locally: the birth and death of stars, etc.; however, if we went back in time, we would have seen a universe identical to our own, as far back as we go, and there was no reason to think that it would change in the future.

When we apply the GR equations, the scale of distances in the Universe must vary over time: either continuously increasing or continuously decreasing. The observation of the escape of distant galaxies, which was not very precise at first, soon enabled us to choose the "branch", which is that of expansion. as the Universe evolves, distances increase, and consequently, density and temperature decrease.

When we return in time, the distances become infinitely small, and the density and temperature are infinitely large. This happens not for an infinite time in the past but for a very specific time in the past, today determined by measurements to date to 13.8 billion years. This point in the past, beyond which it is impossible to return, is the Big Bang. At the beginning of time, the Universe is infinitely dense and infinitely hot and begins to expand. Note that this expansion takes place "everywhere in itself" and not from any particular point.

11.2.2. *The cosmological constant*

The "cosmological constant" has had a famous fate, surprisingly facetious for a physical quantity. To begin with, Einstein did not write it into his equation. He noted that this first approach described a universe that varied with time: expansion or contraction. At the time, the constancy of the Universe over time was taken for granted, and a serious theory had to respect this criterion.

In 1917, Einstein realized that he could add a cosmological constant term to his equation and thus describe a static Universe (actually, it was later realized that this "stabilization" was, in fact, unstable). However, Hubble's observations and the work of Friedmann and Lemaître showed as early as 1927 that distant galaxies were moving away from us and that this distance corresponded to an expanding Universe, as predicted by Einstein's original equation, with no cosmological constant.

In 1929, Einstein withdrew the cosmological constant, calling its introduction "the biggest blunder of my life". He then solidified the model without the cosmological constant in 1931, together with de Sitter.

11.3. A rapidly expanding universe

After the brief introduction of the cosmological constant and its somewhat shameful withdrawal, astronomical measurements have become dedicated to measuring the distance and velocity of the most distant objects possible to constrain the current rate of expansion and verify its slowdown. Indeed, all objects in the Universe are attracted by gravity, which should lead to a slowdown in expansion. While it is fairly easy to measure the velocities of objects, which give rise to Doppler shifts in the frequencies of the light emitted by their atoms, it is difficult to determine their distances. It was only after slow progress, more accurate measurements and, above all, a judicious choice of the objects measured (in this case, a certain type of supernova) that two teams announced a result in 1998 that turned cosmology on its head: not only was the expansion not slowing down but it was actually accelerating.

Perlmutter, Schmidt and Riess were awarded the Nobel Prize in Physics in 2011 for this discovery. The cosmological constant had to be pulled back. The effect is quite small, resulting in little deviation from the simple expansion without a cosmological constant, but it is now very well measured.

The value of the cosmological constant measured in this way is $\Lambda = 1.1 \times 10^{-52}\ \text{m}^{-2}$, an essential parameter of our Universe.

11.3.1. *Vacuum energy interpretation*

If we leave the term Λ to the left of the equation, we can interpret it purely geometrically and leave it at that for GR theory. However, let us move this term to the right-hand side (with a – sign), factor in the coupling κ; and consider the case where there is no matter ($T_{\mu\nu} = 0$).

We then see another interpretation: this is an energy analogous to that of matter, but in the absence of matter, it is therefore the energy of the "vacuum." : $\Lambda = \kappa\varrho_{vacuum}$.

The GR does not claim to describe what matter is made of, what its elementary particles are and how they interact. Therefore, in GR, there is no problem in having a vacuum energy that is only a parameter, albeit an essential one of the Universe.

For the question "What is the energy of the vacuum?", the GR answers accordingly, with the help of recent measurements:

The energy density of vacuum is $\varrho_{vacuum} = 0.53\ \mathrm{nJ/m^3}$ (nano-Joules per cubic meter).

Is that "a lot"? A nano-Joule is a small but not infinitesimal amount of energy on a human scale: it represents the energy consumed by a microprocessor to perform a thousand additions.

In particle terms, when the formula $E = mc^2$ is applied, it is four times the rest of the energy corresponding to the mass of the proton. It is not enough to have any observable consequences on our scale. However, it is enough to have a very important consequence on the scale of the Universe.

In our current model, the cosmological constant, interpreted as vacuum energy, also known as dark energy, accounts for 70% of the Universe's total energy density, with 25% of the remainder being matter of unknown form, known as "dark matter", and 5% of our ordinary matter. It is therefore very important, and its fraction will continue to increase over time. This is because the density of vacuum energy remains constant, whereas the density of matter, which is subject to expansion, decreases.

Shortly after the Big Bang, matter made up the bulk of the Universe's energy density, and far into the future, vacuum energy will dominate. Some even wonder why "our time", 13.8 billion years after the Big Bang, is the one when both contributions – matter and vacuum energy – are of the same order of magnitude, leading to what is commonly referred to as the "coincidence problem".

11.4. Quantum field theory: the basics

11.4.1. *Non-relativistic quantum theory: ground state*

The beginning of the 20th century also saw the emergence of the second fundamental pillar of so-called modern physics: quantum theory (rather than quantum mechanics, as the theory is not limited to mechanics alone). The original, non-relativistic quantum theory was the subject of numerous experimental confirmations: the photoelectric effect, the spectrum of atoms, black-body radiation etc.

With respect to the subject of vacuum energy, one difference from classical theory is already apparent in this framework: for a system endowed with a potential, the state of minimum energy, known as the "fundamental state", is not equal to the mathematical minimum of the potential, as would be the case in classical physics.

In practice, nothing changes, since nothing can be extracted from this new minimum, but the door is open to a vacuum that is more complex than "nothingness".

11.4.2. *Relativistic quantum theory: the basics, the successes*

The need for a relativistic quantum theory was obvious. Several parallel developments, including the oscillator field representation by Heisenberg, Born and Jordan, Dirac's description of the electron and other work, led to quantum field theory (QFT) in the 1930s. The main change from the original quantum theory is that, during an interaction, the number of particles is not conserved. The creation or annihilation of a particle is subject to quantization similar to that of classical variables in the original quantum theory.

Matter is described not by an assembly of immutable bricks but by a "field", a set of particles that can appear and disappear at any instant and at any point in space in the course of any interaction (as long as energy is conserved). This great flexibility constitutes both the extraordinary power of this theory and its burden: a very large number of degrees of freedom, which nature spontaneously exploits on its own behalf but which physicists have great difficulty mastering.

Despite a rather vague physical and mathematical formalism, physicists were soon able to extract spectacular results from the calculations: the creation of electron–positron pairs from a photon, the bremsstrahlung effect (radiation from a relativistic electron in the vicinity of a nucleus), etc.

11.4.3. *Problems of quantum field theory: infinities, renormalization*

In QFT, we see the importance of "virtual" processes. Figure 11.1 shows some examples of the interaction of two electrons, the QFT equivalent of the Coulombian interaction.

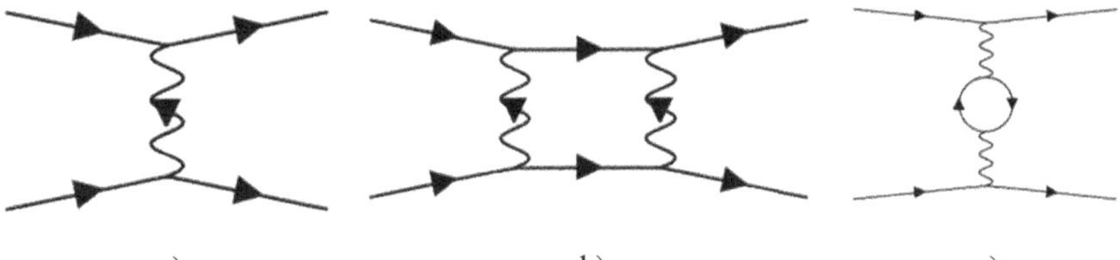

a) b) c)

Figure 11.1. *Examples of Feynmann diagrams for a few of the processes used to describe the scattering of two electrons in Quantum Field Theory*

In this process, a "virtual" photon is emitted by one of the electrons and absorbed by the other (a), or vice versa, or two photons are exchanged (b), or even (c), the virtual photon is transformed into a virtual e+/e– pair that itself annihilates, etc.

The usual simplified explanation for describing virtual processes is as follows: quantum theory "authorizes" an energy fluctuation of magnitude ΔE as long as it is limited to a duration Δt such that ΔE. Δt < ħ. However, owing to relativity, an energy fluctuation ΔE can correspond, for example, to the creation of a particle/antiparticle pair with masses $m = \Delta E/2c^2$. The combination of quantum (fluctuations) + relativity (energy = particles), with a non-conserved number of particles, leads directly to this conception, where all reality is the sum of virtual processes involving all of the particles available in the model.

This great flexibility easily leads to uncontrolled calculations. Even an isolated electron becomes a complex object. Figure 11.2(a) shows an isolated electron, described in the formalism of "Feynman diagrams", invented by Richard Feynman in 1948. This is a very common, intuitive representation of particles and their interactions in the QFT.

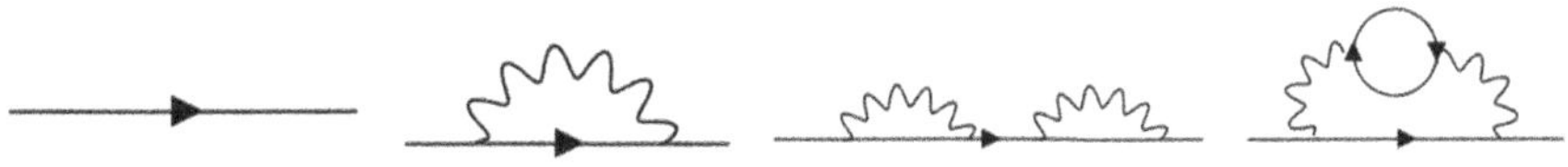

a) b) c)

Figure 11.2. *Example of a few Feynman diagrams describing the propagation of a single electron*

To this simple diagram, an infinite number of diagrams representing virtual processes must be added, such as the emission and reabsorption of a photon (b), and all of the others, for example, those in Figure 11.2(c). In the term corresponding to Figure 11.2(b), the photon momentum is not defined by the initial and final conditions. To calculate the energy of the system, in this case, the mass of the electron, we must therefore sum over all possible states in the quantum sense, which means integrating all photon momenta up to infinity. Unfortunately, this integral does not converge, and the result is infinite. Most QFT calculations involving at least one loop give an infinite result. The same applies to calculations involving an infinite number of particles with infinitely weak momenta, etc.

The solution was found after 1947, using the "renormalization" method initiated by Bethe and developed by numerous authors. Essentially, quantities such as the electron mass cannot be predicted by theory, and the above sum of all virtual processes is fixed at the value of this mass measured experimentally. The same applies to the electric charge, which is fixed at the measured value. To calculate an interaction process, such as a collision between two electrons or the state of an electron around a nucleus, we separate the infinite terms from the calculation and include them in the measured values. We then show that the "interesting" terms, such as an interaction probability or the electron's energy level in an atom, are "finite". The first example of a calculation performed using this procedure was Bethe's famous Lamb shift calculation in 1947. The Lamb shift is a small shift of one level in the hydrogen atom measured against ordinary quantum theory. The method was subsequently extensively studied and improved. In particular, the types of QFTs that can be made predictive by applying this method, referred to as renormalizable theories, have been researched. Quantum electrodynamics (QED), the QFT that applies to electrons and photons, is one of these renormalizable theories, which has ensured its success. For example, the prediction of QED regarding the electron's magnetic moment, agrees with the experimentally measured value over 13 significant digits. This is the model that all theorists dream of for their other favorite theories.

Together, experimentation and theoretical progress led to the development of the standard model (SM) of particle physics, as we know it today. Without going into detail, the main idea of the SM is to describe the other elementary interactions (weak and strong nuclear) with a structure similar to that of quantum electrodynamics, in the hope that they will also be renormalizable. Formal proof of this was provided in 1974 by t'Hooft and Veltman. Describing the electromagnetic and weak interactions with the "right" structure requires them to be generated by a very special process: the Brout–Englert–Higgs mechanism, discovered as early as 1964. The model thus constructed predicts the existence of a particle "witness" to this mechanism, the famous Higgs boson. The discovery of the Higgs boson at CERN in 2012 crowned this theoretical and experimental edifice.

However, the SM does not describe gravitation. We can attempt to make general relativity quantum in a natural way. This can be imagined simply as describing the gravitational interaction by the exchange of "gravitons", thus copying the electromagnetic interaction transmitted by photons. Unfortunately, the resulting theory is not renormalizable. Naively, we can see it like this: because of the quantum, energy fluctuations arise. Through Einstein's equation, these fluctuations manifest themselves as fluctuations in the space–time curvature, which in turn induce energy fluctuations, etc. This vicious cycle is more serious than the fluctuations in the particle fields for the SM and cannot be controlled by a method such as renormalization.

We must not lose sight of the fact that nature manages just fine! Our two descriptions – one quantum and the other geometric – are incompatible. This limitation of our knowledge is one of the fundamental problems of physics today.

11.3. Vacuum energy in a QFT

11.3.1. *Naive estimate*

Here, we focus on the SM, with the exception of gravity. In the QFT, each "particle" included in the theory appears as a set of oscillators everywhere in space, where each oscillator oscillates between 0 particles and 1, 2, 3, etc., particles. The "vacuum" is therefore made up of all of these oscillators for all of the particles when the average values of the number of particles are all equal to zero. However, the energy of these oscillators is not zero! To calculate this value, we have to sum all of the vacuum fluctuation diagrams, such as those shown in Figure 11.3.

Figure 11.3. *Examples of Feynman diagrams describing the fluctuations of the vacuum in Quantum Field Theory*

A priori, the result will once again be infinite. In practice, however, this infinity can be considered irrelevant: since physical processes depend only on energy differences, they ignore the vacuum energy that is present everywhere and is constant over time.

Again, the problem arises when we try to include gravitation in the model, in the form of general relativity. In GR, the space–time curvature depends directly on energy density, not energy differences. The vacuum energy should therefore play its full role, and it becomes essential to estimate its value using our microscopic theory, the SM. For a first estimate, we try to sum the energy of all oscillators. However, up to what limit? We lack a quantum theory of gravitation, but we do know that such a theory would manifest itself for energies of the order of a value $M_P c^2$, where M_P is the Planck mass defined by $M_P = \sqrt{\frac{hc}{G}}$. Planck's energy, $M_P c^2$, equal to 10^{19} GeV, is a gigantic value on the scale of elementary particles, 10^{15} times higher than that of the LHC beams (CERN's largest accelerator), for example, and therefore is completely out of experimental reach. In the absence of any other scale, it is plausible to integrate up to this limit. This first estimate gives the following result for vacuum energy: $M_P^4\, c^5/h^3$, for each particle/field in the theory, that is, the numerical value of:

$$E_{quantum\ vacuum} \sim 10^{122}\ \mathrm{nJ/m^3}$$

This value is compared with the measured value of the Universe expansion acceleration, interpreted in the GR framework: 0.53 nJ/m^3. Even if the above estimate seems naive, it is rare that a simple calculation involving only "dimensional" arguments yields a result thus far removed from reality. Such a large value for vacuum energy would not even allow our Universe to expand.

11.3.2. *Toward a better estimate?*

Figure 11.3, approximately 120 orders of magnitude between theory and reality, is often quoted when discussing the problem of vacuum energy (Martin 2012). However, can we attempt a slightly less naive estimate? It is indeed possible to apply the renormalization method to this calculation in a less "brutal" way than a simple upper bound to the M_P. With more careful treatment, we see that the contribution of each SM particle is on the order of M^4, where M is the mass of that particle, rather than M_P^4. The highest mass SM particles have masses on the order of a hundred GeVs: top quark 173 GeVs, Higgs boson 125 GeVs, Z° 91 GeVs. The contributions to vacuum energy would therefore be on the order of 10^{55} nJ/m^3 rather than 10^{122} nJ/m^3, and we would therefore have "gained" an (enormous) factor of 10^{67}. However, the 10^{55} ratio between calculation and reality remains rather discouraging. The only advantage of this new, more detailed calculation is that the contribution term for each particle has a + or – sign, depending on whether the particle is a boson or a fermion. The vacuum energy therefore contains the following terms: $- M_{top}^4 + M_{Higgs}^4 + M_Z^4 + M_W^4 - M_b^4 \ldots$. In the SM, the masses of bosons and fermions are not equal, as seen from this expression. There is no term-to-term cancellation, and these different signs make little practical difference in the problem. In addition, we know that there are other contributions linked to symmetry breaking in the SM, whose exposition would require further development, but whose order of magnitude is also 100^4 GeV^4.

The confrontation is therefore inevitable and has no solution to date.

11.3.3. *A few leads*

11.3.3.1. *Do quantum fluctuations gravitate?*

Given that the link between quantum theory and gravity, as described in general relativity, is poorly understood, it is legitimate to simply ask whether quantum fluctuations exist in the material sense, in the sense, of having an impact on a macroscopic quantity and, if so, whether they are subject to gravity, as is classical matter.

For the first question, two examples are often proposed: the Casimir effect and the Lamb shift.

11.3.3.2. *The Casimir effect*

We would like to go into more detail about the Casimir effect since it is so frequently cited on this subject. This effect, which was predicted by Casimir in 1948 and has been observed experimentally since then, is often taken as a strong argument in favor of the material reality of vacuum fluctuations. Nevertheless, the discussion remains more open than expected.

The effect is as follows: two parallel conducting plates separated by a distance L, under vacuum, experience an attractive force per unit area:

$$\frac{dF}{dS} = -\frac{\pi^2}{240}\frac{hc}{L^4}$$

This force is very small for macroscopic distances. For example, between two plates with a surface area of 1 cm^2 separated by 1 μm, the force is 10^{-7} N. This force becomes significant for separations in the nm range. It has been measured experimentally for distances L between a few micrometers and a few millimeters, essentially verifying the theoretical calculation.

Casimir's calculation links this force to quantum fluctuations of the vacuum between the plates. More precisely, in this approach, the presence of the conducting plates constrains these fluctuations and suppresses certain modes between the plates, whereas these modes are not suppressed outside the plates. As a result, the pressure of the inner modes is lower than that of the outer modes, giving rise to an attractive force between the plates.

However, Jaffe proposed an alternative interpretation (Jaffe 2005) by directly calculating the mutual influence of the free charges of each plate on those of the other plate in the quantum framework. In practice, the calculation is identical to that of Casimir (in fact, it was at its origin). Jaffe thus rediscovered Casimir's force and formula without reference to the quantum vacuum. He then questioned the material "reality" of vacuum fluctuations, particularly their energy density coupled with gravity.

The Archimedes experiment, which is currently under construction in the Gran Sasso tunnel, is designed to "weigh" a Casimir cavity. It uses the same plates, either close together or far apart. The Casimir force, if integrated from infinity to distance L, corresponds to the energy per unit area:

$$E_{Casimir} = -\frac{\pi^2 \hbar c}{720 L^3}$$

The equivalence principle therefore predicts that the cavity will experience Earth's gravity reduced by $\Delta P = g E_{Casimir}/c^2$, similar to Archimedes' thrust, hence the name of the experiment.

This "thrust" is an extremely weak force, making the measurement very difficult. A result with the required precision would be an experimental tour de force and an interesting measurement. However, as we observed with Jaffe's discussion, it would be risky to deduce from it the material reality of vacuum fluctuations.

11.3.3.3. *Lamb shift*

As mentioned above, one of the first predictions of quantum electrodynamics was Bethe's calculation of the Lamb shift in the hydrogen atom in 1947. This calculation takes into account the "vacuum polarizability" between the proton and electron bound by Coulombic attraction and is therefore presented as proof of the reality of vacuum fluctuations. However, although this has not been done to our knowledge, it is conceivable that the same calculation could be carried out without direct recourse to the quantum vacuum, following the example of Jaffe's calculation for the Casimir effect.

Once again, the successful calculation of the Lamb shift, like that of the Casimir effect, demonstrates the power of quantum field theory in the context of interacting systems. They reveal little about the material reality of the quantum vacuum and, in particular, about the coupling of its fluctuations to gravity.

11.3.3.4. *Atoms, atomic nuclei*

The free fall of matter provides more information on this subject. In the case of nuclei, for example, we can estimate that a few percent of the mass comes from energies linked to "vacuum polarization" corrections and that this fraction depends on the type of nucleus.

Free-fall experiments involving different materials have achieved remarkable precision: in 2022, for example, the "Microscope" satellite experiment demonstrated the validity of the equivalence principle to within 10^{-15} between platinum and titanium masses (Touboul et al. 2022). Clearly, if we interpret a fraction of the masses of these nuclei in terms of vacuum fluctuations, these fluctuations exist and gravitate "normally".

Nevertheless, we do not deal here directly with the energy of the vacuum as an isolated system but rather with the energy differences of an interacting system, where the vacuum can be seen as "connected" to the other elements of the system.

11.3.3.5. *Antimatter*

The very real existence of antimatter and its relationship with ordinary matter are strong arguments in favor of quantum field theory. However, the influence of gravity on antimatter was not measured experimentally until 2023. Even if the equivalence principle, the very basis of general relativity, requires the antimatter to undergo the same gravitational attraction by matter as ordinary matter does, it is not forbidden to wonder whether the antimatter might somehow escape it. Some even hypothesized "antigravity", whereby the gravitational "charge" of an antimatter would be opposed to that of matter, as with the charges of other interactions.

Such a microscopic theory does not exist, as it would give rise to numerous paradoxes. However, if it did exist, its consequences for quantum vacuum gravitation would be radical. Indeed, a matter–antimatter system in the ground state possesses, by definition, the quantum numbers of the vacuum.

In 2023, the ALPHA experiment at CERN obtained the first measurement of terrestrial attraction on antihydrogen atoms and reported an acceleration identical to g (as for matter) to within approximately 20% (Anderson et al. 2023). This experiment, and others such as it, are continuing to improve accuracy.

11.3.4. *Tracks in particle physics*

11.3.4.1. *Fine tuning*

It is conceivable that all of the different contributions to vacuum energy could cancel each other by extraordinarily fine-tuning the parameters of the SM or by a theory that realizes a simple extension of the SM. Note that an additional difficulty would be to maintain this fine-tuning as the Universe evolves.

In general, physicists reject these "coincidence" adjustments, preferring solutions where the exact equality of the amplitude of two large terms of opposite sign is imposed by the model through, for example, a law of symmetry. As a last resort, the multiverse hypothesis, supported by certain theories beyond the SM, authorizes fine adjustments: if there are 10^{500} or even $10^{10^{10^7}}$ universes, as some claim (Linde and Vanchurin 2010), there is indeed one where the adjustment is realized, allowing us to inhabit and think about it.

11.3.4.2. *Symmetry compensation*

The point above which mentioned the sign of the contributions of fermions and bosons, can be developed further: this is the subject of supersymmetry theories. In this framework, there exists a symmetry between bosons and fermions, which is precisely what is needed for compensation.

These theories have several interesting properties for resolving current questions posed by the SM and observations: they can provide a model for the dark matter of the Universe, and they take a step toward solving the "hierarchy problem", a problem linked to that of vacuum energy.

However, no sign of this phenomenon has been detected experimentally: known ordinary particles have no supersymmetric partners, or at least their masses are beyond the reach of our most powerful accelerators, particularly the LHC, that is, greater than one or more TeVs. Sending partner masses to high values means that supersymmetry must be broken, as electroweak symmetry is in the SM. As might be expected, the boson and fermion terms no longer exactly offset each other, and this symmetry breaking at the M_{Susy} mass introduces a contribution to the vacuum energy of the order of M_{Susy}^4, an order of magnitude similar to that previously found.

11.3.5. *General relativity tracks*

Numerous attempts have been made to modify general relativity and make it less sensitive to quantum vacuum fluctuations. We have neither the space nor the competence to go into detail here. Let us mention the names of massive gravity, bimetric gravity, Horndeski gravity, unimodular gravity, non-local gravity, etc. (Bull et al. 2016). Although theoretically rich, none of them has yet produced a perfect solution. Moreover, it is very difficult to modify the GR without negative consequences for accuracy tests. A theorem due to Lovelock states that GR (Einstein's equations) is the only theory that can be derived from an action that contains only the second derivatives of the metric of a four-dimensional space–time. If we remain within the framework of a "traditional" action, the only possibilities for a sensible modification are therefore additional dimensions or fields other than the metric. Even more radical are the use of higher order derivatives, non-locality, or even abandoning the assumption that the equations of motion derive from an action ("emergent" gravity).

11.3.5.1. *Additional dimensions*

In the 1920s, shortly after the invention of general relativity, Einstein himself and others attempted to unify this relativistic theory of gravitation with Maxwell's equally relativistic theory of electromagnetic interactions. Kaluza put forward the revolutionary idea that space could have an extra dimension. The extended GR to five dimensions of space–time reveals a very attractive structure: it contains the normal four-dimensional GR with Einstein's equations plus electromagnetism with Maxwell's equations (and in addition, an extra scalar field, the "radion" or "dilaton"). To account for the fact that the fifth dimension is not observable at our scale and to introduce the quantum character, Klein proposed that the extra dimension, the fifth dimension, should be wrapped around itself within a very small radius. The size of this radius is the scale at which gravitation would become quantum, that is, $(1/M_{Planck})(\hbar/c)$, often noted as $1/M_{Planck}$ in so-called natural units where $\hbar = c = 1$.

The Kaluza–Klein theory, the only way to unify gravitation and electromagnetism, ran up against reality: according to it, particles such as the electron would have masses of the order of M_{Planck}. After a certain popularity in the 1930s, it was left in abeyance until the 1980s. At this point, "superstring" theories appeared, again in an attempt to unify gravitation with other interactions. These theories find their framework in multidimensional space–time (for example, 10 or 11 dimensions), and the "Kaluza–Klein miracle" is recalled to the fore. In the original superstring theories, the scale of additional dimensions remains at the value of $1/M_{Planck}$, which is totally unobservable, and superstring theories appear as a purely mathematical edifice. However, in 1998, Arkani-Ahmed, Dimopoulos and Dvali realized that the extra dimensions need not be so small. Indeed, they could be of the order of $1/M_{SM}$, where M_{SM} represents the mass scale typical of the SM, some hundred GeVs (Arkani-Hamed et al. 2000). Bringing the scale of quantum gravity down to the scale of the SM opens a new avenue for resolving the many problems between gravity and quantum gravity. The idea has since been applied in a wide variety of ways: number of extra dimensions, presence of standard fields throughout space–time ("bulk") or on a lower dimensional variety ("brane").

We can sketch out how this configuration might help with the vacuum energy problem. Suppose, as is the case in most of these models, that gravity has access to all dimensions, for example, six, but that the SM is restricted to a four-dimensional "brane", which is our ordinary spacetime. This explains the weakness of gravity on our scale, as it is "diluted" in a much larger space. By the same token, the fact that SM fields have no access to extra dimensions "decouples" their fluctuations from most gravity, and the energy of these fluctuations is no longer directly equivalent to a cosmological constant. This hypothesis is therefore attractive, but to date, it has not been experimentally confirmed. As the scale of quantum gravity is reduced to that of the SM, which is on the order of a hundred GeVs, most of these models predict important effects observable at the LHC, for example, a series of new particles analogous to ordinary particles but with higher masses ("Kaluza-Klein towers"), missing energy-momentum in the final states of collisions (particles disappearing into extra dimensions), or even the famous creation of microscopic black holes. None of these effects have yet been reported.

11.4. Conclusion

Our descriptions of nature must necessarily admit some "quantum" aspect: observed quantities do not behave as our senses indicate, like classical objects. Quantum theory has been able to account for this "reality" and apply it in many fields, such as atoms, electronics and chemistry. Quantum field theory has pushed the quantum limit even further to the point where the number and identity of elementary particles are no longer immutable but can be exchanged or multiplied in the course of interactions. The extraordinary flexibility of this theoretical ensemble has led to numerous successes in understanding and describing a wide variety of phenomena, from superconductivity to the high energies of the latest accelerators. However, it has also opened up a Pandora box of theoretical problems that remain unsolved today. The "vacuum" state, which should be the simplest, is also the least constrained state, and its description, its interactions with the material world and other theories, are at the heart of these problems. Its interaction with the other great theory, general relativity, poses a problem quantitatively, with a rarely observed incoherence between two different descriptions of nature. In this particular case, the problem can be circumscribed to a precise question: why does the vacuum gravitate normally when it has other interactions with matter, for example, in atoms or nuclei, and gravitate only very weakly when it is subject only to gravity, as this is the case at the scale of the cosmos?

It is tempting to think that this problem is just one of the consequences of our lack of a quantum theory of gravitation. Countless attempts to formulate such a theory have been proposed and are still being explored. However, it is not certain that all current problems – the hierarchy problem (the difference of several orders of magnitude between the energy scales of the weak nuclear interaction and Planck's), the nature of dark matter, matter–antimatter asymmetry, etc. – can be solved by a single theory. This is why creative and ambitious experimental programs are underway or planned to address the problem from different angles: new-generation accelerators, observations of the distant Universe, detection of gravitational waves in space, tests of the equivalence principle, antimatter experiments, precision measurements in the laboratory, etc.

The energy of the vacuum, now more than ever, is a fundamental problem and a source of inspiration for physicists.

11.5. References

Abbott, T.M.C., Allam, S., Andersen, P., Angus, C., Asorey, J., Avelino, A., Avila, S., Bassett, B.A., Bechtol, K., Bernstein, G.M. et al. (2019). First cosmology results using type Ia supernovae from the dark energy survey: Constraints on cosmological parameters. *arXiv*. doi: 10.48550/arXiv.1811.02374.

Anderson, E.K., Baker, C.J., Bertsche, W., Bhatt, N.M., Bonomi, G., Capra, A., Carli, I., Cesar, C.L., Charlton, M., Christensen, A. et al. (2023). Observation of the effect of gravity on the motion of antimatter. *Nature*, 621, 716–722. doi: 10.1038/s41586-023-06527-1.

Arkani-Hamed, N., Dimopoulos, S., Kaloper, N., Sundrum, R. (2000). A small cosmological constant from a large extra dimension. *Physics Letter B*, 480, 193.

Bull, P., Akrami, Y., Adamek, J., Baker, T., Bellini, E., Jiménez, J.B., Bentivegna, E., Camera, S., Clesse, S., Davis, J.H. et al. (2016). Beyond ΛCDM: Problems, solutions, and the road ahead. *Physics of the Dark Universe*, 12, 56–99.

Jaffe, R.L. (2005). Casimir effect and the quantum vacuum. *Physical Review D*, 72, 021301.

Linde, A. and Vanchurin, V. (2010). How many universes are in the multiverse? *Physical Review*, D81, 083525.

Martin, J. (2012). Everything you always wanted to know about the cosmological constant problem (but were afraid to ask). *arXiv*. doi: 10.1016/j.crhy.2012.04.008.

Touboul, P., Métris, G., Rodrigues, M., Bergé, J., Robert, A., Baghi, Q., André, Y., Bedouet, J., Boulanger, D., Bremer, S. et al. (2022). MICROSCOPE mission: Final results of the test of the equivalence principle. *Physical Review Letters*, 129, 121102.

12

Temporality in Relativistic Space–time

Marc Lachièze-Rey[1,2]
[1]*CNRS, Astroparticle and Cosmology Laboratory (APC), Paris, France*
[2]*SPHERE, Paris, France*

12.1. Relativistic temporal effects

The birth of Einstein's relativistic theories stemmed from the contradiction between the existence of time and the constancy of the speed of light: confirmation of the latter implied that time could not exist in physics. Einstein's first merit was to account for this disappearance and to propose a formulation of physics devoid of the notion of time, in favor of that of space–time: special relativity.

Although this nonexistence of time is not always taken seriously to this day, it is not, as some think, a kind of philosophical abstraction without concrete influence. In contrast, it manifests itself (among other things) in very concrete relativistic temporal effects (RTEs). This is the name that will be given to any measurement result obtained by clocks that differs from what Newtonian physics would have predicted, while still being consistent with relativistic theories. These effects are the subject of very precise predictions by relativistic theories, and these predictions have now been fully confirmed observationally and/or experimentally, many times over and with excellent precision. They are very well evaluated, both on Earth and in near and far space, in astrophysics and cosmology. Since this theory was confirmed, they have been used as tools for exploring the cosmos in astrophysics and cosmology, as well as on Earth for tracking, altimetry and prospecting, geophysics, etc.

Quantum Physics and Cosmology,
coordinated by Hervé Zwirn.

We are obliged to describe these effects as "temporal", even though they mark the disappearance of time: a paradoxical and unfortunate situation that underlines the extent to which our language remains immersed in the pre-relativistic conception of time. I repeat that not only do they have nothing to do with time or "a time", but they mark its very disappearance!

Some of these effects are quantitative. This is the case for *time shifts* (which include *spectral shifts*). They resemble the usual (nonrelativistic) Doppler effect but are given by different formulas. *Time delays* are also quantitative, but they are also qualitative. This makes them even more spectacular, as they have no equivalent in nonrelativistic physics. Shifts and delays are direct manifestations of the geometric properties of space–time. To explain their nature, we need to take a brief look at the notion of time as it operates in Newtonian physics.

12.2. From time to space–time

12.2.1. *Aspects of Newtonian time*

To understand why the notion of time cannot be retained in the real world, the first step is to analyze its introduction into physics, or rather its construction by Newton. There is no doubt that it was a fruitful idea, as demonstrated by the success of his theory well into the 20th century. This introduction corresponds to a kind of natural "ansatz", far from being a logical necessity.

The starting point is the personal impression that "time passes" and the evidence that this impression is shared by everyone. This led to the hypothesis that this personal impression could be interpreted as the manifestation of a common universal process: it would not, in fact, be so personal but rather reflect a universal phenomenon that everyone calls the "passing of time". Although plausible and clear, this interpretation is not a logical obligation, and would appear to be largely contradicted by observations. Nevertheless, it was a pillar of the Newtonian conception of the physical world.

I defend the idea[1] that time is not a "primitive" notion but a composite construction. It synthesizes several basic notions, such as *duration*, *dating* and *chronology*, and *causality*. In Newton's Day, this construction was justified by the intimate relationships between the notions mentioned, which made the synthesis

1 Lachièze-Rey, M. (2013). *Voyager dans le temps : la physique moderne et la temporalité*. Editions du Seuil, Paris.

natural and harmonious. Recent experiments and observations have since shown that this task was impossible.

There is no such thing as "measuring time". A clock is a physical device that measures *durations* (a chronometer).[2] Our personal feelings, our experiences, our measurements are always about *durations*. Every history has its own duration! There's no doubt about it, but the Newtonian Ansatz would assume that the indication of our clock *here* concerns the *whole Universe*: that its reading allows me, for example, to conclude that it is 12 o'clock on the planet Mars or in the most distant galaxy. *Duration*, on the other hand, is a "primitive" notion. In addition, it remains perfectly operational and measurable in space–time, as we shall see.

The second notion, *dating*, does not concern *histories* but rather *events* in the Universe: it assigns a unique number to each one, called its "date". It is consubstantial with the notion of time. By comparing dates, dating immediately and unambiguously define *chronology* (including *simultaneity*). The third primary notion of interest is that of *causality*, which Newtonian physics identifies with chronology.

These three primitive notions form the basis of the construction of Newtonian time. They are assumed to be harmoniously linked in the physical world: durations are obtained from dating, defined as the differences between the beginning and end dates of a history; causality is identified with chronology. Details of Newton's ansatz and its failure can be found in the book *Voyager dans le temps* by Lachièze-Rey (op. cit.).

I will focus here on a few RTEs and their analysis.

Einstein emphasized that it is impossible to define physical simultaneity if the speed of light is constant. No simultaneity means no chronology, no dating. In addition, no notion of time to bring them together. This was his motivation for developing a physical theory in which time is absent: *special relativity*. His greatest achievement was to understand this logical impossibility and the consequent disappearance of time, even without any of the numerous experimental or observational results that followed.

Durations, or *proper durations*, do exist in the real world, and they play an absolutely fundamental role. However, neither chronology nor simultaneity are relevant to physics. This in no way prevents us from *constructing* many different "artificial" chronologies and calling them "time-something". They are arbitrary and incompatible with each other, but we adopt one or the other for reasons of social convenience or other motivations. They are not related to measures of duration.

2 idem.

can be *kinematic* or *gravitational*, and the two aspects are often intertwined. *Kinematic* RTEs are predicted and described by both special and general relativity, and *gravitational* RTEs are predicted by general relativity only since special relativity does not consider gravitation.

For many, RTEs remain a mystery. Some speak of an "acceleration" or "slowing down of time" of "contractions or dilations of durations". This language was admissible in Einstein's day, when physics was still bathed in Newtonian concepts and it was difficult to talk about temporal effects without referring to time (even here, I am still obliged to use the expression "temporal effect"!). In fact, it makes no sense at all. In addition, there is no paradox, no mystery, no strangeness if we interpret things correctly in space–time: a shift or delay always concerns a situation that involves *two different histories* and merely expresses the difference between the proper durations of these two histories (whereas Newtonian physics would imply equality between these durations). There is nothing more mysterious.

The common consequence of these effects is that two clocks never give the same reading (except in the very special case where they permanently occupy the same spatial position). This prediction ("clock hypothesis") was certainly surprising, even shocking. This was impossible to verify on Einstein's day, given the limited precision of the instruments available at the time. Today, this is the reason why "official" clocks – those that are supposed to indicate *universal time* (see below) – have to be corrected on a regular basis. The impossibility of any lasting synchronization is well known to all "time" specialists and is at the root of much of their work.

We have long been in the stage where these effects had yet to be proven. They manifest themselves daily in a wide variety of phenomena. Desynchronization between clocks, which has itself been very well measured, now serves as the basis for a variety of practical engineering devices.[5]

12.3. Time shifts

A shift is defined as the (relative) difference between the duration of an observation of a history and the duration of the history itself.

Every history has its own perfectly defined duration, but it can be measured only by an observer (a clock) who takes part in this history and who *travels* through it.

5 Lachièze-Rey, M. (2022). La phénoménotechnique [bachelardienne] du temps. In *Actes du Colloque Phénoménotechnique du temps*, Besançon.

No one else can measure it. I can, however, *observe* a history other than my own. However, it is essential to understand that *my observation of history* is another history than the *history being observed*.

My friend Hélène, for example, travels by plane from Paris to Tokyo. Her history begins with take-off (D) and ends with landing (A). Only Hélène can know (by reading her own on-board clock) the proper duration d(h) of her flight, a history between D and A.

I cannot take the same measurement because neither I nor my clock are involved in history h. However, nothing stops me from *observing this history*, for example, by receiving a radio or TV signal that Hélène sends me continuously. My observation of history h, which is my reception of this signal, constitutes another history h'. It begins at my observation D' of event D and ends at my observation A' of A. This other history h' differs from h. It is lived by me, not by Hélène. I can measure its own duration d(h') with my clock. It differs from d(h). This defines our *time shift* $z = d(h')/d(h) - 1$.

The time shift therefore measures the difference between the proper duration of a history and the proper duration of the observation of that history. If the clock (belonging to Hélène) sends me a beep (a signal) every second, I receive a beep every 1+z s.

12.3.1. *Spectral shifts*

A history often corresponds to the occurrence of an oscillation of a periodic process (e.g. light or electromagnetic radiation emitted by the source). Its duration is exactly what we call the *period*. Applied to such a history, the time shift is called the "spectral shift": a special case of the time shift. It is defined as $z = (T_r - T_e)/T_e$, where the *emission* period T_e is measured by the emitting clock (that of the source), and the *received* period T_r is measured by the clock of the observer receiving the signal.

The notion is relatively familiar since it manifests itself in classical (nonrelativistic) physics in the form of the usual *Doppler effect*. The ambulance siren is often used to describe it: as it moves away (respectively, toward us), we perceive its sound signal with a period (i.e. the inverse of its frequency, which itself is identified with the "pitch" of the sound) greater (respectively, *less*) than that of the transmitted signal. In this example, the shift (period received/period emitted) is given by the *nonrelativistic* formula $z_{nr} = v/v_{sound} - 1$, which involves the speed v of the ambulance and that of the sound v_{sound}. The formula for real-time shifts is quite different.

A shift can be kinematic or gravitational in origin. In both cases, it results directly from the geometry of space–time.

12.3.2. *Kinematic shift*

This is caused by relative motion. This is expressed by a relative velocity V (which I always express in units of the speed of light c[6]) between the source and observer, which is positive or negative depending on whether they are moving toward or away from each other. The shift is expressed by the formula 1 + z = sqrt((1+ V)/(1 − V)). For low speeds (on Earth, speeds rarely exceed the speed of sound, i.e., 10^{-6}), they develop approximately according to $z = V + V^2/2 + \dots$ The first term is linear and corresponds to the *nonrelativistic* Doppler shift formula. The quadratic terms, and those that follow, are generally negligible, except at high speeds. However, the difference is crucial because the contribution of the quadratic term does not cancel out on a round trip, which is the origin of the *time-delay* phenomenon (described below) that is nonexistent in Newtonian physics. The difference can also be seen in the fact that the tangential (and not just radial, as in the Doppler effect) component of the relative velocity is also involved.

The space–time formulation suggests the use of a more fundamental and natural (as well as more convenient) geometric notion than that of relative speed: *Celerity* θ, which geometrically is identified with the *hyperbolic angle* between world lines. It advantageously replaces velocity, which is expressed by the *hyperbolic tangent* function $V = \tanh \theta$. Note that neither the angle θ nor the speed (always expressed in units of c) has any dimension here.

Hyperbolic trigonometry in space–time indicates that when the *Celerity* θ varies from minus infinity to plus infinity, the velocity v varies between −1 and 1. This shows that what we consider to be the limiting velocity (i.e. unit 1 in the correct units) is merely an artifact resulting from the use of an ill-adapted quantity, *velocity* rather than *Celerity*, which does vary without limit. The kinematic shift is written as $1 + z = \exp \theta = 1 + \theta + \theta^2/2 + \dots$ (which can be developed as $z = v + v^2/2 + \dots$). Notably, unlike speeds, Celerities are simply added together.

Box 12.1. *Hyperbolic geometry*

6 This is equivalent to measuring lengths in seconds (generally speaking, light-seconds), which corresponds (apart from one constant) to the legal definition of the meter. To restore more familiar formulas, simply replace V everywhere with V/c.

The high velocities of particles in accelerators and cosmic rays generate intense kinematic shifts. However, the relativistic, quadratic component is generally masked by the nonrelativistic component. It has only recently been detected in the vicinity of a black hole, as described below. The existence and properties of these time shifts (notably spectral shifts) have been widely confirmed and measured.

12.3.3. *Gravitational shifts*

Known as the *Einstein effect*, a gravitational shift is generated by the *curvature* of space–time. It has no counterpart in Newtonian physics. Under conditions of weak gravitation (e.g. on Earth or in the Solar System), its effect is conveniently described using the *gravitational potential* ϕ (expressed in units of c^2): the shift is then simply $z = \delta\phi$, which is the difference between the values of the potential at the source and observer positions.

In the Earth's gravitational potential, for example, this is manifested between clocks at different altitudes: for a difference in altitude δh, general relativity predicts $z = g\ \delta h$ (where g is the acceleration of the Earth's gravity), or approximately 10^{-16} per meter: a clock at higher altitude *seems to beat* faster than one at lower altitude (remember that a clock always beats at the same rate whatever happens). Remember also that the precision of today's atomic clocks reaches approximately 10^{-18}.

12.3.4. *Pound and Rebka's experiment*

The first measurement of gravitational shift (also known as the Einstein effect) was carried out in 1959 on the tower of Harvard University by American physicists Robert Vivian Pound (1919–2010) and his student Glen Anderson Rebka, Jr. (1931–2015). This *Harvard Tower Test*, considered as the third classical test of general relativity, exploits the difference in the Earth's gravitational potential over the 22.6 m separating the receiver (on the ground) from the top of the tower.

The measured shift value (a few 10^{-15}) was well within the predictions of general relativity. The accuracy was improved in 1976 by the American *Gravity Probe A* (also known as the Vessot and Levine experiment) and even further improved in 2014 by a Soviet rocket. Today, the precision of atomic clocks (10^{-18} for an optical clock) means that the Einstein effect can be measured over differences in altitude as small as 1 cm, with an accuracy of 10^{-5}.

The obvious result is that two clocks, even terrestrial clocks, cannot remain synchronized, especially if they do not reside at the same altitude: the shift reaches three microseconds per year for an altitude difference of 1 km. This direct manifestation of the clock hypothesis makes it impossible to support the idea that these clocks measure the same quantity, a "time", for example.

12.3.5. *At the Tokyo Tower*

In 2020, a group from the University of Tokyo connected optical clocks via fiber optics to measure the gravitational shift corresponding to 450 m of the Tokyo Tower, that is, $z = 5 \times 10^{-14}$. The main aim was to demonstrate the sensitivity, accuracy and robustness of these measurements. As a result, these time-effect measures are beginning to be used as practical tools, not only for *altimetry* but also for *gravimetry*.[7]

Gravimetry uses the difference between the indications of two neighboring clocks to measure a difference in space–time curvature (gravitational potential, if you prefer), which reveals the presence (or absence) of massive bodies: buried masses, deposits, or the occurrence of volcanic or seismic activity, with the potential to predict all of these effects.[8]

If it proves too complicated to link two clocks by an optical fiber, the recently launched ACES (Atomic Clock Ensemble in Space) system can be used. This *European Space Agency* (ESA) mission involves installing a set of atomic clocks on the International Space Station. Comparing their indications with those of clocks on the ground will provide the difference between gravitational potentials, enabling us, for example, to determine the Earth's geoid (relativistic geodesy) with a sensitivity of a few centimeters in altitude.[9]

12.3.6. *Astrophysical gravitational shifts*

Gravitational shifts are now well known in astrophysics. They are all the more intense when they concern compact objects.

---

7 See Lachièze-Rey, M. (2022). La phénoménotechnique [bachelardienne] du temps. In *Actes du Colloque. Phénoménotechnique du temps*, Besançon; and Lachièze-Rey, M. (2023). *Gravitation*. Flammarion, Paris.

8 For more details, see *Gravitation* (Lachièze-Rey, op. cit); and Roura, A., Schubert, C., Schlippert, D., Rasel, E.M. (2020). Measuring gravitational time dilation with delocalized quantum superpositions. *arXiv:2010.11156*.

9 See: https://www.researchgate.net/publication/340467276_Test_of_general_relativity_by_a_pair_of_transportable_optical_lattice_clocks.

As early as 1908, Einstein predicted that a gravitational shift on the order of 2×10^{-6} would be generated by the difference in gravitational potential between the surface of the Sun and that of our planet. The most recent and precise confirmation of this prediction of general relativity is accurate to within 1%.[10]

A white dwarf is a star in a late stage of evolution. It is not very massive (less than one solar mass) and is very small in size. Therefore, it is very compact. A first attempt to measure gravitational shift was made in 1925: from the surface of Sirius B, the invisible companion of the familiar star Sirius, which was recognized as a white dwarf in 1915. Unfortunately, the signal proved unusable, but it was later measured by the HST at a value close to 2×10^{-4}. Other white dwarf shifts (of the same order) were subsequently measured.

Neutron stars are the most compact objects known (apart from black holes). They must therefore generate a significant Einstein effect. On the other hand, a pulsar, a fast-rotating neutron star, emits very regular radiation, making it an excellent natural clock. The situation is ideal for highlighting RTEs. A number of such shifts have been measured since the beginning of the century, with varying degrees of precision. The case of a binary system involving a pulsar and another compact star in mutual orbit is exemplary: the signal emitted by the pulsar ("the clock") passes close to its compact companion before reaching us. This causes a delay in the arrival of the pulsar's signal of up to 10 μs. Recent results[11] (detailed in *Gravitation*, Lachièze-Rey op. cit.) represent a fine example of gravitational astronomy, where pure (relativistic) gravitation provides information on the behavior of matter.

12.4. Time delays

Time delays are defined as the difference between the durations of two different stories but two particular stories that share the same beginning and the same end (it is sometimes improperly described as the *contraction of durations*, the *dilation of time*, etc.). If duration could be calculated as a difference between temporal dates, as

---

10 González Hernández, J.I., Rebolo, R., Pasquini, L., Pasquini, L., Lo Curto, G., Molaro, P., Caffau, E., Ludwig, H.-G., Steffen, M., Esposito, M., Suárez Mascareño, A. et al. (2020). The solar gravitational redshift from HARPS-LFC Moon spectra. A test of the General Theory of Relativity. *Astronomy & Astrophysics (A&A)*, 643, A146.

11 Stairs, I.H., Arzoumanian, Z., Camilo, F., Lyne, A.G., Nice, D.J., Taylor, J.H., Thorsett, S.E., Wolszczan, A. (1998). Measurement of relativistic orbital decay in the PSR B1534+12 binary system. *The Astrophysical Journal*, 505(1), 352–357. doi: 10.1086/306151; Weisberg, J.M. and Huang, Y. (2016). Relativistic measurements from timing the binary pulsar PSR B1913+16. *The Astrophysical Journal*, 829(1), 55. doi: 10.3847/0004-637X/829/1/55.

Newtonian physics assumes, this difference would be zero. This is why observations of delays demonstrate very directly that durations cannot be a manifestation of time, regardless of the definition of the latter. Of course, there is no such thing as a temporal delay in Newtonian physics.

In the first decades of the 20th century, the reality of delays was difficult for the community to accept. It was popularized by the story of the *Langevin twins.*[12] One of them sets off into space, whereas his brother remains on Earth. Their separation constitutes the common beginning of their different histories. Then, their reunion on Earth constitutes the common end of both histories, after one has traveled, but not the other. If time existed (as it does in Newtonian physics), both histories would obviously have the same duration, equal to the lapse of time between the end and the beginning. In such a case, the (negative) kinematic shifts during the return journey compensate for those (positive) during the outward journey.

This is not the case in the real world: kinematic shifts in approach do not compensate for shifts during the departure, as they would in the Newtonian case. *Linear* components (the only ones present in the Newtonian case) do compensate for each other but not *nonlinear* contributions, which are always positive in both directions.

Much debate has focused on this story, which was obviously unfeasible in Einstein's day and even in the decades that followed. The effect has since been perfectly verified in principle, as described below. The closest thing to it is probably the few microseconds "lost" by a space station astronaut, compared with his terrestrial peers.

12.4.1. *Hafele–Keating experiment: shifts and delays*

The Langevin twins' situation was reproduced for the first time in an experiment carried out in 1971 by physicist Joseph C. Hafele and astronomer Richard E. Keating. The twins were replaced by atomic clocks aboard airliners (altitude of approximately 9 km). The first of these flew west around the world. Its speed of 830 km/h more or less compensates for the Earth's rotation, bringing it close to rest in relation to the fixed stars. A second describes a similar journey eastward, giving it twice the speed of the Earth's rotation. The third remains stationary in relation to the *US Naval Observatory*.

12 The latter is most often reported in the purely kinematic case, but as with shifts, it can also have a gravitational origin. See, for example, During, É. (2014). Langevin ou le paradoxe introuvable. *Revue de métaphysique et de morale*, 84, 513–527.

The three clocks are perfectly synchronized at embarkation, which is the common initial event. They describe three different histories, and their reunion constitutes the final common event F. Once reunited, the clocks appear to be out of sync. The observed delays are on the order of 100 ns, in perfect agreement with relativistic predictions at the measurement accuracy. The situation combines kinematic and gravitational effects.

A variant of the experiment with reduced speeds enabled us to focus on the *gravitational* delay, which appears to be approximately 50 ns. By sending periodic laser pulses, which were then received on the ground after reflection by the aircraft, the authors were able to test the shifts during flight, confirming that their noncompensation was at the origin of the delays.[13]

12.4.2. *Shifts, delays and GPS*

The most familiar manifestation of RTEs undoubtedly affects the *global positioning system* (GPS): the 24 satellites in the NAVSTAR constellation describe circular orbits (12-hour period, altitude of 20 200 km). Each carries a cesium atomic clock and sends out radio signals incorporating information about its position.

Their velocities ($1.3 \times 10^{-5}\, c$) give them relativistic kinematic shifts. The nonlinear kinetic component, on the order of 8.3×10^{-11}, adds up to delays of 7 μs/day. Those combine with the 45 μs/day gravitational delay, resulting in approximately 38 μs/day. As a result, "forgetting" to take these "relativistic effects" into account (i.e. the assumption that time elapses both on Earth and in satellites) would lead to a positioning error of approximately 15 km/day (a difference of one microsecond corresponds to 300 m in position).

12.4.3. *Close to a black hole*

The most extreme manifestations of these effects are expected close to black holes. The 2020 Nobel Prize in Physics was awarded for observations of stars orbiting *SagX**, which is the central black hole in our galaxy. Their velocities, reaching hundreds of a *c*, generated intense kinematic effects. For the first time, the

13 With relative velocities approximately 10^{-6}, the quadratic component is approximately 10^{-12}. Given the duration of the flights (approximately 40 hours), this corresponds to the measured values.

nonlinear (i.e. purely relativistic) component of the kinematic shift and the gravitational shift could be demonstrated separately. For the So-2 star, one of the closest to the black hole, 100 astronomical units from its horizon, with a velocity of 0.03 *c*.

12.4.4. *The Shapiro delay*

I will conclude by mentioning a more original effect of space–time curvature: the *Shapiro delay* (sometimes referred to as the *fourth test of general relativity*[14]). This concerns a light ray emitted by an astronomical source (a planet), which propagates to the terrestrial observer by grazing a massive object such as the Sun, in other words, a *gravitational lensing* situation. The local space–time curvature imparted by the latter delays the arrival of the signal at the observer compared with what it would be in the absence of the massive object.

A radar signal emitted from the Earth and reflected by a planet can pass close to the Sun. It then experiences a delay of approximately 200 μs (nonexistent in Newtonian physics). The effect was confirmed as early as the 1960s, and its accuracy improved a few decades later with the Viking and Cassini probes.[15]

Since this is a deformation of space–time, the Shapiro effect must apply to neutrinos as well as photons. Hence, during the explosion of supernova SN1987A, the simultaneous arrival of photons and emitted neutrinos indicated that they had indeed undergone the same gravitational delay, estimated at between one and six months. This finding further confirmed the *equivalence principle* of general relativity.

The same applies to gravitational waves. Those of different frequencies from source GW150914 arrived simultaneously: they underwent the same Shapiro delay caused by the mass distribution along the line of sight. These delays are estimated at approximately 1,800 days on a journey of approximately one billion years. The measurements indicate that they are identical within 0.2 s, confirming the principle of equivalence.

14 Shapiro, I.I. (1964). Fourth test of general relativity. *Physical Review Letters*, 13(26), 789–791. doi: 10.1103/PhysRevLett.13.789.

15 Bertotti, B., Iess, L., Tortora, P. (2003). A test of general relativity using radio links with the Cassini spacecraft. *Nature*, 425, 374–376.

12.5. What to do with time

12.5.1. *Time and the speed of time*

On Earth, velocities (always estimated in units of *c*) rarely exceed those of sound, namely, approximately 10^{-6} (this is also the order of magnitude of the Earth's speed of revolution on the surface at the equator). Shifts of this order are common, with a quadratic component on the order of 10^{-12}. At the same time, the points on our planet's surface are at different altitudes, and they are driven at approximately 10^{-4} by orbital motion around the Sun. All of these factors generate shifts and delays, making it impossible to synchronize clocks.

All that remains is to construct an operational notion that corresponds to what we want to call "time". This can be done in a variety of ways, depending on the intended use. Cosmic time, sidereal time, universal time, international time, ephemeris time, etc., are all incompatible with each other. None of them has any real relevance in the physical world, but each one may prove more convenient depending on the circumstances. Each definition is marked by the choice of a collection of clocks entrusted with indicating said "time" approximately. There are two opposing options: at one extreme, "cosmic time" corresponds, for example, to the desire to define a suitable (albeit fictitious) clock at *every point in* space–time, albeit with very limited precision; at the other extreme, the construction of *universal time* is based on a limited collection of clocks (a few hundred), chosen for a particular (socially motivated) local use only, and requires greater precision (albeit much less than that of the best clocks themselves).

The idea of "cosmic time" was first developed by the German mathematician and physicist Hermann Weyl in the early 20th century. The construction procedure[16] begins with the choice of a "congruence of observers" (the term here is synonymous with "clocks") in free fall. In other words, the choice of a clock at each point in space–time: at this point, the value of "cosmic time" is defined as the indication of this fictitious clock. Weyl proposes methods for choosing these clocks within the framework of a cosmological model. However, these are purely fictitious, and we would not be able to build one, not in our own environment, nor in any precise location in space–time. Cosmic time is by no means a measurable quantity (see *L'Âge de l'Univers* by Lachièze-Rey (op. cit.)). It can be estimated only if we are fully aware of all of the characteristics of the cosmological model that describes our Universe. Thus, the value we give for the age of the Universe depends, for example (with an uncertainty of approximately 10%), on how the current controversy over

16 See, for example, Lachièze-Rey, M. (2021). *L'âge de l'univers*. Éditions humenSciences, Paris.

the Hubble–Lemaître constant will be resolved and would probably have to be modified even further if, for example, it were discovered that "dark energy" really does exist (see *L'Âge de l'Univers* by Lachièze-Rey (op. cit.)). The same applies to the value of cosmic time attributed to a given distant galaxy (at the moment we observe it). This value is generally calculated from its (*supposedly* cosmological) *redshift*, assuming the standard Universe model.

Cosmic time, a fictitious, approximate and relatively arbitrary reconstruction, has no real physical meaning. Nevertheless, it has the merit of providing a reference for cosmological events. However, it remains unmeasurable and unrelated to the durations. The duration of any astrophysical processes (a measurable quantity) is not equal to the difference between the values of cosmic time at the beginning and at the end.

For our social life, it is better to construct something slightly more precise, at least on the scale of our planet, even if it means losing "universality" (which is probably why it is called universal!). The *universal time* (UT) is defined as a quantity whose value does not differ much from what a terrestrial clock indicates. Here again, the choice is vast, and there are many options to choose from. The one most commonly used (also called *International Atomic Time* (TAI), which we may call "legal" time) is defined as a sort of average (which is far from being a simple process) of the indications of some 500 atomic clocks spread around the globe. The precision of each is far better than that of the definition of "time" itself. These clocks are chosen so that the duration of a "TAI second" coincides as closely as possible with the physical second, which is locally measured by each clock in the network. However, this requires regular "corrections" by inserting leap seconds.

Notably, no clock (including those in the chosen collection) gives, or can give, the value of any of these "times". The *Bureau international des poids et mesures* (BIPM) is responsible for the complex management of the corrections needed. Several books are needed to detail the procedures involved.

To outline this idea, let us imagine a simplified situation. The construction of a "Parisian" time t_P, which would be defined as the indication of a single clock H_P located in Paris, assumed to be perfect: Parisian time would be defined as the proper time of the reference clock H_P.

Of course, this "Parisian time" can be read only on the spot. No other clock indicates its value (except for fortuitous, fleeting and unforeseeable coincidence). On the other hand, optical fibers or electromagnetic waves can perfectly transmit signals emitted by the H_P clock: beeps emitted every second of this Parisian time t_P.

Wherever I am (say, in Rome), I can receive these beeps. My clock gives me the proper time between the reception of successive bits: 1+z seconds, as indicated above. The time shift, a cause of this lack of synchronization, may come from the altitude difference from Paris in terms of the Earth's gravitational potential or from the relative speed between the two capitals due to their different positions on the globe.[17]

The effects we are talking about are generally beyond 10^{-12}, and are quite negligible in terms of the precision of our social life! But they are not to the precision of modern clocks, which can reach 10^{-18}. These require constant corrections of reference clocks.

12.5.2. *The speed of time*

The "true" duration of a transmitted "second of Parisian time" is 1 + z *physical* seconds, measured by the clock. Hence, it is tempting to call 1/(1 + z) the "speed of time" (the time I have chosen, here t_P) at the position I am occupying.[18]

No clock indicates the manufactured time *t*, whatever it may be (except for the unique reference clock). However, each clock indicates a real physical quantity, the *duration of its own history*, which is unrelated to time. However, the reception of the beeps emitted by the reference clock tells me the *speed of time t* at my position, which makes it a perfectly measurable quantity.[19] This speed of time is manifested by the FPRs described above. Its measurement is directly used for altimetry or

17 The Earth's rotation on itself and its revolution around the Sun, impart a relative speed to two points on the Earth's surface that differ in latitude and longitude. An observer at the equator, for example, moves at a speed of approximately 10^{-6} *c* relative to another observer who is diametrically opposite. The *linear* component of their relative displacement is of the same order but offsets itself over time. The *nonlinear* component, on the other hand, is of order 10^{-12} and generates an accumulating desynchronization that is incompatible with the precision of atomic clocks.

18 As the quantity 1 + z is composed by multiplication, it would be more convenient to define the "speed of time" as its logarithm. In the case of kinematic shifts, this coincides with the quantity we call "Celerity". For not so rapid movements, it coincides with ordinary speed.

19 See, for example, Lachièze-Rey, M. (2023). Time and durations in relativistic physics. In *Time and Science: Physical Sciences and Cosmology*, volume 3, Lestienne, R and Harris, P.A. (eds). World Scientific Publishing, Singapore. doi: 10.1142/q0405-vol3.

gravimetry[20]. For more details, see, for example, Lachièze-Rey's chapter on "Time and durations in relativistic physics"[21] or *Gravitation* (Lachièze-Rey op. cit.).

12.6. Conclusion

All of these effects directly demonstrate the relevance of the space–time framework for describing the real world. They are perfectly consistent with our relativistic theories. Time is not a physical quantity, but this does not prevent us from *constructing* notions of "time-this" or "time-that" that are operational for one purpose or another.

The best comparison that comes to mind is that of the *monetary value* of an object. It varies from one day to the next, from one place to another, depending on needs, abundance, the facilitator involved, the commission of a possible intermediary, etc. There is no a precisely defined monetary value for an object as much as there is time in physics.

The idea of a physical time that elapses in our Universe, or even on the scale of our planet (or even, for a precision physicist, on the scale of his laboratory), has become obsolete. This disintegration is no longer an incidental curiosity of relativistic theory. It is a fact of our concrete world is exploited by today's technologies. It should not remain without consequences for metaphysical or philosophical reflections and questions about the nature of the physical world. The "illusion of time", even if it works perfectly well in everyday life, is largely out of date for reflections on the constitution of the real world, and it seems to me that philosophical reflection should take more account of it.

20 See Lachièze-Rey, M. (2022). La phénoménotechnique [bachelardienne] du temps. In *Actes du Colloque Phénoménotechnique du temps*, Besançon.

21 Lachièze-Rey, M. (2023). Time and durations in relativistic physics. In *Time and Science: Physical Sciences and Cosmology*, volume 3, Lestienne, R and Harris, P.A. (eds). World Scientific Publishing, Singapore. doi: 10.1142/q0405-vol3.

13

A History of Thought Experiments in Quantum Gravity

Thiago HARTZ[1] and Olival FREIRE JUNIOR[2]
[1]*Instituto de Matemática, Universidade Federal do Rio de Janeiro, Brazil*
[2]*Instituto de Física, Universidade Federal da Bahia, Salvador, Brazil*

13.1. Introduction

The development of quantum theory was led by several remarkable experiments, ranging from Rubens' measurement of the spectral energy density of black body radiation to Clauser's, Aspect's, and Zeilinger's experiments, which demonstrated the violation of Bell's inequalities. Several of these experiments were initially conceived as thought experiments, that is, hypothetical situations designed to test the consistency of the quantum mechanical formalism – with no expectation of ever being realized in a laboratory. While some of these thought experiments eventually became real experiments, others remained purely theoretical, such as Einstein's box from the 1930 Solvay conference and Heisenberg's gamma ray microscope (from 1927).

The need for consistency tests arose from the rather counterintuitive aspects of quantum mechanical formalism, which challenged physicists' intuitions. The situation was even more dramatic in certain quantum systems where no real experiments were expected to be feasible in the foreseeable future. Thus, when physicists began to introduce the quantum of action into general relativity, leading to the development of what would later be called quantum gravity theories, the need for thought experiments became paramount.

Quantum Physics and Cosmology,
coordinated by Hervé ZWIRN.

In this chapter, we will discuss the history of thought experiments in the development of quantum theories of the gravitational field. We will show that the role of thought experiments in quantum gravity has changed significantly over the past 90 years. Initially conceived to test whether the situation in quantum gravity was analogous to quantum electrodynamics, these thought experiments later served both to support and to challenge the need to quantize the gravitational field. We will contextualize these efforts within the changing attitudes toward the foundations of quantum mechanics in the second half of the 20th century.

Quite surprisingly, several research groups have suggested the possibility that quantum measurements on massive quantum systems may become real lab experiments and will soon possibly shed light on the quantum nature of the gravitational field. Therefore, this chapter also hopes to provide some historical perspective on those endeavors.

The chapter is structured as follows. In section 13.2, we review the role of thought experiments in physics prior to the advent of quantum mechanics. In section 13.3, we discuss notable experiments in quantum theory. In section 13.4, we discuss how thought experiments in quantum mechanics became real experiments. In section 13.5, we outline the main approaches physicists have taken in the search for a quantum theory of the gravitational field. In section 13.6, we examine the experiments proposed to test the quantum nature of gravity. In section 13.7, we briefly describe the potential of real experiments in quantum gravity and explain their connection to the thought experiments discussed in the previous section. Finally, in section 13.8, we present our conclusions.

13.2. Thoughts experiments before quantum theory

The expression "thought experiment" was coined in 1811 by the Danish chemist and physicist Hans Christian Ørsted. However, as Stuart et al. (2018) noted, it was not until the 1980s that philosophers began to pay more systematic attention to the concept. In the everyday practice of science and philosophy, however, thought experiments have long been widespread, so it is difficult to pinpoint their origin.

Thought experiments seem to be as old as rational thought itself, as paradoxes among the Greeks (such as Zeno's paradox and Anaximander's demonstration of the stability of the Earth, for instance) and medieval criticism toward Aristotelian physics may illustrate. They continued to be used in the transition from natural philosophy to current physics from the 16th century on ward. Historians have long debated whether Galileo Galilei indeed performed real experiments with the inclined plane. Alexander Koyré famously claimed that Galileo's work was possible only because of his use of thought experiments (Koyré 1966)[1].

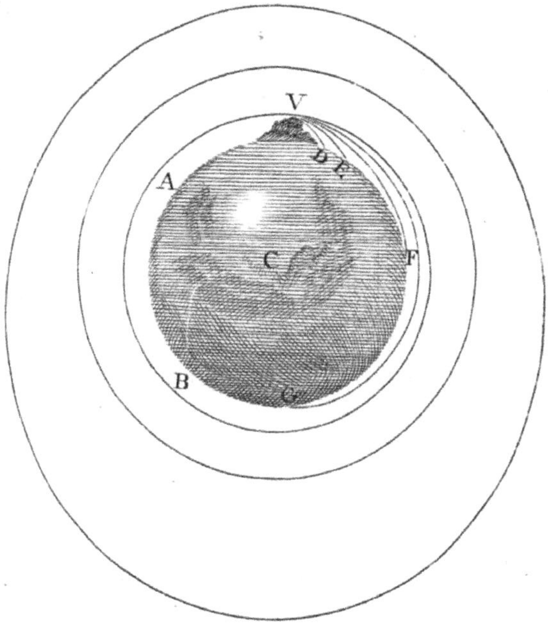

Figure 13.1. *Newton's thought experiment connecting parabolic and elliptical movements. Source: Newton (1728, p. 6)*

Thought experiments also played an important role in Isaac Newton's work. In Principia's Book III, Proposition IV, Newton presented his thought experiment on the Moon's fall, which was crucial in his unification of celestial and terrestrial physics (Newton 1999, p. 803). For Newton, this was a rhetorical and argumentative strategy because he could avoid conjecturing about the cause of gravity, which faced opposition from his contemporaries. This experiment still has pedagogical value (Freire et al. 2004). Densmore (1996) considered it the most telling and moving demonstration in *Principia,* and Bernard Cohen considered it "[one] of the great climaxes of the Principia" (Newton 1999, p. 64). The French poet and philosopher Paul Valéry (1957, p. 384) stated that "one had to be Newton to perceive that the moon falls, when everyone else clearly sees that it does not" ("il fallait être Newton pour apercevoir que la lune tombe, quand tout le monde voit bien qu'elle ne tombe pas"). In the *scholium* to Book I, Newton proposed his rotating bucket argument –

1 Koyré's views were challenged by Stillman Drake (Galilei and Drake 1974).

a thought experiment that aimed at proving the existence of absolute space (Newton 1999, pp. 106–108, 412). In his posthumous *A Treatise of the System of the World*, Newton suggested a thought experiment where a stone is thrown horizontally with different velocities, thus connecting parabolic and elliptical movements, as shown in Figure 13.1.

The development of Laplacian determinism also relied on a hypothetical situation: an intelligence that knew the initial state of all particles in the world, all forces that act on those particles, and how to solve all of the dynamical equations of the world would also know how every single particle would move in the future (Laplace 1814, pp. 2–3). Recently, the role of determinism in the history of physics has been an interesting point of discussion among historians of science (e.g. Del Santo 2021; van Strien 2021).

Thought experiments were also used to test the limits of existing theories and to guide the development of new theories. James Clerk Maxwell imagined a being – later called "Maxwell's demon" by William Thomson – intended to disprove the universality of the second law of thermodynamics. The development of Einstein's theory of relativity began with a thought experiment testing the consistency of the explanations of the appearance of an electric current in a wire that approaches a magnet. Later, several other thought experiments were proposed, such as the twin paradox by Paul Langevin and the rotating disk paradox by Paul Ehrenfest. The construction of the general theory of relativity was also guided by thought experiments (Norton 1991).

The philosophy of the science community has long debated whether thought experiments surpass empiricism. John Norton defended that "thought experiments in science are merely picturesque arguments. Their epistemic reach can always be replicated by an argument, and this is best explained by their merely being arguments" (Norton 2004, p. 64). As we aim to show in this chapter, experiments have played multiple roles in testing the consistency of quantum theoretical descriptions of physical systems – particularly the gravitational field. On the one hand, they have sometimes sought to transcend the bounds of empiricism; on the other hand, more modestly, they have served to probe the consistency and limits of existing theories.

13.3. Thought experiments in quantum theory

In addition to all of this noble lineage in the whole history of physics, the debates on the foundations of quantum mechanics also indicate that thought experiments occupied the main stage (see Jammer 1974). The uncertainty principle was proposed by Werner Heisenberg, as part of his dispute with Erwin Schrödinger about new quantum mechanics, as a way of expressing what he considered to be visualizable (*anschaulich*). A central idea was the so-called "gamma ray microscope" – a thought experiment in which the position of an electron could be measured (Heisenberg 1927).

The debates between Niels Bohr and Albert Einstein on the consistency and completeness of quantum theory took the form of analyses of thought experiments. Often, the victory of a physicist came by showing that a given thought experiment would not lead to the result intended by the other physicist. Thought experiments assumed a central role in those debates, primarily because the theory had to grapple, from its inception, with the limitations of its mathematical formalism in providing a descriptive framework for nature.

Einstein conceived a thought experiment with a box to weight a photon as a challenge to the description of quantum systems provided by quantum mechanics. Einstein also conceived another thought experiment with a pair of correlated particles (the EPR experiment) to suggest first the inconsistency and then the incompleteness of quantum mechanics. Bohr rebuked Einstein, analyzing these experiments and showing that, once the proper experimental conditions were taken into consideration, they would corroborate quantum mechanics and its putative completeness (Bacciagaluppi and Valentini 2009; Bacciagaluppi and Crull 2024).

Bohr and Heisenberg put thought experiments at the center of quantum mechanics, although for different reasons (Rosenfeld 1971, pp. 87–88; Tanona 2004). Their works gave rise to the idea of a measurement process, which later – in the 1950s – became the measurement problem. Jørgen Kalckar, Bohr's last assistant, expressed well the role of thought experiments according to Bohr, even though not all quantum physicists adhered to this project:

> Measurability problems in the sense of quantum theory are consistency problems. The discussion of imaginary experiments provides a testing ground for the compatibility of concepts in situations which are simple enough to allow comprehensive analysis. The rules to which the manipulation of the apparatus and other material bodies are subjected are not imposed by experience of what is "actually" possible. The whole argumentation must be regarded as a symbolic illustration of the rules embodied in the abstract quantal algorithm, and each step in the analysis corresponds to a definite formal operation within the mathematical framework of quantum theory. It is the purpose of the analysis to demonstrate the equivalence

between definability and measurability of physical quantities. (Kalckar 1971, p. 127)

The most famous thought experiment of quantum theory was introduced by Erwin Schrödinger in 1935 to argue that the use of quantum mechanics to address the coupling of microsystems and macrosystems would lead to the absurdity of a cat being dead and alive at the same time. As we will see, the proposal of a macroscopic superposition will be important to us later, in sections 13.6 and 13.7.

However, experiments in quantum mechanics have also appeared in pedagogical contexts. This was the case with Werner Heisenberg (1930, p. 31), who used in his Chicago lectures his gamma ray microscope to illustrate his uncertainty principle as well as to convince the readers of the inevitability of this principle. This was also the case with Richard Feynman, who used the double-slit experiment with single electrons or photons in his quantum mechanics lectures and stated that it contained the full mystery of quantum mechanics (Feynman et al. 1966).

13.4. Thought experiments turned real

There has often been tension between considering thought experiments as simply hypothetical situations and considering thought experiments as a proposal of an experimental setup that is momentarily unfeasible for technological reasons. Therefore, a century of history of quantum mechanics led to an interesting twist for some thought experiments: they became real experiments. This was possible owing to a combination of theoretical work, such as the Bell theorem, which contrasts local realism and quantum results; the prediction of entanglement and the decoherence effect; and technical developments that allowed the manipulation of single quantum systems such as photons, electrons and atoms (Bromberg 2006; Freire 2015, 2022; Kaiser 2022). The novelty here was not only that those real experiments corroborated quantum predictions, but also that they allowed physicists to trust that new physical features – such as entanglement and decoherence – were real effects and that physics could be developed, including in its applications, taking them for granted.

Some recent Nobel prizes were awarded to the physicists who carried out laboratory experiments previously considered idealized experiments. Illustrations of this were the 2022 prize for John Clauser, Alain Aspect and Anton Zeilinger with pairs of entangled photons; and the 2012 prize for Serge Haroche and David Wineland, which manipulated quantum single systems, with Haroche making with quantum cavities equivalent to Schrödinger's cat experiment and measuring the time of loss of coherence. Thus, experiments that were once a matter for debates among physicists to check consistency and the foundations of theory became part of the physics full enterprise, entering the laboratories and revealing new physics phenomena.

13.5. Approaches to quantum gravity

The development of quantum field theory began soon after the development of quantum mechanics. As early as 1925, Max Born and Pascual Jordan conjectured that the electromagnetic field could be treated using the new matrix mechanics (Born and Jordan 1925, p. 885). What was a very clever, although purely formal, anticipation would one year later become quantum field theory. The same two physicists, now in collaboration with Heisenberg, published a paper in 1926 whose last section presented the quantization of a one-dimensional scalar field under Dirichlet's boundary conditions – physically, that corresponds to quantizing the motion of a string with fixed endpoints. This section was written by Jordan Schroer (2011). His goal was to show that such a system would exhibit fluctuations in accordance with Einstein's 1909 fluctuation formula (Kojevnikov 2002). This was still quite a formal study.

In 1927, however, the application of quantum mechanics to the electromagnetic field was finally able to account for unexplained phenomena. Paul Dirac was able to show that the spontaneous emission of the hydrogen atom, which was suggested by Einstein in 1917 and was not predicted by the new quantum mechanics of Heisenberg and Schrödinger, could be explained if the quantum aspect of the electromagnetic field surrounding the atom was considered. In 1929, Heisenberg and Pauli further developed the theory, inventing the gauge fixing process and establishing the connection between Lagrangian and Hamiltonian descriptions of electromagnetic theory.

Quantum electrodynamics began to present numerous conceptual problems circa 1930, with divergent improper integrals in scattering processes (Rueger 1992). Several physicists, such as Lev Landau and Rudolf Peierls, have attempted to identify the genesis of the problem (Hartz 2022). The theory was in a state of crisis until the late 1940s, when the application of renormalization processes provided a method to address divergent integrals. Hans Bethe was able to explain the Lamb shift of the hydrogen atom. Sin'itiro Tomonaga, Julian Schwinger and Richard Feynman independently developed general formalisms to implement renormalization, which were later proven to be equivalent to that by Freeman Dyson. Despite some remaining criticisms by physicists such as Landau and Dirac, the theory was largely successful in accounting for existing experimental data (Schweber 1994).

The historical development of quantum theory in the first half of the 20th century presents all of the features that Thomas Kuhn later identified in the development of scientific theories. First, there was a crisis in classical theory and old quantum theory. Then, quantum mechanics became a successful paradigm. However, the divergent quantities that appeared when the quantum mechanical formalism was applied to fields created a new crisis, which was finally solved when renormalization was able to accommodate the experimental evidence in the existing paradigm. The historian Alexei Kojevnikov suggested a compelling conjecture: since Kuhn was educated as a physicist in the 1940s and 1950s, he probably drew his historical concepts from his practice as a young physicist at that time; therefore, it is no coincidence that his historical concepts fit perfectly to explain the development of quantum electrodynamics (Kojevnikov 2015). Naturally, those concepts are quite limited when accounting for other areas of physics, such as the ongoing controversy about the foundations of quantum mechanics (Freire 2014).

The Kuhnian perspective is enlightening when we try to understand the history of nuclear physics and quantum gravity in the 1930s. The successful and promising developments of quantum electrodynamics set a collection of puzzle-solving methods, which were then applied to other interactions. This situation is analogous to the core of the early developments of nuclear physics (Darrigol 1988). Analogical reasoning was also systematically applied in the development of quantum gravity (Hartz and Freire 2015; Rickles 2020). That was so evident at that time that Bryce DeWitt remembered having the following idea as a graduate student in 1947: "the thought occurred to me, why don't I just go over [Schwinger's] lectures on quantum electrodynamics and in place of the electromagnetic field use the gravitational field [and] try to do the same thing as Schwinger was doing[?]" (DeWitt-Morette 2011, p. 51). However, doing for gravity the same as Schwinger, what had been done for electromagnetism was far from trivial.

The first to apply Heisenberg and Pauli's formalism to the gravitational field was Léon Rosenfeld in 1929. At that time, this was merely a formal exploration of the new framework, since there was no reason – theoretical or experimental – to believe that the gravitational field should be quantized. The formal explorations were further developed by Matvei Bronstein and Jacques Solomon in the 1930s, as we will mention in the next section. In the 1950s, discussions about the quantum aspects of the gravitational field gained new momentum (Blum et al. 2015).

Following Schwinger's approach to quantum electrodynamics, DeWitt revisited Rosenfeld's results in an explicitly covariant way (Seligman 1950). The covariant approach to quantum gravity evolved quite slowly. Suraj Gupta explored the Lagrangian of a spin 2 particle (Gupta 1952). Ten years later, Feynman presented, through a perturbative approach, the shocking result that the Einstein–Hilbert Lagrangian led to a non-unitary S matrix, that is, to a theory whose probabilities do not add 100% (Feynman 1963). The solution was found by Feynman himself: to restore unitarity, it was necessary to add a new fermionic field – later called Faddeev–Popov ghost field – to the Lagrangian. This field did not have asymptotic states, so it could not be detected in a laboratory. DeWitt was able to obtain the Feynman diagrammatic rules for that new field up to second order, in 1964, and to all orders, in 1967 (DeWitt 1964, 1967b). In the same year, Ludwig Faddeev and Victor Popov, working on the Yang–Mills field, arrived independently at the same rules and provided a beautiful geometrical interpretation (Faddeev and Popov 1967). The main criticism of the covariant approach was that it relied on the linearized approximation and overlooked a fundamental aspect of Einstein's field equations – its nonlinearity.

In the late 1940s, Peter Bergmann and collaborators developed new methods to address constrained Hamiltonian systems, such as Maxwell electromagnetism and general relativity (Salisbury 2008). The same methods were discovered one year later by Dirac, who used them to obtain a Hamiltonian formalism for general relativity (Dirac 1950, 1958). Hamiltonian methods became the core of the so-called canonical quantization approach to quantum gravity, which was developed by several physicists – including Richard Arnowitt, Stanley Deser, Charles Misner, DeWitt, and Wheeler – and soon became a competing project against the covariant approach (Arnowitt et al. 1959; DeWitt 1967a; Deser 2021). The main criticism of the canonical approach was that it separated space and time, breaking explicit covariance and thus contradicting the very essence of the theory of relativity. Nevertheless, canonical formalism had the advantage of keeping the full nonlinear aspect of general relativity.

In the mid-1950s, general relativity went through a phase called the "renaissance" period (Blum et al. 2015). There is a rich historiographical debate about whether the word Renaissance adequately characterizes that period (Groener 2017). Clearly, after 1955, a community of physicists and mathematicians working on general relativity coalesced (Einsenstaedt 1986; Lalli 2017; Lalli et al. 2020a, 2020b). That community met on several occasions: first at the 1955 conference organized by André Mercier in Berna, Switzerland; then at the 1957 conference organized by DeWitt and Cécile DeWitt-Morette in Chapel Hill, NC, United States; later at the 1959 conference organized by André Lichnerowicz and Marie-Antoinette Tonnelat in Royaumont, France; and finally at the 1962 conference organized by Leopold Infeld in Jabłonna, Poland. Several textbooks or lecture notes were written in the 1950s and 1960s, including those by Christian Møller (1955), John Synge (1960), Vladimir Fock (1958), Feynman (2018[1962]).

Among the many debated topics in the general relativity community in the 1950s, a central topic was the quantization of the gravitational field. Should the gravitational field be quantified, and if so, for what reasons? Every physicist working in the field had to address criticisms on this issue. DeWitt blatantly stated:

> No apology will be made for this [desire to attack the problem of developing a quantum theory of gravitation], although needless to say, recent experiments have nothing to do with it! In the author's opinion, it is sufficient that the problem is there, like the alpinist's mountain (DeWitt 1957).

DeWitt later sought to provide stronger justifications, such as adopting Heisenberg's (1938) project, which aimed to derive the masses of elementary particles from a theory incorporating the constants $\hbar$, c and G.

At the Chapel Hill conference, in January 1957, it was stated that:

> Bergmann first asked the question, "why quantize?". His reply was that physical nature is an organic whole and that various parts of physical theory must not be expected to endure in "peaceful coexistence". An attempt should be made to force separate branches of theory together to see if they can be made to merge, and if they cannot be united, to try to understand why they clash (DeWitt-Morette and Rickles 2011, p. 165).

Nevertheless, it was still true that experiments had nothing to say about the quantization of the gravitational field. In the absence of such empirical evidence, physicists proposed thought experiments, which could tell whether the quantization of the gravitational field would be consistent within the existing mathematical formalisms.

13.6. Thought experiments in quantum gravity

13.6.1. *Analogy between gravitation and electromagnetism*

The most influential thought experiment in quantum electrodynamics was the measurement process of a quantum electromagnetic field – that is, the problem of devising a thought experiment in which the quantum electromagnetic field could be measured while ensuring that its uncertainties remained consistent with the commutation relations prescribed by the mathematical formalism (Hartz and Freire 2015; Hartz 2022).

The first physicist to address that problem was Werner Heisenberg, in his 1929 Chicago lectures (Heisenberg 1930). His attempt was a natural extension of his 1927 discussion of the uncertainty principle, which relied on the gamma ray microscope to explain what he meant by visualizability (*Anschaulichkeit*), and of his 1929 work with Wolfgang Pauli on the Lagrangian formalism of quantum electrodynamics. Soon after, Landau and Peierls approached the same problem and concluded that the measurement analysis led to more restrictive field uncertainties than those predicted by the mathematical formalism of quantum electrodynamics, making the theory inconsistent (Landau and Peierls 1931). The problem was settled in 1933, when Bohr and Rosenfeld presented their famous measurement analysis of the quantum electromagnetic field, arriving at the conclusion that Landau and Peierls were wrong and that the theory is consistent (Bohr and Rosenfeld 1933; Darrigol 1991; Hartz and Freire 2015).

The analogical reasoning that shaped the development of quantum gravity led physicists to attempt a similar thought experiment for the gravitational field. Therefore, throughout the history of quantum gravity, several physicists have attempted to devise a thought experiment capable of measuring the quantum gravitational field. Bronstein, in 1934, obtained an uncertainty for the measurement of a single component for the Christoffel symbol, suggesting that the mathematical formalism of quantum gravity, which did not yet exist, should predict such a single component uncertainty that had no analogue in the electromagnetic case (Hartz and Freire 2015, p. 405). At about the same time, Walter Heitler, Lothar Nordheim and Edward Teller conceived a thought experiment to measure the quantum gravitational field, which was supported by George Gamow and Jean Mariani but criticized by Jacques Solomon (Hartz and Freire 2015, p. 406).

13.6.2. *Proper order between formalism and measurement analysis*

Many years later, in 1957, the same problem was addressed again. The attitudes of physicists were quite diverse. Some of them believed that a measurement analysis should be pursued only after mathematical formalism was fully developed.

At the Chapel Hill conference in January 1957:

> Wheeler suggested that perhaps one should simply forget about the measurement problem and proceed with other aspects of theory. The history of electrodynamics shows that it is always ticklish business to conclude too early that there are certain limitations on measurement. He would propose rather to emphasize the organic unity of nature, to

> develop the theory (i.e., quantum gravidynamics) first and then to return later to the measurement problem (DeWitt-Morette and Rickles 2011, p. 179).

We will return soon to the Chapel Hill conference. A few months later, in July, a few physicists who had gathered in Copenhagen approached the problem again (Blum and Hartz 2017).

According to DeWitt:

> The question of measurability was raised in Copenhagen by Professor Møller, who had some lengthy discussions on it just a few weeks previously with Professor E. P. Wigner of Princeton. In approaching the problem of measurement, Wigner has begun by trying to construct the most efficient measuring devices which the principles of quantum mechanics allow. […] In Professor Møller's view it will be equally necessary [as had happened in quantum electrodynamics], in the case of quantum gravidynamics, to have a valid formalism available, before one can make any reliable statements about the measurement problem (DeWitt 2017, pp. 161–162).

Therefore, both Wheeler and Møller believed that it would be advisable to have a quantum gravity formalism before approaching the measurement process in quantum gravity. Tulio Regge shared the same view in an undated manuscript from the 1950s (indicated as possibly from 1956): "The problem of the measurability of gravitational field averages […] from the point of view of quantum mechanics seems to be a hard and premature one" (APS, John A. Wheeler Papers, Series VI, Box 115). Interestingly, a few years later, DeWitt subverted that position, using a field measurement, that is, a thought experiment, to arrive at a mathematical formalism for quantum gravity:

> [Our] indebtedness [to Bohr and Rosenfeld's article] may seem in one respect surprising, not, to be sure, because of any present-day diminution in the importance of this classic work, but because its content, as Bohr and Rosenfeld have repeatedly indicated, was guided in every way by the existence of an already developed formalism, whereas here we are trying to 'put the cart before the horse'—to develop the formalism itself with the aid of the ideas of the theory of measurability (DeWitt 1962a).

As we have discussed in a previous work, DeWitt's attitude toward this thought experiment was consistent with the pragmatic attitude that prevailed in quantum theory after the Second World War (Hartz and Freire 2015; see also Schweber 1986). His thought experiment is shown in Figure 13.2.

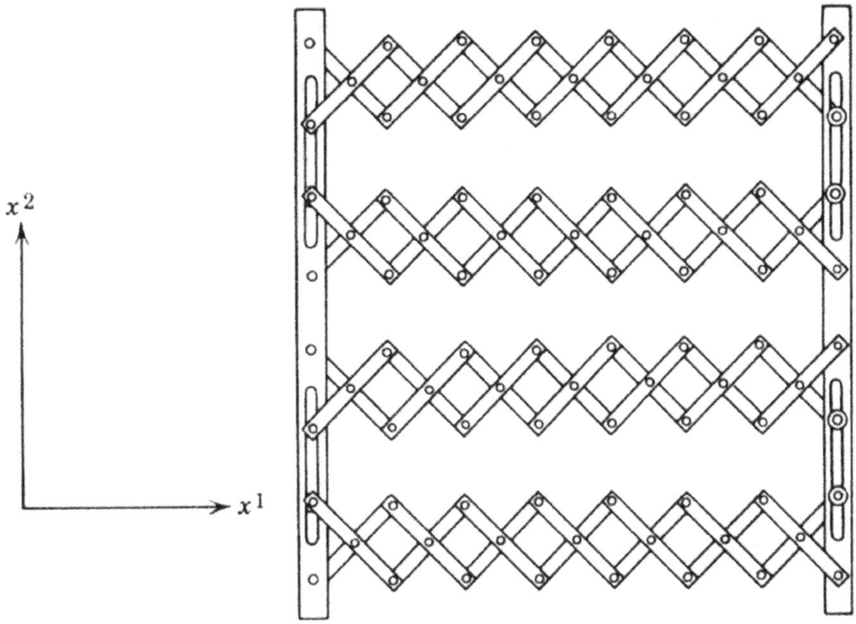

Figure 13.2. *Bryce DeWitt's drawing of an apparatus for measuring the gravitational field. Source: DeWitt (1962a, p. 344). Reproduction authorized by Wiley. Copyright © 1962 by John Wiley & Sons, Inc.*

DeWitt's analysis not only provided a path to a quantum theory of the gravitational field – that later would develop into his second most cited work (DeWitt 1967b) – but also led him to a mathematical theorem, which stated that any classical system coupled to a quantum system would become a quantum system (DeWitt 1962b). Coupled systems transfer quantization.

This theorem was severely criticized by Rosenfeld (Infeld 1964, pp. 144–145, 219–222; Rosenfeld 1963, 1978) and is a matter of discussion until nowadays (see, for example, Albers et al. 2008).

DeWitt's thought experiment reflected several interesting theoretical attitudes toward the general theory of relativity. One of the most notable was the emphasis on the motion of extended bodies, exemplified by his measuring device. In his words, after a talk by Dirac at the Jabłonna conference in 1962:

> I should like to make a comment about a matter of principle [...]. I think the example Prof. Dirac has shown us is a very excellent example of the intimate relation which exists between the physical description of the geometry of space-time and the dynamical behavior of bodies which occupy space-time. I should like to suggest that [...] we should study more, perhaps, the actual description of material objects which occupy space-time. This, for example, I found very useful in the analysis of the Bohr-Rosenfeld problem: to really describe the elastic test bodies that one uses. One learns very interesting new things this way. After all, our original ideas of distance and Riemannian geometry, are based on our experience with physical objects like rods and clocks. And the effort to describe these objects in manifestly covariant language and to learn how they behave is worth it, I think (DeWitt as quoted in Infeld 1964, p. 174).

Dirac replied to DeWitt with the following succinct remark: "I agree completely". Until the 1980s, a few other physicists also discussed the measurement of a quantum gravitational field, but later, this problem became less relevant, as the community turned to new problems (e.g. Borzeszkowski and Treder 1988, pp. 24–42; see also Hartz and Freire 2015; Hartz 2022).

13.6.3. *The search for inconsistencies and contradictions*

The works cited above illustrate how thought experiments were used over several decades to explore the analogy between quantum gravity and quantum electrodynamics.

The Chapel Hill conference, held in January 1957, marked a particularly important moment in this sequence of historical events, as it featured the proposal and discussion of a few thought experiments in quantum gravity, to which we now turn. Zeh (2011) and Aspelmeyer (2022) have already analyzed the debates on thought experiments at Chapel Hill; however, their focus is almost exclusively on Feynman's contributions, whereas our aim here is to offer a broader perspective.

The use of thought experiments in quantum gravity was a recurrent topic in the second half of the Chapel Hill conference and was mentioned by several participants.

The question was raised first by Peter Bergmann, who gave the first talk about the quantization of the gravitational field. He stated that "a study should be made of the extent to which arguments based on the uncertainty principle force one to the conclusion that the gravitational field must be subject to quantum laws" (CH,

p. 165).[2] Although it was not a new problem – having been discussed in the physics community since Bronstein's work – new points were presented at this conference.

After Bergmann's discussion, the following question was asked: "What are the limitations imposed by the quantum theory on the measurements of space–time distances and curvature?" (CH, p. 167). The conference report does not record who asked that question. In the sequence, Helmut Salecker discussed his work in collaboration with Eugene Wigner, in which they proposed a thought experiment to measure curvature using clocks. They expected to obtain information about the quantum aspects of space–time from the quantum aspect of the clock used in such a measurement. In the discussion after Bergmann's and Salecker's talks, Rosenfeld "took the floor to make some remarks about the difficulties of trying to extend the theory of measurement to the gravitational field in the manner in which he and Bohr had done for the electromagnetic field" (CH, p. 178). He simply expressed the old problems that had been known since the 1930s: the existence of positive and negative electric charges and the possibility of controlling the charge–mass ratio were used by Bohr and Rosenfeld in their analysis of the measurement of the quantum electromagnetic field; no similar situation existed in the measurement of the gravitational field, limiting the possibility of proceeding by strict analogy. Therefore, Rosenfeld "agreed with Salecker that there will be a fundamental limitation on the accuracy with which one can measure the gravitational field" (CH, p. 179).

Later in the discussion, Rosenfeld "took the floor to express some second thoughts which he had had on the question of measurability. [...] [He said] that perhaps his original pessimism in regard to the measurability of gravitational fields, as compared with the electromagnetic case, might be unjustified" (CH, p. 184). Rosenfeld argued that only after carrying out the calculations would it be possible to determine how the uncertainties in the test particle might influence the gravitational potential. It was conceivable that the result would reveal reciprocal pair uncertainties, as had occurred in the case of the quantum electromagnetic field. Interestingly, Bronstein's work was not mentioned at Chapel Hill and was apparently unknown to most participants, including Rosenfeld himself, even though he had known Brontein in his 1934 travel to the USSR (Gorelik 2005).

2 The Chapel Hill report was written by Cécile DeWitt-Morette in March 1957 to the Wright Air Development Center (WADC). The indicated pages refer to the 2011 edition (DeWitt-Morette and Rickles 2011), henceforth indicated as CH. It is important to point out that the report was written by DeWitt-Morette based on audio recordings and written notes taken by her during the event. As a result, the report employs indirect speech and refers to the participants in the third person. On the activities of the WADC, see Goldberg (1992).

The fact that Rosenfeld – who had worked on field measurements for decades, possessed extensive knowledge of general relativity, and was usually quite assertive in his remarks in conferences – expressed doubts about the issue suggests that the connection between measurement analysis and quantization was far from clear to the conference participants. In particular, it was uncertain whether the interaction between a classical field and quantized matter would lead to inconsistencies – an issue that lay at the heart of the debate on field measurement since the early 1930s. The same problem was raised later in the conference by Thomas Gold and James L. Anderson:

> GOLD said the theory of electrodynamics required quantization because there was a broad range of phenomena that had to be accounted for; for example, the infrared catastrophe had to be somehow explained. With gravitation, one could say that either it must be quantized because otherwise one might get into a contradiction with the logical structure of quantum theory, or else that there exists a broad range of phenomena which the classical theory (without quantization) is unable to explain. But this latter range of phenomena seems to be missing [...]. ANDERSON asked if there was anything wrong with the argument that if it were possible to measure all gravitational fields accurately, that one would have enough information to violate the uncertainty principle. [...] GOLD said that he wanted still to be convinced that one gets into a contradiction by not quantizing the gravitational field (CH, p. 243–244).

There was a usual confusion at that time between the field uncertainties obtained in the field measurement using a test particle and the field uncertainties obtained when the field source was considered a quantum system. As Feynman observed, "Salecker was trying to measure the gravitational field produced by a small mass whereas Rosenfeld was here considering the response of that mass to a given gravitational field" (CH, p. 184). This confusion had some interesting consequences since physicists were concomitantly discussing both problems.

After Gold's remark, DeWitt raised a subtle point, which would later be much discussed. If matter is quantized and the gravitational field is not, we must decide what to place on the right-hand side of Einstein's field equations. One possibility is to consider the expectation value of the stress–energy tensor. He hints that this problem is an experimental one. However, from a purely theoretical perspective, he argues that the expectation value depends on the quantum state of matter and, therefore, a measurement may suddenly change the gravitational field. Thus, it is possible that "classical gravitation theory works only because the experimental fluctuations are so small on the scale at which gravitational effects become noticeable." (CH, pp. 243–244). That was not the first time DeWitt was reflecting on the possibility of having the expected value of the stress–energy tensor on the right-hand side of Einstein's field equations.

In his 1953 essay to the Gravity Research Foundation, he stated, "The gravitational field [in Einstein's equations] could be produced, not by the quantum stress tensor of all the other fields in nature, but rather by the quantum mean value of this tensor" (DeWitt 1953). His claim that classical gravitational theory could suffice because experimental fluctuations are too small prompted Feynman to propose a thought experiment to evaluate under which circumstances such an assumption would hold.

Feynman imagined a situation in which there was a superposition of states of a massive particle such that the gravitational potential produced by the particle would have an uncertainty. This uncertainty in the measurement of the gravitational potential is related to the uncertainty in the measurement of the particle's mass. This could be done, for example, with a double-slit experiment with a mass indicator near one slit. A back-of-the-envelope calculation led him to a mass uncertainty of 10^{-5} g. Therefore, he argued that "either gravity must be quantized because a logical difficulty would arise if one did the experiment with a mass of order 10^{-4} grams, or else that quantum mechanics fails with masses as big as 10^{-5} grams" (CH, p. 244).

The discussion then moved to the interpretation of the metric. Ryoyu Utiyama observed that $g_{\mu\nu}$ can be considered the gravitational potential or the space–time metric tensor. Historian David Kaiser discussed how these two views were under dispute in the 1950s and 1960s (Kaiser 1998). Utiyama argued that "contrary to Einstein's original intention, the separation of this geometrical concept from the interpretation of $g_{\mu\nu}$ seems convenient for the quantization of the gravitational field" (CH, p. 245). This is an important mathematical issue if we want to investigate whether it would make sense to have the expectation value of the stress–energy tensor. After some discussion on the field variables and quantization prescriptions, the discussion returned to the thought experiments.

Salecker suggested a thought experiment to investigate whether the equivalence principle would be consistent with quantum mechanics. A stream of particles falls on a diffraction grid. The particles' wavelength depends on their masses and, therefore, the diffraction pattern depends on the masses, contradicting the equivalence principle. A discussion followed, with objections by Wheeler and Bergmann (CH, p. 249). Feynman partially supported Salecker: "the point which

Salecker had raised [was] an interesting point and a true point, but not necessarily a paradoxical one", as he then explained. Salecker expressed dissatisfaction, asking "again the question why the gravitational field needs to be quantized at all". Proceeding in analogy with the electromagnetic case, he stated that "charged quantized particles already serve as sources for a Coulomb field which is not quantized". Salecker's intended meaning remains unclear, as the editor noted.

13.6.4. *The experiment suggested by Belinfante and Feynman*

Frederik Belinfante responded to Salecker with the following observation. If, as mentioned by DeWitt, a measurement may cause a sudden change in the matter state and, consequently, in the expectation value of the stress–energy tensor, then "the laws of nature […] unfold continuously only as long as the observer does not bring extra knowledge of his own into the picture" (CH, p. 250). While the stress–energy tensor evolves continuously as an operator that obeys certain ordinary differential equations, its expectation value does not. He claims that the only way to avoid a sudden change in the gravitational field is to treat it, too, as an operator that evolves continuously. He argues that this conclusion certainly applies to the static part of the gravitational field and may also apply to gravitational radiation.

This was the longstanding problem that DeWitt had already addressed in 1953, as we mentioned, and which had also been discussed earlier during the conference.

However, Belinfante's phrasing was so clear that it prompted Feynman to propose a second thought experiment, aimed at implementing Belinfante's ideas in a concrete measurement setting. Feynman's speech was transcribed verbatim by DeWitt-Morette: "I'd like to repeat just what Belinfante said with an example – because it seems clear to me that we're in trouble if we believe in quantum mechanics but don't quantize gravitational theory" (CH, p. 250). He suggested the following thought experiment. A beam of spin ½ particles passes through a Stern–Gerlach experimental setup. At the end, there are two counters, one for each path of the split beam. Each counter, by means of a mechanical setting, moves a small ball up or down. Therefore,

> We have an amplitude that the ball is up, and an amplitude that the ball is down. […] And if we imagine that the ball can be analyzed through the interconnections up to this dimension (cm) by the quantum mechanics, then before we make an observation we still have to give an amplitude that the ball is up and an amplitude that the ball is down. Now, since the ball is big enough to produce a real gravitational field […] we could use that gravitational field to move

> another ball, and amplify that, and use the connections to the second ball as the measuring equipment.

This is a kind of Schrödinger's cat thought experiment – that is, a superposition involving a macroscopic system. Interestingly, this experiment envisages a situation where a quantum gravity effect can be observed at low energies. However, it was far from clear, at the time, whether such a setup could be realized in a laboratory. Following Feynman's explanation, Louis Witten raised the following question: "What prevents this from becoming a practical experiment?". The challenges posed by this thought experiment at Chapel Hill – including the role of decoherence and the cut between classical and quantum systems – and its feasibility as a real experiment were carefully discussed by Zeh (2011) and Aspelmeyer (2022).

13.6.5. *General relativity as a theory valid only for small fluctuations*

Five years later, at the conference on gravitation organized by Leopold Infeld and held in Jabłonna, Poland, the same discussion was raised again. In the discussion session after a talk by DeWitt, which claimed that the gravitational field "must" be quantized, Rosenfeld made some lengthy remarks explaining that thought experiments could not show the necessity of quantization.

DeWitt did not respond, but Belinfante, who was also present during the discussion, remarked: "I still feel after Rosenfeld's remarks that there is almost a 'must' for quantization, certainly if one believes in Einstein's equation and in the quantization of the matter fields going into the $T_{\mu\nu}$, just like it is hard to assume that in $div\, \boldsymbol{E} = 4\pi\rho$ the $\boldsymbol{E}$ would be a c-number while the ψ-field in ρ is second-quantized". Given the significance of this exchange, we follow their arguments step by step (see Infeld 1964, pp. 219–222). Rosenfeld pondered that "one could use the expectation value of $T_{\mu\nu}$ instead of $T_{\mu\nu}$ itself", and Belinfante answered:

> This would be most unusual to equate a c-number to an expectation value of a q-number, and I think this violates the principles of quantum theory. Suppose we had a situation which where a superposition of states, say 40% probability for one state of the matter and 60% probability for a different state. Suppose one did a million experiments; then one would expect a correlation between the measured actual state of the matter and the surrounding gravitational field. That is, one would in 400,000 cases find one gravitational field and in 600,000 cases a different gravitational field. If, however, the gravitational field were given by the expectation value, one should in

> all 1,000,000 cases find the same gravitational field, obtained by a 40–60 average.

Rosenfeld retorted: "The case you consider here is rather extreme. One seldom would have fluctuations so large that their gravitational effects would be measurable". Belinfante answered, "As you want a c-number gravitational field, I can discuss it classically, and by a thought experiment, I can then measure the field as accurately as I want". Rosenfeld repeated his argument: "It is thinkable, however, that Einstein's equation has the character of a thermodynamic equation which breaks down and is not valid in case of extreme fluctuations".

Belinfante concluded:

> This is the reason why I feel that quantization is almost a "must", as I would feel very unsatisfied if Einstein's equation were a thermodynamic equation, while Maxwell's equations would be q-number relations, though we obtain both by varying one single Lagrangian, of which we would prefer to see all terms treated alike as q-numbers.

Rosenfeld ended the conversation as follows: "On the other hand, since we know the validity of the gravitational equations only for macroscopic systems, it is perhaps not so impossible to look at them as a kind of thermodynamic equation". The idea that general relativity might resemble thermodynamics was somewhat ironic – as Rosenfeld was undoubtedly aware – given that it was precisely thermodynamics that served as a model for Einstein in formulating the special theory of relativity (see Brown 2005; Dieks 2009).

Rosenfeld's skepticism about the quantum nature of gravitation seemed surprising to many physicists, as he was regarded as the first to approach quantum gravity from a modern perspective using quantum field theory, back in 1930 (see Hartz and Freire 2015; Blum and Lessel 2022; Hartz 2022). A year later, expanding his arguments presented in Jabłonna, he published an article opposing the use of field measurement analysis to justify the need for quantizing fields, which ended with the harsh words:

> The preceding hints may suffice to expose the inanity of a reasoning which amounts to putting physics upside down. It is nice to have at one's disposal such exquisite mathematical tools as the present methods of quantum field theory, but one should not forget that these methods have been elaborated in order to describe definite empirical situations, in which they find their only justification. Any question as

to their range of application can only be answered by experience, not by formal argumentation (Rosenfeld 1963, pp. 355–356).

At the Einstein Symposium organized by Hans-Jürgen Treder in East Berlin in 1965, Rosenfeld reiterated his arguments in even greater detail, concluding that "no logical compulsion exists for quantizing the gravitational field; rather the question of whether and how such a quantization is to be carried out can only be decided empirically" (Rosenfeld 1978, p. 606).

In the discussion session, Helmut Hönl suggested a thought experiment that is a variation of the diffraction experiment put forward by Salecker in 1957. He claimed that such an experiment would show that the gravitational field does not have a quantum nature. Consider a particle that falls through two holes, forming an interference pattern in the screen under the holes. A small mass is then placed near one of the two holes, attracting gravitationally – according to classical general relativity – the particle if it passes through that hole. The expectation would be that we could use this mass to investigate which hole the particle has passed through. Hönl, in Rosenfeld's words, arrived at the following conclusion: "Placing the mass next to a hole will thus necessarily result in a complete blurring of the diffraction pattern, and a contradiction can never arise between the consistent application of quantum theory on one hand and of classical gravitational theory on the other" (Rosenfeld 1978, pp. 607–608).

13.7. Promises of experimental quantum gravity

Currently, with experiments concerning gravity in domains where the quantum effects cannot be discarded, we may be making the same turn we have had with thought experiments in quantum mechanics at large, namely, leading previously idealized experiments to real labs. In fact, in recent years, many physicists have suggested that quantum measurements of massive quantum systems both test quantum mechanics in domains far beyond where it is already well tested and shed light on the quantum nature of the gravitational field. Among these teams, the laboratory led by Markus Aspelmeyer in Vienna is noticeable.

It is beyond the scope of this chapter to present and analyze the details of these intended experiments. However, we want to highlight the line of conceptual continuity between the current experiments and those once thought of as idealized experiments. Thus, Bengyat, Di Biagio, Aspelmeyer and Christodoulou, in a paper dedicated to the study of gravity-mediated entanglement, that is, gravity producing the entanglement, the quintessential quantum effect, framed their work in the conceptual and chronological line remounting to Feynman and the 1957 Chapel Hill conference, where gravitational thought experiments were debated. According to the authors:

> Feynman argued [at the Chapel Hill conference, in 1957] that detecting the gravitational pull of a mass in superposition would be direct evidence of the quantum nature of gravity. Due to the weakness of gravity this has remained a *gedankenexperiment* the past 65 years. Partly motivated by experimental progress there has been a revival of this fundamental question and several proposals have been put forward that analyze gravity-mediated entanglement (GME) in more detail. Given the increase in quantum control of mesoscopic solid masses and advances in gravity measurements at small scales, these experiments are expected to become feasible in our times (Bengyat et al. 2024).

Nevertheless, the current trend in real experiments concerning gravity and quantum effects is unclear. Unlike the experiments on the Bell theorem, where the validity of entanglement was at stake, here, we have different questions and distinct possible answers. Following Bengyat et al. (2024), for example, such experiments may lead to the eventual detection of GME. This would have the consequence of "viewing gravity as a communication channel", and this "would imply that gravity cannot be a classical channel"; however, "the validity of employing such a model is uncertain". For these physicists, "a related interpretation is that detection can be taken as evidence of the spacetime metric being in a quantum superposition", but "another related perspective is to argue that the Newtonian potential must possess the ability to convey quantum information and, to prevent violations of causality, vacuum fluctuations must be inherent to gravity". These options do not exhaust the spectrum of possibilities as "using perturbation theory, it has been argued that off-shell (virtual) gravitons are the cause of entanglement"; finally, "some have argued that the entanglement arises from a constraint rather than a degree of freedom and that detecting entanglement does not say anything about the quantum nature of gravity". This should not be seen as a discouragement because we know that science, as stated by the French physicist and philosopher Pierre Duhem, has no crucial experiments.

13.8. Conclusions

During the 20th century, physicists tried to investigate the gravitational field by strict analogy with the electromagnetic case. Sometimes, the analogy proved excellent, for example, in the expectation of gravitational waves, whereas in other cases, the analogy failed, owing to the nonlinearity of general relativity, the fixed ratio between gravitational charge and mass, and the absence of a negative mass.

A guiding tool in exploring those analogies was the use of thought experiments, which presented hypothetical situations that allowed physicists to explore the consistency and limitations of the quantum description of the gravitational field. In the absence of experimental evidence in support of the quantization of the gravitational field, thought experiments were sometimes conceived in the quantum theory of gravitation. They were, in fact, quite popular, as we observed in the preceding sections.

Experiments were used with opposite purposes to support the necessity of quantizing the gravitational field and to support that such a field should remain classical. The discussion did not reach a decisive conclusion but was paramount in pushing the development of quantum gravity from the 1930s until the 1960s, when a more formalistic period began, with the exploration of the canonical and the covariant approaches, leaving the thought experiments aside.

Currently, owing to the development of experimental methods to create quantum superpositions of macroscopic objects, real experiments are promising for bringing life to those old thought experiments about the gravitational field, reaching perhaps an empirical answer to the question of whether the gravitational field has a quantum nature. From a matter of perspective, it is quite interesting to observe that, once again, as had happened before in the history of quantum mechanics, experiments that were envisaged as purely hypothetical situations are about to become real experiments, showing that the development of physical theories is far from linear or predictable.

13.9. Acknowledgements

We thank Reinaldo de Melo e Souza for his comments on an earlier version of this chapter. Olival Freire Jr. acknowledges the Brazilian National Council for Scientific and Technological Development (CNPq) for a fellowship that partially funded this research.

13.10. References

Albers, M., Kiefer, C., Reginatto, M. (2008). Measurement analysis and quantum gravity. *Physical Review Part D*, 78, 064051.

Arnowitt, R., Deser, S., Misner, C.W. (1959). Dynamical structure and definition of energy in general relativity. *Physical Review*, 116(5), 1322–1330.

Aspelmeyer, M. (2022). When Zeh meets Feynman: How to avoid the appearance of a classical world in gravity experiments. In *From Quantum to Classical: Essays in Honour of H.-Dieter Zeh*, Kiefer, C. (ed.). Springer, Berlin.

Bacciagaluppi, G. and Crull, E. (2024). *The Einstein Paradox – The Debate on Nonlocality and Incompleteness in 1935*. Cambridge University Press, Cambridge.

Bacciagaluppi, G. and Valentini, A. (2009). *Quantum Theory at the Crossroads: Reconsidering the 1927 Solvay Conference*. Cambridge University Press, Cambridge.

Bengyat, O., Di Biagio, A., Aspelmeyer, M., Christodoulou, M. (2024). Gravity-mediated entanglement between oscillators as quantum superposition of geometries. *Physical Review Part D*, 110, 056046.

Blum, A. and Hartz, T. (2017). The 1957 quantum gravity meeting in Copenhagen: An analysis of Bryce S. DeWitt's report. *European Physical Journal H*, 42, 107–157.

Blum, A. and Lessel, B. (2022). The interpretation debate and quantum gravity. In *The Oxford Handbook of the History of Quantum Interpretations*, 1st edition, Freire, O. (ed.). Oxford University Press, Oxford.

Blum, A., Lalli, R., Renn, M.J. (2015). The reinvention of general relativity: A historiographical framework for assessing one hundred years of curved space-time. *Isis*, 106(3), 598–620.

Bohr, N. and Rosenfeld, L. (1933). Zur Frage der Messbarkeit der elektromagnetischen Feldgrössen. *Mat.-Fys. Medd. Dan. Vidensk. Selsk.*, 12, 1–65.

Born, M. and Jordan, P. (1925). Zur quantenmechanik. *Zeitschrift für Physik*, 34, 858–888.

Borzeszkowski, H.-H. and Treder, H.-J. (1988). *The Meaning of Quantum Gravity*. D. Reidel, Dordrecht.

Bromberg, J.L. (2006). Device physics vis-a-vis fundamental physics in Cold War America: The case of quantum optics. *Isis*, 97(2), 237–259.

Brown, H.R. (2005). *Physical Relativity Space-Time Structure from a Dynamical Perspective*. Clarendon Press, Oxford.

Darrigol, O. (1988). The quantum electrodynamical analogy in early nuclear theory or the roots of Yukawa's theory. *Revue d'histoire des sciences,* 41, 225–297.

Darrigol, O. (1991). Cohérence et complétude de la mécanique quantique : l'exemple de 'Bohr-Rosenfeld'. *Revue d'histoire des sciences*, 44(2), 137–179.

Del Santo, F. (2021). Indeterminism, causality and information: Has physics ever been deterministic? In *Undecidability, Uncomputability, and Unpredictability*, Aguirre, A., Merali, Z., Sloan, D. (eds). Springer, Cham.

Densmore, D. (1996). *Newton's Principia: The Central Argument – Translation, Notes, and Expanded Proofs, Translations and Diagrams by W.H.* Green Lion Press, Santa Fe, Donahue.

Deser, S. (2021). The ADM version of GR at Sixty: A brief account for historians. *European Physical Journal H*, 46, 14.

DeWitt, B.S. (1953). *New Directions for Research in the Theory of Gravitation*. Gravity Research Foundation, Wellesley Hills, MA.

DeWitt, B.S. (1962a). The quantization of geometry. In *Gravitation: An Introduction to Current Research*, Witten, L (ed.). John Wiley & Sons, New York.

Dewitt, B.S. (1962b). Definition of commutators via the uncertainty principle. *Journal of Mathematical Physics*, 3(4), 619–624.

DeWitt, B.S. (1962c). The quantization of geometry. In *Gravitation: An Introduction to Current Research*, Witten, L. (ed.). John Wiley & Sons, New York.

Dewitt, B.S. (1964). Theory of radiative corrections for nonabelian gauge fields. *Physical Review Letters*, 12(26), 742–746.

DeWitt, B.S. (1967a). Quantum theory of gravity. I. The canonical theory. *Physical Review*, 160(5), 1113–1148.

DeWitt, B.S. (1967b). Quantum theory of gravity. II. The manifestly covariant theory. *Physical Review*, 162(5), 1195–1239.

DeWitt, B.S. (2017). Exploratory research session on the quantization of the gravitational field. *European Physical Journal H*, 42, 159–176.

DeWitt-Morette, C. (2011). *The Pursuit of Quantum Gravity: Memoirs of Bryce DeWitt from 1946 to 2004*. Springer, Berlin.

DeWitt-Morette, C. and Rickles, D. (eds) (2011). *The Role of Gravitation in Physics: Report From the 1957 Chapel Hill Conference*. Max-Planck-Gesellschaft zur Förderung der Wissenschaften, Berlin.

Dieks, D. (2009). Understanding in physics: Bottom-up versus top-down. In *Scientific Understanding: Philosophical Perspectives*, de Regt, H.W., Leonelli, S., Eigner, K. (eds). University of Pittsburgh Press, Pittsburgh.

Dirac, P.A.M. (1950). Generalized Hamiltonian dynamics. *Canadian Journal of Mathematics*, 2, 129–148.

Dirac, P.A.M. (1958). The theory of gravitation in Hamiltonian form. *Proceedings of the Royal Society of London Series A*, 246(1246), 333–343.

Einstein, A. (1909). Zum gegenwärtigen Stand des Strahlungsproblems. *Physikalische Zeitschrift*, 10, 185–193.

Eisenstaedt, J. (1986). La relativité générale à l'étiage : 1925–1955. *Archive for History of Exact Sciences*, 35(2), 115–185.

Faddeev, L.D. and Popov, V.N. (1967). Feynman diagrams for the Yang–Mills field. *Physics Letters B*, 25(1), 29–30.

Feynman, R.P. (1963). Quantum theory of gravitation. *Acta Physiologica Polonica*, 24.

Feynman, R.P. (2018 [1962]). *Feynman Lectures on Gravitation.* CRC Press, Boca Raton.

Feynman, R.P., Leighton, R.B., Sands, M. (1966). *The Feynman Lectures on Physics*, volume 3. Addison–Wesley, Boston.

Fock, V. (1958). *The Theory of Space Time and Gravitation.* Pergamon Press, London.

Freire, O. Jr. (2014). On the influence of science milestones on the history and philosophy of science. In *Toward a History of the History of Science: 50 Years Since "Structure"*, Blum, A., Gavroglu, K., Renn, J. (eds). Edition Open Access, Berlin.

Freire, O. Jr. (2015). *The Quantum Dissidents – Rebuilding the Foundations of Quantum Mechanics 1950–1990.* Springer, Berlin.

Freire, O. Jr. (2022). Alain aspect's experiments on Bell's theorem: A turning point in the history of the research on the foundations of quantum mechanics. *European Physical Journal D: Atomic, Molecular, Optical and Plasma Physics*, 76, 248.

Freire, O. Jr., Matos Filho, M., Valle, A.L. (2004). Uma exposição didática de como Newton apresentou a Força Gravitacional. *Física na Escola*, 5, 25–31.

Galilei, G. and Drake, S. (1974). *Two New Sciences: Including Centers of Gravity & Force of Percussion.* University of Wisconsin Press, Madison.

Goenner, H. (2017). A golden age of general relativity? Some remarks on the history of general relativity. *General Relativity and Gravitation*, 49, 42.

Goldberg, J. (1992). US Air Force support of general relativity, 1956–1972. In *Studies in the History of General Relativity*, Eisenstaedt, J. and Kox, A. (eds). Birkhäuser, Boston.

Gorelik, G.E. (2005). Matvei Bronstein and quantum gravity: 70th anniversary of the unsolved problem. *Physics-Uspekhi*, 48(10), 1039–1053.

Gupta, S.N. (1952). Quantization of Einstein's gravitational field: General treatment. *Proceedings of the Physical Society. Section A*, 65(8), 608–619.

Hartz, T. (2022). The measuring process in quantum field theory. In *The Oxford Handbook of the History of Quantum Interpretations*, Freire, O. Jr. (ed.). Oxford University Press, Oxford.

Hartz, T. and Freire, O. Jr. (2015). Uses and appropriations of Niels Bohr's ideas about quantum field measurement, 1930–1965. In *One Hundred Years of the Bohr Atom*, Aaserud, F. and Kragh, H. (eds). Royal Danish Academy of Science and Letters, Copenhagen.

Heisenberg, W. (1927). Über den anschaulichen Inhalt der quantentheoretischen Kinematik und Mechanik. *Zeitschrift für Physik*, 43, 172–198.

Heisenberg, W. (1930). *The Physical Principles of the Quantum Theory.* Chicago University Press, Chicago.

Heisenberg, W. (1938). Über die in der Theorie der Elementarteilchen auftretende universelle Länge. *Annalen der Physik*, 32, 20–33.

Infeld, L. (ed.) (1964). *Proceedings on the Theory of Gravitation, Conference in Warszawa and Jablonna*. Editions Scientifiques de Pologne, Warszawa.

Jammer, M. (1974). *The Philosophy of Quantum Mechanics: The Interpretations of Quantum Mechanics in Historical Perspective*. John Wiley & Sons, New York.

Kaiser, D. (1998). A ψ is just a ψ? Pedagogy, practice, and the reconstitution of general relativity, 1942–1975. *Studies in History and Philosophy of Science Part B: Studies in History and Philosophy of Modern Physics*, 29(3), 321–338.

Kaiser, D. (2022). Tackling loopholes in experimental tests of Bell's inequality. In *The Oxford Handbook of the History of Quantum Interpretations*, Freire, O. Jr. (ed.). Oxford University Press, Oxford.

Kalckar, J. (1971). Measurability problems in the quantum theory of fields. In *Fondamenti di meccanica quantistica: Rendiconti della scuola internazionale di fisica Enrico Fermi*, d'Espagnat, B. (ed.). Academic Press, New York.

Kojevnikov, A. (2002). Einstein's fluctuation formula and the wave–particle duality. In *Einstein Studies in Russia*, Balashov, Y. and Vizgin, V. (eds). Birkhäuser, Boston, MA.

Kojevnikov, A. (2015). Quantum electrodynamics and the specter of a "Permanent Revolution". In *Talk at the Fourth History of the Quantum Conference (HQ4)*, San Sebastián.

Koyré, A. (1966). *Études galiléennes*. Hermann, Paris.

Lalli, R. (2017). *Building the General Relativity and Gravitation Community During the Cold War*. Springer, Cham.

Lalli, R. (2020). The multiple lives of the general relativity community, 1955–1974. In *Biographies in the History of Physics: Actors, Objects. Institutions*, Forstner, C. and Walker, M (eds) Springer, Cham.

Lalli, R., Howey, T.R., Wintergrün, D. (2020a). The dynamics of collaboration networks and the history of general relativity, 1925–1970. *Scientometrics*, 122(2), 1129–1170.

Lalli, R., Howey, T.R., Wintergrün, D. (2020b). The socio-epistemic networks of general relativity, 1925–1970. In *The Renaissance of General Relativity in Context*, Blum, A., Lalli, R., Renn, J. (eds). Birkhäuser, Cham.

Landau, L. and Peierls, R. (1931). Erweiterung des Unbestimmtheitsprinzips für die relativistische Quantentheorie. *Zeitschrift für Physik*, 69(1–2), 56–69.

Laplace, P.-S. (1814). *Essai philosophique sur les probabilités*. Courcier, Paris.

Møller, C. (1955). *The Theory of Relativity*. Clarendon Press, Oxford.

Newton, I. (1728). *A Treatise of the System of the World*. F. Fayram, London.

Newton, I. (1999). *The Principia: Mathematical Principles of Natural Philosophy*. University of California Press, Berkeley, CA.

Norton, J. (1991). Thought experiments in Einstein's work. In *Thought Experiments in Science and Philosophy*, Horowitz, T. and Massey, G. J. (eds). Rowman & Littlefield Publishing Group, Savage, MD.

Norton, J. (2004). Why thought experiments do not transcend empiricism. In *Contemporary Debates in Philosophy of Science*, Hitchcock, C. (ed.). Blackwell Publishing, Malden.

Rickles, D. (2020). *Covered with Deep Mist: The Development of Quantum Gravity* (1916–1956). Oxford University Press, Oxford.

Rosenfeld, L. (1963). On quantization of fields. *Nuclear Physics*, 40, 353–356.

Rosenfeld, L. (1971). Men and ideas in the history of atomic theory. *Archive for History of Exact Sciences*, 7, 69–90.

Rosenfeld, L. (1978). Quantum theory and gravitation. In *Selected Papers of Léon Rosenfeld*, Cohen, R.S. and Stachel, J. (eds). D. Reidel, Dordrecht.

Rueger, A. (1992) Attitudes toward infinities: Responses to anomalies in quantum electrodynamics. *Historical Studies in the Physical and Biological Sciences*, 22, 309–337.

Salisbury, D.C. (2008). Rosenfeld, Bergmann, Dirac and the invention of constrained Hamiltonian dynamics. In *The Eleventh Marcel Grossmann Meeting: On Recent Developments in Theoretical and Experimental General Relativity, Gravitation and Relativistic Field Theories*, Kleinert, H., Jantzen, R., Ruffini, R. (eds). World Scientific, Singapore.

Schroer, B. (2011). Pascual Jordan's legacy and the ongoing research in quantum field theory. *European Physical Journal H*, 35, 377–434.

Schweber, S.S. (1986). The empiricist temper regnant: Theoretical physics in the United States, 1920–1950. *Historical Studies in the Physical and Biological Sciences*, 17, 55–98.

Schweber, S.S. (1994). *QED and the Men Who Made It: Feynman, Dyson, Schwinger, and Tomonaga*. Princeton University Press, Princeton.

Seligman, C.B. (1950). I: The theory of gravitational interactions. II: The interaction of gravitation with light. PhD Thesis, Harvard University, Cambridge, MA.

van Strien, M. (2021). Was physics ever deterministic? The historical basis of determinism and the image of classical physics. *European Physical Journal H*, 46(8), 202.

Stuart, M.T., Fehige, Y., Brown, J.R. (eds) (2018). *The Routledge Companion to Thought Experiments*. Routledge, New York.

Synge, J. (1960). *Relativity: The General Theory*. North-Holland, Amsterdam.

Tanona, S. (2004). Uncertainty in Bohr's response to the Heisenberg microscope. *Studies in History and Philosophy of Modern Physics*, 35, 483–507.

Valéry, P. (1957). *Œuvres, I*. Gallimard, Paris.

Zeh, H.D. (2011). Feynman's interpretation of quantum theory. *European Physical Journal H*, 36, 63–74.

14

Possibility in Physics

Baptiste LE BIHAN, Emilia MARGONI and Annica VIESER
Department of Philosophy, University of Geneva, Switzerland

14.1. Introduction

What is it about physical reality that allows us to walk from our flat to our favorite library in the city center and choose a book to read in the park, but prevents us from getting there instantaneously or through an admittedly thrilling flight path? Although we have shared intuitions about what is physically possible in everyday situations, it is generally accepted that science plays a key role in exploring and ultimately defining the domain of physical possibilities – possibilities that align with the way our universe is structured and functions.

A central starting point for the philosophical investigation of possibility is the observation that we can distinguish different notions. Are perpetual motion machines possible? In one clear sense, they are not: they would violate the first law of thermodynamics. At the same time, the concept of a perpetual motion machine does not appear self-contradictory: we can imagine a world with different physical laws in which perpetual motion machines exist. Perpetual motion machines thus seem impossible in one sense and possible in another. For this reason, it is customary to distinguish between different kinds of possibility that can be ordered by strength: logical possibility stands as the broadest notion, whereas physical possibility, that is the set of possibilities compatible with the physical laws of our

Quantum Physics and Cosmology,
coordinated by Hervé ZWIRN. © ISTE Ltd 2026.

world, satisfies more restrictive requirements.[1] Perpetual motion machines could thus be construed as being logically possible yet physically impossible. Another notion of possibility, which cannot as easily be cast in the inclusion ordering of "objective" possibility spaces, is that of epistemic possibility. What is epistemically possible for a person or group is what is compatible with their knowledge. For instance, before the development of modern theories of thermodynamics, perpetual motion machines represented an epistemic possibility.

This chapter focuses on physical possibility and on a feature of physical frameworks that might complicate the way we understand it: physical theories rely on two kinds of solution spaces associated with the kinematical and dynamical models of the theory. In general, the kinematical models are consistent with the theoretical framework under consideration before dynamical equations are added, whereas the dynamic models further obey the latter constraints.

The kinematical–dynamical distinction invites a number of questions. Are kinematical and dynamical models really representing distinct spaces of possibilities, or are kinematical solutions better thought of as a mere mathematical device? Are there two kinds of physical possibility, or is one of the two sorts of possibility to be identified with logical or another type of possibility?

To address these questions, this chapter analyzes how the distinction between kinematics and dynamics is carried out in different physical theories. We show how the mechanism through which dynamical models are generated from kinematical models varies in different theoretical settings. In classical mechanics and general relativity (GR), dynamical solutions correspond to a subset of the kinematical space, contrary to non-relativistic quantum mechanics, where dynamical models are built out of weighted sums of kinematical solutions. The lack of universality for the subset approach suggests that kinematical possibilities should be construed as resulting from the application of a preliminary layer of constraints on the way to even more constrained dynamical models. These dynamical models ultimately define the genuine space of physical possibilities. We thus conclude that kinematical solutions are not genuine physical possibilities and should be seen as epistemic, logical or linguistic possibilities (we set aside the question of their exact nature).

---

1 Another notion, popular in contemporary metaphysics but less so in philosophy of physics, is that of *metaphysical possibility*, born from the work of Kripke (1980). The notion has become popular in a priori approaches to modality, and its status remains a subject of ongoing debates (see Divers 2018; Norton 2022). We will discuss it in the following.

Note also that here, we focus on a limited number of theoretical contexts.[2] We do so because they represent the foundational milestones for a proper understanding of the distinction between kinematical and dynamical solutions. In addition, GR and non-relativistic quantum mechanics embody the two major challenges for the tenability of the distinction between kinematics and dynamics: the former calls into question the very idea of temporal evolution that dynamic solutions typically incorporate, whereas a certain understanding of the latter questions the ordering relationships between kinematical and dynamical possibilities.

The chapter is structured as follows. We first articulate how dynamical models are selected from kinematic models in the Lagrangian and Hamiltonian formulations of classical mechanics (section 14.2). On the basis of our findings, we then discuss different possible interpretations of the kinematic–dynamical distinction (section 14.3). We suggest that kinematical possibilities should be understood as epistemic, logical or linguistic preconditions for the formulation of a theory's dynamics, which then defines the space of genuine physical possibilities. Section 14.4 discusses how the selection mechanism of dynamical solutions from kinematical ones should be generalized in the context of GR. We turn to non-relativistic quantum mechanics in section 14.5, examining the nature of the kinematic–dynamical distinction there, and find that kinematical models appear to be highly idealized, not representing genuine physical possibilities. We conclude by discussing potential implications from speculative approaches to quantum gravity for the nature of physical possibilities and the distinction between kinematical and dynamical models (section 14.6).

14.2. Classical mechanics

The original formulation of classical mechanics is Newtonian mechanics, which determines the evolution of a system's position with respect to a temporal parameter through the system's *equation of motion*. This equation of motion – referred to as Newton's second law – is a second-order differential equation that fixes the relationship between the total force acting on a system and the associated acceleration of the system itself.

However, the exact range of solutions in Newtonian mechanics can be quite difficult to evaluate explicitly. For example, consider the simple case of the classical pendulum's equation of motion: solutions are exactly defined for small angular

2 We will not examine special relativity and relativistic quantum mechanics. Although these theoretical frameworks would require an in-depth evaluation, we believe that our analysis will naturally extend to them. We leave this task for future research.

values only. This is why more powerful formalisms with broader application scopes, *Lagrangian* and *Hamiltonian mechanics*, were introduced in the 18th and 19th centuries, respectively. They help drastically ease the calculation in many contexts. For this reason, before comparing how the distinction between kinematical and dynamical possibilities is carried out, a preliminary discussion of the Lagrangian and Hamiltonian frameworks is necessary. This will allow us, in the next section, to provide a more careful characterization of the general distinction between kinematical and dynamical possibilities.

Central to Lagrangian mechanics is the Lagrangian *L* of a system:[3]

$$L = K - V$$

where *K* (often denoted *T* in the Lagrangian framework) refers to kinetic energy and *V* refers to potential energy. The kinematical solutions of the system are evaluated via an action *S*, itself defined as the integral of the Lagrangian function between two points in the configuration space, at two successive instants t_1 and t_2:

$$S = \int_{t_1}^{t_2} L(q(t), \dot{q}(t), t)dt$$

where q is the position and $\dot{q}(t)$ is the associated velocity. The formalism is thus defined in terms of configuration-like and velocity-like quantities. The Lagrangian can be used to calculate equilibrium states satisfying an extremization principle for the paths a system can undergo. This can be done by evaluating the so-called Euler–Lagrange equations, a system of second-order differential equations whose solutions correspond to stationary points of the action *S*:

$$\frac{\partial L}{\partial q} - \frac{d}{dt}\frac{\partial L}{\partial \dot{q}} = 0$$

In other words, the Lagrangian formalism shows that the solutions of the equation of motion for a system of interest are obtained by extremizing a quantity, in this case, the Lagrangian, in the space of possible trajectories a system can undergo while moving through spacetime. This procedure is called the *principle of least action*.[4] For systems that can be cast in this form, the Euler–Lagrange equations are equivalent to Newton's equation of motion. The difference with respect to the latter

3 For a more detailed, philosophical presentation, see North (2009, pp. 68–70).

4 The variational principle is typically referred to as the principle of least action as dynamical solutions often correspond not just to stationary points, but more specifically to minima of the action (for a more detailed discussion see, Arnold 1989, Chapter 3).

is that the focus of Lagrangian mechanics is on the space of solutions, which can generally be interpreted as a space of possible "worlds", out of which the dynamical solutions are selected via the extremization principle. Contrary to the picture captured by Newton's vectorial mechanics, which fixes initial states and allows temporal evolution via the application of dynamical laws (as encoded in Newton's second law of motion), the Lagrangian formalism hence ranges over the *temporally extended paths* a system can undergo, or *constrained histories*.

Turning to Hamiltonian mechanics, we recover the Newtonian idea of describing the evolution of states. The formalism uses the Hamiltonian function associated with a system, which reads:

$$H = K + V$$

where, again, K refers to the kinetic energy and V refers to the potential energy of the system. The variables involved are the position of the system and its associated momentum (that is, the product of the system's velocity with its mass m). The Hamiltonian is then used to move from Newton's equation of motion to Hamilton's equation of motion.

These equations correspond to a pair of first-order differential equations encoding the time derivative of the position and momentum of the system under analysis:

$$\dot{q} = \frac{\partial H}{\partial p}$$

$$\dot{p} = -\frac{\partial H}{\partial q}$$

The Newtonian and Hamiltonian equations are generally regarded as equivalent.[5] The focus of Hamiltonian mechanics is on the space of initial conditions. The underlying assumption is that, provided with initial data plus the function representing the energy of the system, past and future states will follow deterministically.

The upshot is that the Lagrangian and Hamiltonian formulations of classical mechanics differ in that the former deals with constrained *histories*, whereas the latter describes the evolution of *states*. Let us unpack this difference more thoroughly.

The Lagrangian formulation focuses on the space of four-dimensional solutions, namely, possible histories or trajectories of the system at hand. Provided the space

5 However, on the equivalence between the various formulations of classical mechanics, see Curiel (2014).

of kinematically possible trajectories, dynamical solutions follow from the application of the principle of least action. Kinematical solutions can be viewed as a space of functions between independent variables (typically, spatial, temporal or spatiotemporal coordinates) and dependent variables (field values for a quantity of interest, including spatial coordinates). Dynamical solutions correspond to the space of functions between independent and dependent variables which respect the additional constraint set by the Euler–Lagrange equations. In the Lagrangian formalism, dynamical possibilities thus correspond to a subset of the broader class of kinematically possible solutions. To provide an intuitive example (Belot 2007, p. 154), take a particle in Euclidean space on which a position-dependent force $F\big(x(t)\big)$ is applied. The independent variable acts as the temporal parameter, whereas the dependent variables are associated with the possible positions of the particle. In this case, an arbitrary continuous function $x(t)$ of the kind $t \in \mathbb{R} \rightarrow x(t) \in \mathbb{R}^3$ sets a kinematically possible trajectory for the particle, whereas the space of dynamically possible trajectories is defined by the subset that satisfies the Newtonian equation $\ddot{x} = F\big(x(t)\big)$. Within the Lagrangian framework, the role of differential equations, or laws of motion, is thus to select dynamically possible solutions among the broader class of kinematically possible ones.

On the other hand, the Hamiltonian formulation defines the space of initial data, or instantaneous states, and evaluates the flow of this space via the application of Hamilton's equations. These equations fix, when plugging in initial conditions, the space of dynamically possible models of the theory. The dynamics is thus provided by an action over the space of initial data, which implements a temporal evolution. This means that the history of the system under investigation corresponds to a trajectory through the space of initial values. As the emphasis is on instantaneous states – namely, a specification of the value of the field at each point of space, together with the associated temporal rate of change – the Hamiltonian formulation requires a global decomposition of spacetime into space and time (see Belot 2007, p. 165). This partitioning of the spacetime manifold presupposes a notion of absolute simultaneity, defined by a preferred family of observers. This $3 + 1$ splitting can be uniquely defined for Newtonian spacetime without causing any particular issue. However, as we discuss later, the need to partition spacetime will become a burdensome challenge in the context of relativistic physics.

14.3. Kinematical and dynamical possibilities

In the last section, we have shown that the Lagrangian and Hamiltonian approaches concur in distinguishing between two spaces of models: kinematical models construed in terms of their compatibility with the general setting of the theory and dynamical models satisfying additional constraints given by the equation of motion. The different spaces of solutions are constructed differently in the two formalisms: kinematical solutions are instantaneous states in Hamiltonian mechanics, and four-dimensional histories in Lagrangian mechanics. Dynamical models are selected by the Euler–Lagrange equations for Lagrangian mechanics and by Hamilton's equations for Hamiltonian mechanics.

With this initial grasp of the kinematical–dynamical distinction, we can return to the broader philosophical questions concerning how this distinction shapes our picture of physical modality. A first question concerns how to read off possibilities from the formal apparatus of a given theory. We will mostly sidestep this question and speak - admittedly, somewhat loosely - of both solutions and models as (representing) kinematical or dynamical possibilities.

As explained in the introduction, the existence of two realms of physical possibilities, suggested by the existence of two kinds of models, seems to clash with the traditional understanding of physical possibility as confineable to a single "layer" (Ruyant and Guay 2024, p. 2). That there could be two types of possibilities in physics where one would expect to find only one leads to a number of intriguing questions. Is the distinction between kinematical and dynamical models latching onto a real separation between two types of possibilities, or is it a mere artifact of the methodology used in physics? If two types of possibilities exist in physics, are they both physical possibilities in the traditional sense? In particular, are kinematical models genuine possibilities at all?

Classical mechanics supports a view called the *subset view*, on which the set of dynamical possibilities is a subset of the set of kinematical possibilities. Given the subset view, Ruyant and Guay (2024) entertain three different options for conceptualizing the modal status of kinematical and dynamical possibilities. In the first option, kinematical models are identified with *metaphysical possibilities* – a distinct sort of possibilities situated between logical and physical ones – whereas dynamical models are identified with physical possibilities.[6] The nature (and existence) of these metaphysical possibilities remains debated, but they are supposed to capture the idea of metaphysical laws going beyond physical ones, conceptualized for example as associated with the natures or essences of the entities constituting the world (Fine 1994). In the second option, kinematical models are associated with a new sort of possibilities situated between metaphysical and physical possibilities, the latter being again identified with dynamical possibilities. Finally, the third

6 This kind of view is defended by Hirèche et al. (2021), who argue that the kinematical space fixes metaphysical necessity, while the dynamical space determines nomic necessity.

alternative holds that kinematical and dynamical models each form a subspecies of physical possibility and which we will substantiate in the remainder of the paper.

The only view we take on this is that the third option is incorrect: purely kinematical models do not represent physical possibilities. Importantly, this position remains neutral on whether the subset view holds in other physical theories than classical mechanics - and, as we shall see, there are reasons to doubt its viability in certain theoretical contexts. Our position resonates well with Wallace's (2022) distinction between the kinematics and dynamics of a theory, according to whom the former represents the ensemble of models with the proper mathematical features to represent a physically possible world, whereas the latter represents the ensemble of models that actually represent a physically possible world. In other words, kinematics fixes the space of solutions or data that display suitable features (such as continuity, four-dimensionality, differentiability and so forth) to properly represent a physical state of the universe, whereas dynamics defines the space of solutions or data that actually represent a physical state of the universe.

In our view, we can think of the kinematical space as the space that defines the preconditions for the application of what are often described as *dynamical laws*. This means that the kinematical space somehow fixes the space of possibilities in a general sense, whereas the dynamical space adds dynamical laws as a second layer of constraints on the range of what is physically possible.

Before turning to physical settings other than classical mechanics, let us highlight another interesting problem concerning physical possibility that arises from the coexistence of multiple physical frameworks. Different physical theories – while all of them simultaneously accepted and in use by physicists; consider, for example, GR and quantum mechanics – yield different possibilities, which makes it far from clear how to delineate the border between the physically possible and the physically impossible. Theoretical physics comprises a collection of frameworks forming a complex web that provides different answers to these questions: special and GR, non-relativistic quantum mechanics, quantum field theories, thermodynamics and statistical mechanics to mention the most fundamental building blocks of our best empirically confirmed physics. What seems bewildering is that a certain phenomenon may be possible according to one framework but impossible according to another (for a discussion of this issue, see Baron et al. 2024). For example, signals propagating faster than the speed of light are forbidden by special and GR but appear to be possible, at least in the sense of not being clearly ruled out, on the basis of non-relativistic quantum mechanics. Despite being an important point, we will not focus on this aspect in the present context. Let us simply mention that one solution would be to accord a privileged status to some theories, and perhaps to one only – a final theory of everything – in discerning what is, and what is not, physically possible.

14.4. General relativity

What happens to the kinematical–dynamical distinction in GR, and how are physical possibilities represented within the theory? A defining feature of GR, which distinguishes it from other theoretical contexts, is that it follows the *principle of general covariance*. It states that physical laws need to be written in an invariant form under arbitrary coordinate transformations. The terminology is thus somewhat confusing as the term of art, *co*variance, expresses a form of *in*variance of the laws under arbitrary coordinate transformations. As we will discuss, the notion of general covariance, which has sparked much interpretative debate (see, e.g., Belot and Earman 2001), calls for a generalization of the selection mechanism of dynamical solutions from kinematically allowed models. Despite this complication, we will show that in GR as well, many kinematical solutions can be regarded as valuable methodological tools, guiding the construction of genuine physical possibilities via dynamical models. To illustrate this, we will begin with a semitechnical overview.

GR is currently our best empirically established theory of spacetime. In GR, a spacetime can be described as a differentiable manifold M defined by an ensemble of points with specific topological properties (such as being four-dimensional and continuous) and a metric tensor field g (with a Lorentzian signature $-+++$). Matter and energy are represented by the stress–energy tensor field T. The Einstein field equation (EFE) connects the metric structure (especially the local spacetime curvature) with the matter distribution. The formal expression is as follows:

$$R_{ab} - \frac{1}{2} R g_{ab} + \Lambda g_{ab} = 8\pi T_{ab}$$

where R_{ab} is the Ricci tensor (which specifies how curvature infinitesimally varies in all directions with respect to every other direction), R is the Ricci scalar (which describes the local curvature), g_{ab} is the metric field (encoding the lightcone structure of spacetime), Λ is the cosmological constant (a coefficient providing a counterbalancing effect to gravity) and T_{ab} is the stress–energy tensor (where natural units are adopted, namely, $c = G = 1$). A model of GR is then a triple $\langle M, g_{ab}, T_{ab} \rangle$ satisfying the EFE. The solution of this tensor equation describes how the metric evolves given a specific distribution of matter fields. Note, however, that the stress–energy tensor does not simply operate as a source for the spacetime curvature because its physical meaning is also contingent upon knowledge of the associated metric. As a result, the EFE requires solving the spacetime metric and the matter distribution concomitantly. This is why exact solutions are only available for specific, highly symmetric cases.

In the context of GR, it is then necessary to outline which specific class of solutions one is considering, as different sectors, that is, classes of solutions of the theory, will have different implications for the availability of the Hamiltonian and Lagrangian formalism. To this end, an important consideration is to evaluate whether the specific sector under examination allows us to define a *globally hyperbolic manifold*, which is required for the adoption of the Hamiltonian formulation: only globally hyperbolic solutions can be cast in the Hamiltonian form. Let us note that if one takes non-globally hyperbolic solutions seriously, this seems to provide a compelling argument for the superiority of the Lagrangian formalism over the Hamiltonian formalism, at least in GR.

Global hyperbolicity encodes a particular condition on the causal structure of the spacetime manifold. A fruitful way to identify a globally hyperbolic manifold is via the existence of Cauchy surfaces. Given (M, g), a set of points $\Sigma \subset M$ is a Cauchy surface of (M, g) if every inextendible timelike curve in (M, g) intersects Σ just once. A manifold M is globally hyperbolic if it can be foliated into Cauchy surfaces, namely, decomposed in the form of $\Sigma \times \mathbb{R}$, where Σ stands for a stack of $3d$ (for a four-dimensional manifold) space, ordered via a universal time parameterized by $\mathbb{R}$. Within this class of solutions, both the Lagrangian and the Hamiltonian formulations are available.[7] Intuitively, the Hamiltonian approach is applicable within this sector of the theory because a Cauchy surface can be described as a spatial state at a certain time, whereas successive Cauchy surfaces can be linked together to form a history.

On the one hand, the Lagrangian formulation of GR is defined via the application of the corresponding extremization principle to the so-called Einstein–Hilbert action. The peculiarity of the formalism in this context comes from the fact that, as already discussed, GR must obey general covariance. This implies that, in this context, certain transformations do not change the physical content. They represent different ways of describing the same physical situation by smoothly shifting the spacetime coordinates without altering the underlying geometry. This is why solutions can be grouped into so-called *gauge orbits*. Two metrics, called g and g^1,

7 For a detailed discussion of the Hamiltonian and Lagrangian formulation of GR, and the more specific conditions required for the Hamiltonian approach to be applicable (e.g. orientability), see Wald (1984, Appendix E).

belong to the same gauge orbit if there exists a diffeomorphism $d: M \to M$ such that $g^1 = d^*g$. The idea, roughly, is that if one metric can be transformed into the other by smoothly changing the coordinates, then the two metrics belong to the same gauge orbit. We then define the reduced space of solutions as the one that divides the solutions into gauge orbits. This is the starting point for the application of Lagrangian machinery to GR.

On the other hand, the Hamiltonian formulation operates analogously to the classical case, where a $3 + 1$ slicing of the four-dimensional manifold M in terms of $\Sigma \times \mathbb{R}$ for some 3-dimensional manifold Σ is realized to define the space of initial data and successively evaluate Hamilton's equations. In this context, the main issue in applying the procedure is that GR does not allow a solution-independent decomposition of the manifold M. Even if one provides a procedure to address this additional complication (see Belot 2007, p. 201), the more dramatic aspect of the Hamiltonian formalism in GR is that it directly leads to a frozen formalism. This is the so-called *problem of time in GR*. When we apply the standard dynamical framework to analyze how the relevant variables evolve, the dynamical space is made up of those paths in the space of initial data that remain within the same gauge orbit. In other words, the evolution we track only occurs along gauge-related curves (such as diffeomorphisms), which are typically interpreted as encoding unphysical information (yet, on this basis, see Belot and Earman 2001). The general result is that, within this sector of the theory, no genuine notion of time, evolution, or change seems available because the physical variables remain confined within the gauge orbit, where the physical content, by definition, remains unchanged.[8]

Clearly, the broader set of solutions, in which some of the so-far assumptions are abandoned, renders the problem even more dramatic. If we drop the global-hyperbolicity condition, the manifold M cannot be sliced into Cauchy hypersurfaces, and the Hamiltonian formalism is no longer applicable. For this sector, only the Lagrangian formalism is available. Let us then focus on the latter with the intent of spelling out how the construction of the kinematical and the dynamical spaces is realized and what implications should be drawn concerning the status of physical possibilities.

---

8 While *prima facie* a possible way out of the conundrum would be to focus on the Lagrangian formalism, which does not require the $3 + 1$ splitting, it can be shown that, under the so-far introduced assumptions concerning the kinematical space, an isomorphism between the reduced space of (Lagrangian) solutions and the corresponding reduced space of (Hamiltonian) initial data should be expected. This means that the problem of time figures in this sector of GR, no matter what formalism one aims to adopt (see Belot 2007, pp. 202–203 and 209–211).

Starting from the Lagrangian,[9] the solution space is evaluated from the action S, which in classical mechanics is defined as the integral of the Lagrangian between two points in the configuration space at two successive instants t_1 and t_2, as illustrated above. More specifically, it was shown that the actual physical path is the one that minimizes the action S and that the minimization procedure is obtained by evaluating the Euler–Lagrange equations.

Analogously, in the context of GR, the Lagrangian formalism can provide the starting basis for the definition of an action principle that allows the evaluation of dynamical solutions. The main issue arising in the context of GR is how to provide a generalization of the concept of temporal integration typical of the Lagrangian formalism in classical and quantum mechanics. This issue can be framed in terms of establishing how to define a volume element, called $d\mu_g$. A possibility would be to justify its definition on the basis that the integral:

$$\int_M f \, d\mu_g$$

for a function f defined on the manifold M and a measure $d\mu_g$, which may be dependent upon the metric, is independent of the choice of coordinates (as prescribed by general covariance). This requires the integral to be invariant under the smooth coordinate change $\{x^i\} \rightarrow \{\hat{x}^i\}$. Without entering too many technical details, it can be shown that this implies imposing $d\mu_g = \sqrt{|g(x)|}d^4x$ (where $g(x)$ stands for the determinant of the metric tensor). The volume element $d\mu_g$ is typically called the measure corresponding to the metric g.

This last condition teaches us that the Lagrangian formalism of GR represents a generalization of the one employed in the context of classical mechanics, whereby no unique notion of temporal evolution is assumed, let alone adopted. For the Lorentzian metric, one typically employs $\sqrt{-g}$ instead of $\sqrt{|g(x)|}$. As the volume element is contingent upon the metric, it should figure in the variation as long as the metric represents the field variable the variational principle applies to. From this, one arrives at the definition of the Einstein–Hilbert action:

$$S_{EH} = \int R d\mu_g = \int \sqrt{-g} R d^4x$$

9 Note that although the Lagrangian formalism is also available for more complex sectors of GR, here we will mention, for a matter of simplicity, the manageable scenario of vacuum solutions, whereby the only available field is the metric field.

where R is the Ricci scalar and the integral is performed over a compact region D with smooth boundary conditions. From the subsequent application of the variational principle, it is then possible to determine the solutions that minimize the Einstein–Hilbert action, namely, those corresponding to dynamical solutions.

From the above discussion, three aspects emerge that are of particular interest for the kinematics–dynamics distinction. First, as GR allows no solution-independent decomposition of the manifold M in a global space Σ and a global time $t \in \mathbb{R}$, the Lagrangian formalism is more apt to track the *solution space* of the theory. Second, GR calls for a redefinition of the notion of evolution, requiring that the volume element used in integrating the Lagrangian reflect the principle of general covariance. This means that one should allow for a generalization of the selection mechanism of dynamical solutions with respect to the space of kinematically allowed models. Third, the subset view still applies as long as the just mentioned generalization of the Lagrangian formalism remains available: the Einstein–Hilbert action encodes kinematical possibilities, whereas the successive application of the variational principle generates the dynamical space, in our interpretation, the one that captures genuine physical possibilities.

14.5. Nonrelativistic quantum mechanics

Is the characterization of the distinction between kinematical and dynamical possibilities from above also suitable for non-relativistic quantum mechanics? The situation is quite complex, as quantum mechanics can be formulated in different ways.[10] Historically, the three major methods are the *matrix formulation*, the *wavefunction formulation* and the *path integral approach*, advanced, respectively, by Heisenberg, Schrödinger and Feynman.

Here, we restrict our attention to the path-integral approach for several reasons. First, it is a natural and fruitful method to convey the main insights of quantum mechanics. Second, the classical limit is relatively straightforward in this approach (MacKenzie 2000, p. 2). Third, while it might be more complicated than the matrix and wavefunction formulations, it becomes extremely useful in various contexts, such as classical and quantum field theory, statistical mechanics and condensed matter physics (MacKenzie 2000, p. 1; Styer et al. 2002, p. 290), so that the conceptual results endorsed here can be straightforwardly generalized to those settings. Fourth, many approaches to quantum gravity embody the spirit of the path integral formalism.

10 See Styer et al. (2002), which provides no less than nine formulations.

The path integral formulation addresses the transition probabilities between initial and final states. The idea is analogous to what happens in the Lagrangian formulation of classical mechanics. Suppose that we have a particle located at x_1 at time t_1 and assume that we want to calculate the probability of finding it at x_2 at time t_2. This is what the path integral approach prescribes:

– evaluate all the possible classical paths from x_1 to x_2;

– determine the action for each classical path (recall that the action is provided by the temporal integral of the associated Lagrangian);

– ascribe to each classical path a transition amplitude, which can be shown to be proportional to $e^{\frac{iS}{\hbar}}$;

– sum the amplitude over all possible classical paths by using the summing rule provided by quantum mechanics;

– the sum, which assumes that the possible paths vary continuously, corresponds to a path integral, provides the transition amplitude from x_1 to x_2 and its square magnitude corresponds to the transition probability.

The total probability that a certain particle, located at x_1 at t_1, reaches x_2 at t_2, is calculated as a weighted sum of all the possible paths the particle can follow from the initial to the final state. If one assumes, for the sake of the present argument, that such an approach to quantum mechanics is not a mere instrumental way of expressing transition probabilities directly measurable in laboratories, the picture of reality we obtain is quite peculiar. Indeed, the lesson appears to be that quantum particles take all viable paths from initial to final states, not only one of them.

What is the relationship between the classical and quantum versions of Lagrangian mechanics? A transition amplitude is associated with each of the possible paths the particle can follow, and the total transition amplitude is given by summing them. How are we then to combine this with the fact that, classically, objects appear to follow clearly defined, and arguably unique, trajectories? By evaluating interferences between adjacent paths, it can be shown that the closer the paths are to the classical path, the more they interfere constructively. The net effect is that, on average, paths close to the classical path will interfere constructively, whereas arbitrary paths will interfere destructively (MacKenzie 2000, p. 13).

From classical mechanics, we know that stationary states correspond to dynamical solutions. In that context, we showed that the latter can thus be regarded as a subset of the kinematical space. Correspondingly, one may wonder whether what we dubbed the subset view is still available in the context of non-relativistic quantum mechanics. At first glance, the path integral approach to quantum mechanics does not seem to differ from the classical case in this respect: among the various possible paths, the particle can follow a subset that maximizes, in the classical limit, constructive interference. This subset should accord with the status of dynamical solutions.

However, Ruyant and Guay (2024) challenge this view. They argue that quantum mechanics raises an issue for the metaphysics of modality. Their argument is as follows: classical mechanics and quantum mechanics frame the relationship between kinematical and dynamical solutions differently. While in classical physics, dynamical possibilities are selected from kinematical possibilities (which thus turn out to be both kinematical and dynamical possibilities), in quantum mechanics, kinematical solutions are *not* dynamical solutions, suggesting that dynamical possibilities are not kinematical possibilities. This is because the kinematical space is modeled as continuous, whereas the transition amplitudes (which stand for the quantum analogue of classical dynamical possibilities) correspond to discontinuous jumps. As a result, kinematical possibilities never become realized in nature: we need them to construct dynamical possibilities, but they are "impossible histories" (Ruyant and Guay 2024, p. 11).

We agree with Ruyant and Guay that the subset approach no longer works in the quantum context or, at the very least, requires substantive adjustments. Nevertheless, we believe that a broader approach to the relationship between kinematical and dynamical solutions applies equally to the classical and quantum contexts. This broader view is that dynamic models are built from kinematical ones and that we should not take the latter to indicate genuine physical possibilities.

An immediate challenge against this deflationist attitude toward kinematical possibilities in quantum mechanics arises from the fact that dynamical models appear to be constructed from weighted sums of kinematical models, making the latter seem physically possible. How could kinematical models fail to represent genuine physical possibilities if they are, in some important sense, constitutive of dynamical possibilities? To address this, it is helpful to distinguish between two opposing interpretations of the path integral: one that treats it as a mere mathematical tool and another that regards it as ontologically significant.

The first approach suggests that the path integral should not be taken too seriously, ontologically. This perspective may involve adopting an alternative formulation of quantum mechanics and viewing the path integral as a mere mathematical representation, useful for empirical predictions but without ontological bearing. Under this view, dynamical possibilities would not literally be constituted by kinematical possibilities. Instead, the kinematical–dynamical structure would be an artifact of the mathematical framework employed.

The second approach adopts the view that quantum particles do not follow *one* path but rather *all* possible paths from initial to final states. This perspective aligns naturally with Everettian quantum mechanics (EQM), the many-worlds interpretation of quantum mechanics (Wallace 2012). According to the EQM, many worlds or branches are grounded in the fundamental universal quantum state. This implies that all trajectories are realized in nature. Kinematical models, therefore, can be regarded as describing the branches of an extraordinarily vast multiverse. If one considers the branches to represent physical possibilities rather than proper parts of such a multiverse, it follows that kinematical models do represent physical possibilities, after all.[11] However, even if one does not endorse the modal interpretation of the EQM, the kinematical models still represent physical possibilities since the particles actually follow them. Indeed, what is actually happening in our world should also, a fortiori, be physically possible.

Does this mean that the EQM implies that branches are kinematical possibilities, thereby undermining our claim that kinematical models do not represent genuine physical possibilities? Not quite.

First, it is important to note that in an important sense, a particle cannot, in this interpretation, follow only one trajectory. It must also follow other trajectories – or, alternatively, other particles must do so, depending on how one individuates entities across the quantum multiverse. For example, if the EQM is the correct interpretation of quantum mechanics, the fundamentality of the universal quantum state rules out the possibility that each physically possible world coincides exactly with one classical world only. This is because the branches are emergent, that is, less fundamental than the universal quantum state. According to EQM, physical branches are a package deal – they always come in large numbers, never individually. Therefore, if the branches correspond to physical possibilities, those are of a peculiar sort, requiring the existence of other coexisting, entrenched possibilities. In our view, this speaks against taking branches as genuine classical possibilities and thus against the view that purely kinematical models can be interpreted as genuine physical possibilities in EQM.

11 The modal interpretation of EQM is endorsed by Wilson (2020).

Second, and relatedly, it is sometimes underappreciated that the quantum multiverse, as described by EQM, is emergent – at least in the standard case of decoherence-based EQM. This implies that there are infinitely many ways to define branches in the quantum multiverse. Consequently, either the branching involves some degree of indeterminacy, or the branches should be excluded from the ontology of EQM, regarded instead as mere heuristics to interpret the universal quantum state (Glick and Le Bihan 2024). This suggests that, in this context, kinematical models should not be viewed as representing classical possibilities rather than quasiclassical ones. They appear classical only if one neglects the subtle residual entanglement with degrees of freedom localized in other branches. Viewing them as truly classical possibilities is thus an idealization. Hence, dynamical models are constructed from classical models, which in turn are understood as approximations of quasiclassical models. Quantum states, although they come after classical states in the order of theory construction, are more fundamental than classical states. Thus, even assuming EQM, or more generally any form of realism about the paths involved in the path integral, we should not take purely kinematical possibilities to represent genuine physical possibilities.

14.6. Concluding remarks

The distinction between kinematical and dynamical models reflects a specific methodology used by physicists: the kinematics provide the stage setting from which dynamical models, corresponding to physical possibilities, are constructed. A central theme of our investigation has been the coexistence of two distinct mathematical frameworks in theoretical physics: the Lagrangian/path integral and Hamiltonian approaches.

Interestingly, GR appears to suggest that the Lagrangian formulation is more fundamental, as certain solutions within the theory cannot be accommodated by the Hamiltonian framework. Furthermore, quantum mechanics suggests that dynamical models are more fundamental than kinematical models. A natural conclusion, then, is that physical possibilities are best represented by dynamical solutions within the Lagrangian or path-integral framework. In contrast, kinematic models do not represent genuine physical possibilities but merely serve as preliminary approximations or rough characterizations of potential candidates for physical possibilities.

Looking ahead to the future of physics, what insights into physical possibility might a theory of *quantum gravity*, going beyond relativistic and quantum physics, offer? While a comprehensive study is necessary, we can highlight one key point.

As extensively discussed in the literature, virtually all approaches to quantum gravity suggest that spacetime is not, in a way to be further specified, fundamental (see, e.g., Oriti 2009; Crowther 2018; Vistarini 2019; Huggett and Wüthrich 2025). Rather, it emerges from a non-spatiotemporal structure.

The possible emergence of spacetime leads to a number of distinct issues, both conceptual and empirical (Le Bihan 2021). The fact that the Lagrangian approach is less temporally flavored than the Hamiltonian approach is suggestive of its potential capacity to describe non-spatiotemporal physical systems and thus to address at least some of those issues. By deconstructing space and time more deeply, spacetime emergence could thereby further underscore the metaphysical superiority of the Lagrangian approach over the Hamiltonian framework.[12] In this way, a Lagrangian formulation may provide our best prospect for identifying physical possibilities within such a non-spatiotemporal theory.

However, the task will not be easy: the Lagrangian formulation is traditionally framed in a spatiotemporal setting, as it involves integrals over paths typically parameterized by spatiotemporal coordinates. In classical physics, action is a function of positions, velocities and time, which are intrinsically spatial and temporal notions. In quantum mechanics, the path integral formulation extends this idea by summing all possible trajectories in spacetime. Thus, in both cases, spacetime plays an essential role in defining the possibilities and dynamics of the system. Our Lagrangian understanding of non-spatiotemporal physics will thus have to be clearly articulated to obtain a sense of the nature of non-spatiotemporal, physical possibilities.

What about the kinematic–dynamical distinction itself in non-spatiotemporal contexts? If the distinction does not hinge on a temporally flavored picture of dynamical laws acting on otherwise non-dynamical systems, but rather on the particular methodological route by which we arrive at physical possibilities, then the challenge posed by spacetime emergence to our understanding of physical possibilities might become conceptually easier to address. Moreover, a reduced emphasis on the modal significance of kinematical models – viewed as mere tools to guide the search for true physical possibilities – suggests that purely kinematical models should not be overvalued when attempting to understand the nature of reality through research programs in quantum gravity. Overall, the lesson seems to be that,

12 Note that this is consistent with the fact that, at the moment, many quantum gravity approaches fruitfully put the Hamiltonian formulation to work. The approach might be highly useful as a way to develop a more fundamental, perspicuous formulation.

when seeking genuine physical possibilities in theoretical physics, the primary focus should be on the maximally constrained solution space of the theory.

14.7. Acknowledgments

Our thanks to Enrico Cinti for his valuable feedback during discussions of the content of this paper, Erik Curiel for a constructive exchange on the kinematical–dynamical distinction broadly conceived and Saakshi Dulani for her insightful comments. This work was supported by the Swiss National Science Foundation via Le Bihan's SNSF Starting Grant *Space, Time and Causation in Quantum Gravity*.

14.8. References

Arnold, V.I. (1989). *Mathematical Methods of Classical Mechanics*. Springer, New York.

Baron, S., Le Bihan, B., Read, J. (2024). Physical theory and physical possibility. *Erkenntnis* [Online]. Available at: https://philsci-archive.pitt.edu/23283/.

Belot, G. (2007). The representation of time and change in mechanics. In *Handbook of the Philosophy of Physics. Philosophy of Physics*, Butterfield, J. and Earman, J. (eds). Elsevier, Amsterdam.

Belot, G. and Earman, J. (2001). Pre-Socratic quantum gravity. In *Physics Meets Philosophy at the Planck Scale*, Callender, C. and Huggett, N. (eds). Cambridge University Press, Cambridge.

Crowther, K. (2018). *Effective Spacetime*. Springer, Cham.

Curiel, E. (2014). Classical mechanics is Lagrangian; it is not Hamiltonian. *British Journal for the Philosophy of Science*, 65(2), 269–321.

Divers, J. (2018). W(h)ither metaphysical necessity? *Aristotelian Society Supplementary Volume*, 92(1), 1–25.

Fine, K. (1994). Essence and modality. *Philosophical Perspectives*, 8, 1–16. doi: 10.2307/2214160.

Glick, D. and Le Bihan, B. (2024). Metaphysical indeterminacy in Everettian quantum mechanics. *European Journal for Philosophy of Science*, 14(1), 3.

Hirèche, S., Linnemann, N., Michels, R., Vogt, L. (2021). The strong arm of the law: A unified account of necessary and contingent laws of nature. *Synthese*, 199(3), 10211–10252.

Huggett, N. and Wüthrich, C. (2025). *Out of Nowhere: The Emergence of Spacetime in Quantum Theories of Gravity*. Oxford University Press, Oxford.

Kripke, S.A. (1980). *Naming and Necessity: Lectures Given to the Princeton University Philosophy Colloquium*. Harvard University Press, Cambridge, MA.

Le Bihan, B. (2021). Spacetime emergence in quantum gravity: Functionalism and the hard problem. *Synthese*, 199(2), 371–393.

MacKenzie, R. (2000). Path integral methods and applications. *arXiv:quant-ph/0004090*. doi: 10.48550/arXiv.quant-ph/0004090.

North, J. (2009). The 'structure' of physics: A case study. *Journal of Philosophy*, 106(2), 57–88.

Norton, J.D. (2022). How to make possibility safe for empiricists. In *Rethinking the Concept of Law of Nature. Jerusalem Studies in Philosophy and History of Science*, Ben-Menahem, Y. (ed.). Springer, Cham.

Oriti, D. (2009). *Approaches to Quantum Gravity: Toward a New Understanding of Space, Time and Matter*. Cambridge University Press, Cambridge.

Ruyant, Q. and Guay, A. (2024). Lagrangian possibilities. *Synthese*, 203(4), 1–22.

Styer, D.F., Balkin, M.S., Becker, K.M., Burns, M.R., Dudley, C.E., Forth, S.T., Gaumer, J.S., Kramer, M.A., Oertel, D.C., Park, L.H. et al. (2002). Nine formulations of quantum mechanics. *American Journal of Physics*, 70(3), 288–297.

Vistarini, T. (2019). *The Emergence of Spacetime in String Theory*. Routledge, Oxford.

Wald, R.M. (1984). *General Relativity*. The University of Chicago Press, Chicago.

Wallace, D. (2012). *The Emergent Multiverse: Quantum Theory According to the Everett Interpretation*. Oxford University Press, Oxford.

Wallace, D. (2022). Observability, redundancy, and modality for dynamical symmetry transformations. In *The Philosophy and Physics of Noether's Theorems: A Centenary Volume*, Read, J. and Teh, N.J. (eds). Cambridge University Press, Cambridge.

Wilson, A. (2020). *The Nature of Contingency: Quantum Physics as Modal Realism*. Oxford University Press, Oxford.

List of Authors

Jeffrey A. BARRETT
University of California
Irvine
USA

Matteo BARSUGLIA
CNRS
Astroparticule et Cosmologie
Paris
France

Françoise COMBES
LUX – Observatoire de Paris – PSL
and
Collège de France, CNRS
Sorbonne University
Paris
France

Olival FREIRE JUNIOR
Instituto de Física
Universidade Federal da Bahia
Salvador
Brazil

Thiago HARTZ
Instituto de Matemática
Universidade Federal do Rio de Janeiro
Brazil

Richard HEALEY
Emeritus of Philosophy
University of Arizona
USA

Marc LACHIÈZE-REY
CNRS, Astroparticle and
Cosmology Laboratory (APC)
and
SPHERE
Paris
France

Baptiste LE BIHAN
Department of Philosophy
University of Geneva
Switzerland

Matthew S. LEIFER
Institute for Quantum Studies
Schmid College of Science
and Technology
Chapman University
Orange
USA

Bruno MANSOULIÉ
IRFU
CEA
Université Paris-Saclay
Gif-sur-Yvette
France

Emilia MARGONI
Department of Philosophy
University of Geneva
Switzerland

Simone MASTROGIOVANNI
INFN Sezione di Roma
Rome
Italy

Nathalie PALANQUE-DELABROUILLE
Lawrence Berkeley National Laboratory
California
USA
and
CEA-Saclay
Paris
France

Thomas RYCKMAN
Stanford University
USA

David SCHMID
Perimeter Institute for
Theoretical Physics
Waterloo
Canada

Lev VAIDMAN
School of Physics and Astronomy
Tel-Aviv University
Israel
and
Institute for Quantum Studies
Chapman University
Orange
USA

Annica VIESER
Department of Philosophy
University of Geneva
Switzerland

Yìlè YĪNG
Perimeter Institute for
Theoretical Physics
Waterloo
and
Department of Physics and Astronomy
University of Waterloo
Canada

Hervé ZWIRN
Institut d'Histoire et de Philosophie des
Sciences et des Techniques
Paris
and
École Normale Supérieure
de Paris-Saclay
Centre Borelli
France

Index

E, F

G, H

I, L

M, N

O, P

Q, R

S, T

U, V

W, Z

Printed and bound by CPI Group (UK) Ltd, Croydon, CR0 4YY
16/08/2026
14942685-0006